T0136734

Connectionism and Meaning:
From Truth Conditions to Weight Representations

Ablex Series in Artificial Intelligence
Yorick Wilks, Series Editor

Chess and Machine Intuition, by George W. Atkinson

A Compendium of Machine Learning, Vol. 1: Symbolic Machine Learning, by Garry Briscoe and Terry Caelli

The Turing Test and The Frame Problem: AI's Mistaken Understanding of Intelligence, by Larry J. Crockett

Connectionism and Meaning: From Truth Conditions to Weight Representations, by Stuart A. Jackson

A Connectionist Language Generator, by Nigel Ward

In Preparation:

Processing Metonymy and Metaphor, by Dan Fass

Computer Language and Vision Across the Pacific, by Yorick Wilks and Naoyuki Okada

Connectionism and Meaning:
From Truth Conditions to Weight Representations

by Stuart A. Jackson
University of Sheffield

Printed in the United States of America

Library of Congress Cataloging-in-Publication Data

Jackson, Stuart A..
Connectionism and meaning : from truth conditions to weight representations / Stuart A. Jackson.
p. cm. -- (Ablex series in artificial intelligence)
Includes bibliographical references and index.
ISBN 1-56750-157-5 (cloth). -- ISBN 1-56750-158-3 (paper)
1. Computational learning theory. 2. Philosophy of mind.
3. Cognitive science. 4. Connectionism--Data processing.
I. title. II. Series.
Q325.7.J37 1996
006.3'5--dc20 96-6194
CIP

Ablex Publishing Corporation
355 Chestnut Street
Norwood, New Jersey 07648

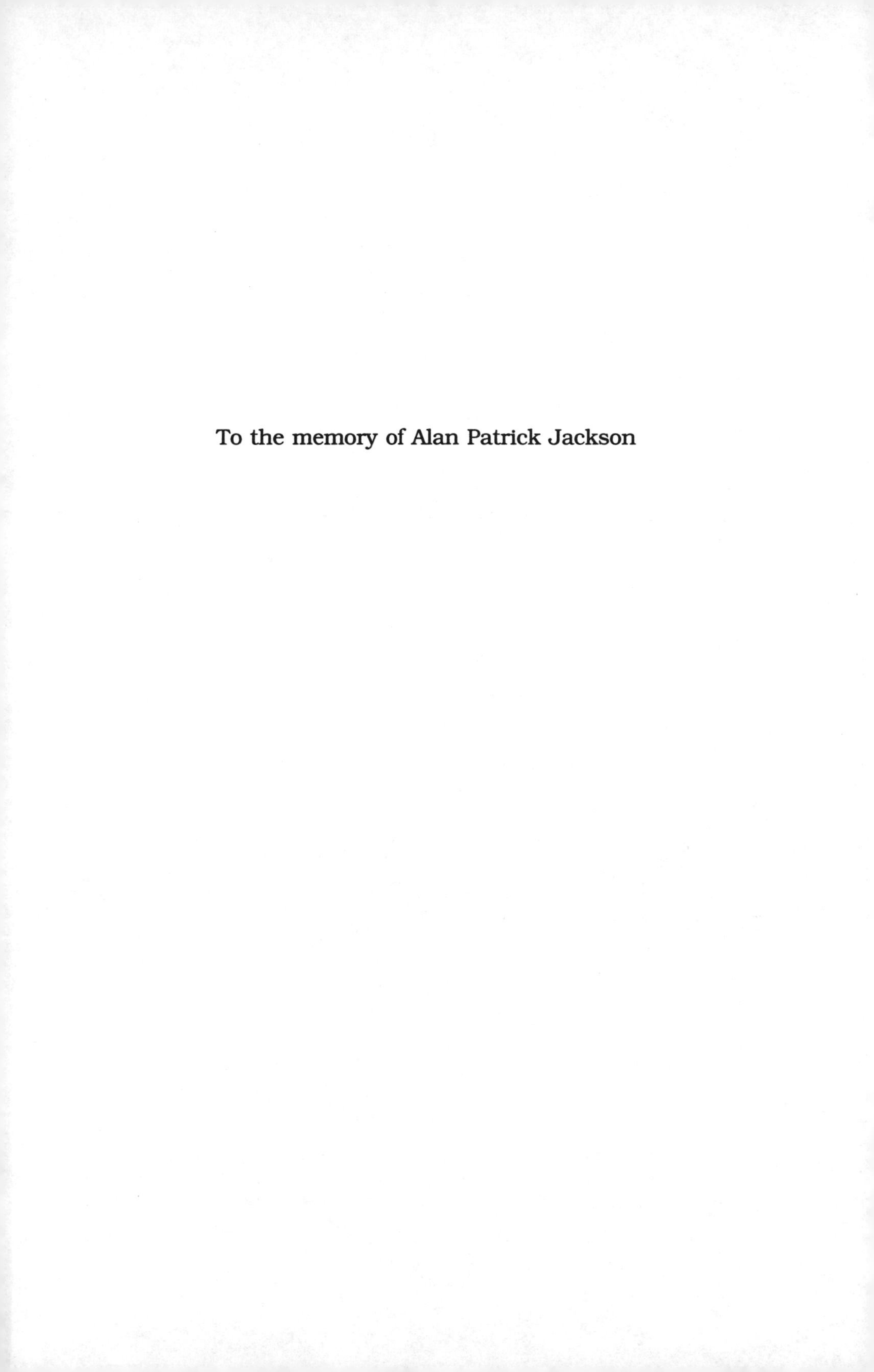

To the memory of Alan Patrick Jackson

Contents

Preface

During the long, hot summer of 1987, I was part of an undergraduate class at the University of Sussex that met once a week to discuss the philosophy of Artificial Intelligence. Enthusiasm made up for ignorance, and the debates were open and lively. There was a sense of coming together, as if each of us taking part was helping the others to open their minds to the fascinating problems before us. One of the topics we discussed later on was a new, messy, exciting field of study called parallel distributed processing. At the time, no one knew anything about it. It was a fledgling area of psychology, just starting to make an impact, with the two ground-breaking PDP books having been published just a year before in 1986. I was intrigued with the little diagrams of simple neural nets, with the nodes and links, and I thought, "Well, of course this is how to do computational psychology. Forget all these clunky, brittle AI models!" I rushed out and bought both volumes of the PDP books and was immediately hooked. At the time, I spent many happy/frustrating/baffling hours reading and thinking and puzzling over what they were trying to tell me, and so (no doubt along with many other connectionists of my generation) I would first like to thank James McClelland and David Rumelhart (and the others of the PDP research group) for writing two such iconoclastic and stimulating books.

Many people have helped me and taught me in the years since that first class on the philosophy of Artificial Intelligence. At Sussex, Pasha Parpia, who took that first class; Alan Garnham, who taught me psycholinguistics; and Pete Clifton, who taught me more than I ever wanted to learn about Neurophysiology. At Exeter, I want to gratefully acknowledge the enthusiasm and support of Noel Sharkey, to whom I owe a great intellectual debt. I want to thank all of the people at the Computer Science Department of the University of Exeter who helped me with arguments and (occasionally scathing) comments to refine my ideas, particularly Adrian Baldwin, Paul Day, Niall Griffith, Ajit Narayanan, Derek Partridge and Amanda Sharkey. I also want to thank all those friends who sat and listened and made suggestions while I variously moaned and conjectured and waved my hands, particularly Lucy Allsopp, John Kinsman, and Jane Compson. I would also like to thank Yorick Wilks, the editor of this Ablex series; and the Production Editor at Ablex, Anne Trowbridge, who took charge of the manuscript and efficiently guided it into print.

—Stuart Jackson,
Brighton, 1996

1

Prolegomenon

Computationally driven, philosophically inspired, and psychologically motivated, this book is a theoretical investigation into *meaning*, examining and combining concepts from formal and psychological semantics, the philosophy of mind, and computational theory within the framework of connectionist cognitive science. The investigation begins by examining the invention of formal calculi, late in the 19th Century, and ends by proposing how the atomic representations of computational engines can be hooked with what they are representations of—a broad sweep of concerns indeed.

The book is largely about ideas: of what semantics is; of what meaning is, in a general sense, of how language signs have meaning; of how the meanings of words and sentences in language might be mentally represented; and, more basically, of how meaning might arise at the ineffable boundary between representations and represented. These, I believe, are fascinating questions, and I hope that the book conveys some of the intellectual excitement generated in considering them.

The book is also, unashamedly, connectionist, drawing on the novel properties of networks of simple computing elements as its computational metaphor. Although connectionist cognitive science is a young discipline, effectively less than 10 years old, it is an exciting and vigorous one, with the potential of actually delivering what symbolic cognitive science has long promised but never really achieved—a comprehensive theory of mental life. This is not, I believe, misplaced optimism. Again and again, connectionists have been forced to recognize seemingly insurmountable problems with their novel class of computing machinery: that they are merely associative devices, that they lack computational power, that they are unable to support structured representations, that they merely implement symbolic architectures, and so on, and so on. In each case, these seemingly insurmountable problems have been overcome with the resultant generation of truly innovative ideas in the process.

The investigations of meaning that I undertake in this book are in this same pioneer spirit. The reader will find established wisdom hand in hand with novel ideas, but at all times anchored to the actual computational properties of connectionist networks. It is my fervent wish that the book provokes reaction, whether it be surprised agreement or irritated disagreement: Both are indications that the interest of the reader has been engaged in the subject matter, which is all that an author can reasonably expect. Much worse than vociferous criticism is bland apathy: Being branded boring is indeed a terrible thing.

FOOLISH QUESTIONS

Since the inception of their discipline some 25 centuries ago, philosophers have had the dubious title of Curators of the Department of Foolish Questions foisted on them. Among their dusty collection of exhibits are assorted ridiculous inquiries into appearance and reality, substance and property, truth and falsity, God, time and consciousness, all notions that lie at the heart of human existence. Some sections of the department are old and moth-eaten, such as the gallery of redundant inquiry, the gallery of abandoned questions, and the gallery of the antithesis. There is one gallery, however, to which the curator is especially attached, one of the oldest, dust-free, kept clean by the feet of many visitors. The foolish questions exhibited here are all concerned, in one way or another, with *meaning* in all of the multifarious ways in which this word can be interpreted.

Meaning is the champion of epistemological intransigence, stubbornly resisting all attempts to make it tractable (John Cleese's portrayal of the Black Knight in Monty Python and the Holy Grail might be a usefully unpompous visualization). In part, such intransigence stems from the fact that meaning inhabits such a vast space of possibility: Literally anything can have a meaning, or be ascribed a meaning, or be described as having a meaning. Philosophers may talk about, for example, the meanings of signs and of concatenations of signs, but they may also talk about the meaning of a raised fist, or the meaning of the texture of tree bark, or the meaning of an ordered sequence of lights. Intuitively, it might seem that in considering meaning, the philosopher is prone to glossing over a fundamental and useful distinction: namely, is the meaning of a word the same kind of thing as the meaning of something that is not a word, like the texture of tree bark? One should always trust intuition.

The first of many distinctions to consider when talking about meaning is indeed between (various species of) *linguistic* meaning, and what (for want of a better term) can be called *environmental* meaning. The former is concerned with the nature of the meaning conveyed by, and associated with, sign systems of various descriptions—most notably the systems of signs constituted by natural and artificial languages—and the latter is concerned with the nature of the meaning conveyed by, and associated with, everything that any given sign system is (roughly) capable of being about. Subsequent usage of the term *meaning* should be taken as shorthand for the term *linguistic meaning.*

WHAT IS MEANING?

This is the most fundamental and intriguing, exhibit in the curator's collection of foolish questions, and it is something of a fractal entity. A close examination of it does not reveal a comprehensible microstructure but rather a recession of nested complexities. The question-exhibit *What is meaning?* hides and contains the more complex question-exhibit *What is the meaning of meaning?*, which in turn contains as part the question-exhibit *What is the meaning of the meaning of meaning?*, and so on. The fractal nature of the curator's most prized question-exhibit highlights a very real concern for the philosopher interested in formulating a theory of linguistic meaning—namely, what question, or set of questions, to ask: From where should the inquiry begin? and how should it begin? Put another way, the gist of the confusion concerns what exactly a theory of meaning is supposed to be about.

To the layperson, it might seem as if this situation is slightly bizarre. Surely, the objection might go, philosophers are being slightly amiss in being uncertain of exactly what it is they are supposed to be philosophizing about. This objection can be considered by looking at a representative selection of the kinds of questions that a theory of meaning would (ideally) answer:

- What is meaning?
- What is the meaning of meaning?

- What is the meaning of a sentence?
- What is the meaning of a word?

Suppose that the philosopher's theory of meaning is intended to explicate the first question, What is meaning? Perhaps some light can be shed by consulting a lexicon (a dictionary gives the meanings of words after all). Unfortunately, the lexicographical flame burns very dim. One finds meaning defined alternatively as that which is in the mind (or thoughts), that which is the sense intended, or that which is significant. Plainly this is not good enough. Without sounding tautological, what is the meaning of *sense* as used in such a definition? or the meaning of *significant*? (Equally, of course, but a little more tangentially, what is the meaning of *is*?)

Now consider the second question, What is the meaning of meaning? Does the question presuppose that meaning is some kind of nonlinguistic entity? Does meaning itself have a meaning? If it doesn't have *a* meaning, then how is the philosopher ever to answer the question? If it does have a meaning, then the question is pointless because it supposes a prior understanding of what meaning is.

Such perplexities over the formulation of questions about meaning derive, in part, from a failure to distinguish the *use* and the *mention* of a term. Fortunately, there is a handy device for handling this problem that involves the theorist adopting the convention that single quotes enclosing a word, phrase, or sentence indicate that what is being talked about is the term itself, not that to which the term refers. This avoids the confusion of the previous paragraph (What is the meaning of the meaning of the meaning of... and so on) by distinguishing, for example, dogs, which have four legs, from "dogs," which has four letters. Thus, to *mention* a term is to treat it as a language object, whereas to *use* a term is to treat it as a language tool. This distinction allows philosophers to recouch the question that they want to ask as follows: What is the meaning of "meaning?" This places their inquiry into the word "meaning" on a similar footing to that of their inquiry into the meaning of any other term.

TRUTH CONDITIONS AND SENTENCES

Earlier I asked what a theory of meaning was supposed to be *about*. There is a large body of opinion that maintains it is the sentence, and specifically the meaning of a sentence, that a theory of meaning is supposed to be about. Sentences, the argument states, are the primary bearers of meaningful information in language; consequently, to ask in what the meaning of a sentence consists is to frame the subject matter of a theory of meaning.

When posed directly, however, my third question, What is the meaning of a sentence?, seems ambiguous, in that it is not at all clear just what is being asked—questions about the meanings of sentences in general, or the meanings of sentences in particular? There is a consensus in the literature that the first of these questions is the most important, and that a theory of meaning should be able to reveal the meaning of any sentence in a clear and perspicuous manner. What is sought, in answering this question, is a way of characterizing the meaning of all sentences using only one metric. It is hard, however, to conceive, intuitively, of the metric that would suffice to encompass the disparate meanings of the following three sentences:

- The matchbox is to the left of the lighter.
- Intelligent front-end methodologies are boring.
- The horse raced past the barn fell.[1]

By common tacit consensus among modern theorists of meaning, the concept of *truth conditions* has become (just about the only viable candidate for) the metric characterizing the meaning of any given sentence of a language. That is, the meaning of a sentence is assumed to be its truth conditions. When the philosopher (or logician) gives the truth conditions of a sentence (that is to say, the conditions both necessary and sufficient for the truth of that sentence), then she is also giving the meaning of that sentence.

This way of viewing meaning is beguiling in its apparent simplicity: An individual who understands the sentence "Intelligent front-end methodologies are boring," must also (quite naturally, the argument states) know and understand what would be the case if what was asserted (that is, the fact that the methodology of intelligent front-ends is boring) was true. Hence, an individual who understands a sentence must also correspondingly know the truth conditions of that sentence. Such a *truth-conditional* view of meaning is very pervasive in the literature. Indeed, the reader is

---

[1] The sentence is an example of a *garden path* sentence. Though "raced" is originally interpreted as the main verb, a re-reading forces the reader to reorganize the sentence. "The horse (that) raced past the barn fell (over)."

asked to go even further; The following is from Michael Dummett (1978):

> Under any theory of meaning whatever . . . we can represent the meaning (sense) of a sentence as given by the condition for it to be true, on some appropriate way of construing `true': the problem is not whether meaning is to be explained in terms of truth conditions, but of what notion of truth is admissible. (p. 67)

On the surface, equating meaning with truth conditions seems an eminently elegant and sensible strategy. Dig a little deeper, however, and the beast within becomes apparent. A series of sample questions will suffice to illustrate the present concern. I have been talking about truth conditions: Accordingly, a sensible first question must be, What is truth? Philosophers make great use of necessary and sufficient conditions in their discussions of truth: Conditions of what exactly, with regard to what? Atmospheric conditions? How is the sentence, "The matchbox is to the left of the lighter" true in any formal sense? What, in point of fact, *are* truth conditions and where are they? Are they immanent in the states of affairs in the world, or are they mental constructions? If a theory of meaning is not to be a vacuous construct, the potentially tendentious nature of the substitution of truth conditions for meaning must not be allowed to occur.

The book's investigations into meaning begin in Chapter 4, which discusses both the origins of the ideas and concepts used in current theories of meaning, and the principles of the technical apparatus employed by meaning theorists to explicate their fiendishly elusive subject matter. As I will explore, the concept of truth conditions, on which many modern theories of meaning are based, arose and evolved from the purely formal tradition of the philosopher and the logician. However, the evolution of truth conditions has been neither simple nor linear, but rather curiously "lumpy," and has resulted in the two sophisticated and powerful notions of *intension* and *extension* becoming commonplace in theories of meaning. These two terms have a central importance for the whole of this book.

THE MEANING OF A WORD

Despite the primacy of the sentence in modern theories of meaning, this book is more concerned with the essence of the fourth of the questions that I posed earlier, What is the meaning of a word? This question is in some ways easier to think about than the previous ones, perhaps because words are generally about things (in some general sense), or simply that they *refer*. The word "Frisbee™," for instance, can be considered to refer to a class of disc-shaped objects made of a semirigid material, usually plastic, combining the lift and flight characteristics of an aerodynamic wing, and the stability characteristics of a gyroscope, used recreationally as an integral part of several kinds of game. Such a specification could go on, and in a technical parlance might be construed as a list of conditions both necessary and sufficient to define the meaning of that word. However, plainly such a specification does not entirely capture the word's full meaning: A Frisbee is still a Frisbee if (as often happens) a dog has mauled it, and its flight characteristics are impaired by a number of annoying teeth marks.

In the case of some other words, it is equally obvious that the philosopher would be ill-advised to equate meaning solely with some (vague) notion of referring. For example, the Native-American Hopi Indians have a word in the vocabulary of their language, *koyaanisqatsi* (pronounced coy-arn-ees-cat-see). To speakers of English, the word is unusual orthographically (there is no "u" after the "q" for example), phonetically, and most importantly, semantically. Consider what the meaning of this word is. Lexicographical illumination would reveal that the word means variously:

- Life in turmoil.
- State of life that calls for another way of being.
- Crazy life.

The conceptual complexity of the word is quite remarkable, and one can only marvel at the sophisticated philosophy on life that made the use and application of the word possible. Anthropology and the study of dead languages is not my concern here, however. Rather, I use the example to show that, although words do undoubtedly refer (in some general sense), it is not at all clear just *how* some words refer.

As this example makes clear, recourse to a vague notion of referring is unlikely to significantly advance an individual's understanding of in what the meaning of a word consists. Without such an understanding, however, the discipline of *lexical semantics* is dead, and with it any hope of understanding how humans manage to understand and use natural language in everyday discourse. The question, What is the meaning of a word? is addressed in Chapter 5, where the challenge of spelling out the state of the art in theories of word meaning is taken up from the perspective of the psychological theorist. A number of theoretical frameworks are explored, with the notion of a *procedural semantics* finding the most favor. In a procedural semantics, the meaning of a word is conceived in terms of how a given computational system processes it—a horribly vague description, open to all kinds of different interpretation. As I will show, the two intension and extension terms find as useful an application in procedural accounts of word meaning as they do in formal accounts of sentence meaning.

In contrast to Chapter 5, Chapter 6 considers how, given that a theorist has an understanding of what the meaning *qua* intension of a word might be, that meaning might be represented computationally. The book begins its departure from accepted doctrine by investigating the computational representation of meaning in the context of the class of complex information processing mechanisms termed *artificial neural networks*. Chapter 6 is the first of the three experimental chapters in the book, and it takes the meanings of spatial terms in simple sentences as the problem domain. The notions of procedure and intension are combined in this chapter to form the basis for a *Microprocedural Proposal* of meaning, in which the weights of a connectionist network are viewed as a novel representational genera (novel to the connectionist community at large, that is) encoding the meaning of spatial terms such as *to the left of*. The details and implications of this proposal for such issues as the context dependency of representations, and the notion of neural computation itself, are also explored.

Whereas Chapter 6 is mainly concerned with the notion of meaning-as-intension, the second of the experimental chapters, Chapter 7, is mainly concerned with the notion of meaning *qua* extension. Specifically, it explores the requirement for an elaboration of the Microprocedural Proposal that, in line with psychological theory, the *discourse representation* that a putative cognitive semantics constructs in the course of comprehension be a *virtual* representation, with an *analogical* structure. In line with this requirement, the *collapse2 strategy*, a method of constructing analogical representations using the standard tools of the neural network theorist, is explained and detailed.

HUMBLING THE CHAMPION

The original aim of the research that forms the basis of this book was to provide some fresh, new answers to the questions, What is the meaning of a word? and, How might such meanings be mentally represented? All of the neural network simulations that I will present I regard as computational illustrations of theoretical points; that is, they are not intended to be complete psychological models, nor should they be merely viewed as toys. Rather, they are intended to perspicuously represent aspects of stated theoretical positions and to provide support for the distinctly connectionist notions of meaning that I will present.

As I have said, Chapter 6 presents a novel conception of the weights of a connectionist network, whereby they are viewed as a legitimate representational resource in and of themselves, readily able to encode the meanings of spatial terms. Chapter 7, moreover, shows that the computational resources of a connectionist network are sufficient to systematically support (virtual) analogical representations. Given my earlier characterization of meaning as epistemologically intransigent, and on the basis of the findings in Chapters 6 and 7, we might describe the Champion as down on one armor clad knee, chastened, but still defiant (and, recalling the Black Knight once more, still threatening to savage the ankles of anyone who crosses him).

In order that the champion be properly defeated, the third experimental chapter of the book, Chapter 8, explores what I consider to be the central and most important problem in the study of meaning: how to explain the relation between a word and what the word refers to, external to the symbol system of which that word is a part. Or to pose the problem differently, how to express the relation between the representation of something and the thing represented? As I will show, this problem has a variety of guises, which find mention throughout the book, one of the most recent and intriguing of which is the *symbol* grounding problem (cf. Harnad, 1989). In Chapter 8, I present arguments for the need to refer to a more expansive variant of this problem, termed *representation* grounding. A solution to this latter problem of representation grounding is presented, called the Radical Connectionist response, in the form of two complementary processes—extensional and intensional grounding. The humbling of the epistemological champion lies, I argue, in recognizing the need for both of these processes.

Building on the insights of Chapter 8, in Chapter 9 I consider extensional and intensional grounding from a slightly different perspective. The usefulness of referring to a *syntactic engine* is motivated, the theoretical mechanism on which "classical" explanations of mental phenomena in cognitive science is based, and contrasted with the more novel *spatial engine*, on which sub-symbolic or "connectionist" theory building in cognitive science is (and will increasingly be, I believe) based. The differences between these two kinds of mechanisms are detailed, drawing on some of the most recent work in the connectionist literature.

Given the equivalence in computational power between the two engines, ascertaining which is more suitable for building theories of cognitive phenomena is a complex question with no obvious solution. However, by considering the kinds of computational and representational requirements of extensional grounding, I show that the syntactic engine displays a relative paucity of (representational) resources, in comparison to those available to the spatial engine. Chapter 9 points out that the inequality is important, because it is precisely these resources that are crucial in "grounding" or "hooking" atomic representations extensionally in the world. That is, although it is possible for both kinds of computational engine to be grounded extensionally, it is only the spatial engine that posseses the requisite resources in and of itself.

The astute reader will have noticed that in this overview of the book's contents, I have not mentioned Chapters 2 and 3. This is not an oversight, but rather a deliberate omission, as these two chapters share a different flavor from the rest of the book. Both are, in a sense, "rough guides." I have included Chapter 2, for instance, as a rough guide to connectionism, and it describes the various different neural network models that are used in the course of the investigations into meaning reported later. These neural networks are the tools of the trade of the connectionist cognitive scientist, comprising architectures and algorithms developed during the course of the last eight years by the likes of Rumelhart, Hinton, Pollack, and Elman (luminaries of the connectionist literature). In addition, Chapter 2 also describes two methods of network analysis. The first, *cluster analysis*, has been ported into connectionism from conventional statistics, although the second, *hyperplane analysis*, is a more recent, and distinctly connectionist, innovation, although its origins lie in the beginnings of neural network research in the 1950s.

In a similar vein, Chapter 3 is included as a rough guide to meaning. It is an attempt to present some of the key notions of the book in as straightforward a fashion as possible. Much of the simulation detail of subsequent experimental chapters is omitted in Chapter 3 for clarity, focusing more on theoretical interpretation rather than the specifics of particular simulation experiments. Three key ideas are introduced: The first, the Microprocedural Proposal, is a conjecture that weights of connectionist networks are a legitimate representational genera in and of themselves. The second is a particular, novel way of thinking about such *weight representation*, as procedural entities. The synthesis of these two notions engenders the third key idea of Chapter 3; namely, a formally grounded *connectionist procedural semantics*, and Chapter 3 presents two versions of it—the first, available to inspection on a symbol surface, and the second, dipping below such a (putative) surface.

Those readers, however, who may wish to skip these rough guides initially, and go straight to reading about meaning, will suffer no loss of continuity. Chapter 3 particularly, I believe, can serve as a useful source of reference for the book's substantive theoretical arguments, and to which the reader may refer back at any point.

2

Artificial Neural Networks

The tools of the trade of a connectionist cognitive scientist are the rich and diverse class of neurally inspired information processing mechanisms, generically termed (artificial) neural networks. On the one hand, neural networks can differ from each other *architecturally*, which is to say, in terms of properties of units and of groups of units, and in terms of the properties of the connections between units. Neural networks can also differ from each other *algorithmically*, which is to say, in terms of how a given architecture learns to perform a complex information processing task: Principally, such *learning algorithms* either require a teacher external to the network, in which case the algorithm is described as *supervised*, or no external teacher is required, in which case the algorithm is described as self-organizing or *unsupervised*.

Since the advent of neural networks, the number of learning algorithms and architectures in use by the neural theorist has blossomed dramatically to include such esoterica as cascade correlation (Fahlman & Lebiere, 1990), radial basis functions (Broomhead & Lowe, 1988), real time recurrent learning (Williams & Zipser, 1989), and many, many others. The purpose of this chapter is not to detail all of this proliferation, however, but more simply to extract from the huge literature the architectures and algorithms that were used in the research for this book on meaning. Specifically, I will look at the architectures of feed-forward networks, simple recurrent networks, and recursive auto-associative memory (RAAM) networks. These three different architectures all learn using the back propagation algorithm (Rumelhart, Hinton, & Williams, 1986), and I will look at this too.

ARCHITECTURES

All neural networks consist of large numbers of simple interconnected computing elements or *units* designed to mimic the first-order characteristics of the biological neuron. Each unit is assumed to receive real-valued activation (either excitatory, inhibitory, or both) along the links that it has to all of the other units to which it is connected. Typically, units do little more than sum this activation and change state as a function of this sum. More precisely, a set of inputs is applied, each representing the output of another unit. The set of inputs is multiplied by an intrinsic but modifiable property associated with each of the connections between any two given units, called a *weight.* A set of inputs x_1, $x_2, \ldots x_n$, collectively referred to as the vector **X**, is applied to the unit. Each input signal is multiplied by an associated weight w_1, $w_2, \ldots w_n$, collectively referred to as the vector **W**. A given unit sums the weighted activation that it receives from all of the units to which it is connected, to determine its own activation level, **net**.

$$\mathbf{net} = \mathbf{XW} \tag{1}$$

The activation of a given unit **net** is often further processed by an activation function f to yield the unit's output signal **out**. The function f is often a threshold function such that:

$$\mathbf{out} = 1 \text{ if } \mathbf{net} > \mathbf{T}$$
$$\mathbf{out} = 0 \text{ otherwise}$$

where **T** is simply a threshold value. Or less simply, **T** can be a complex function that more accurately simulates the non-linear transfer function of the biological neuron. The logistic or sigmoid function is regularly used to achieve such nonlinearity, often referred to as the *squash function*: This function serves, no matter the size of **net**, to compress a given unit's output signal **out** to some value between 0 and 1. The function has the following form.

$$\mathbf{out} = f(\mathbf{net}) = \frac{1}{1+e^{-\mathbf{net}}} \tag{2}$$

The pioneering work of investigating the computational properties associated with arranging such simple computing elements in *layers* was done by McCulloch and Pitts (1943), and later by Minsky and Papert (1969) with their work on perceptrons. A percep-

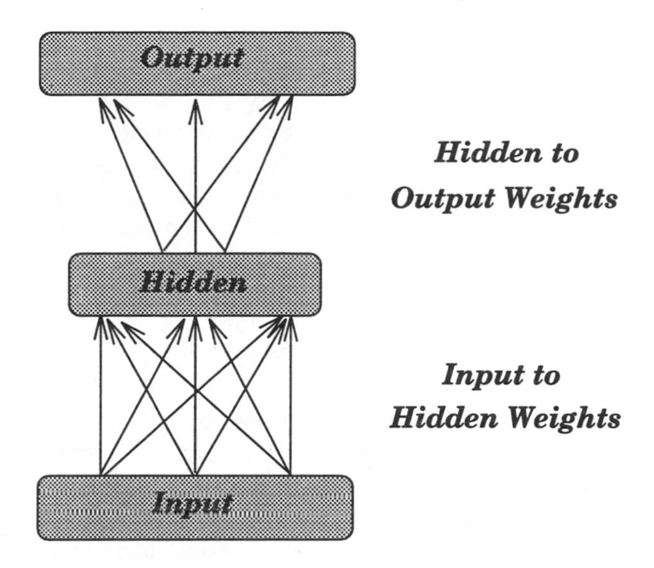

FIGURE 2.1. The architecture of a simple feed-forward network.

tron is a single layer ANN (artificial neural network), and it learns by means of the perceptron convergence theorem (cf. Minsky & Papert, 1969, which is also termed the *delta rule* by Rumelhart, Hinton, & Williams 1986). The startling things that such deceptively simple mechanisms such as single layer networks can do are, unfortunately, equally balanced by their disappointing *inability* to do other things, the most notable of which, historically speaking, is the exclusive or (XOR) problem.

Much more interesting than single layer perceptrons, and considerably more powerful, are multilayer perceptrons. These kind of ANNs have three (or more) layers of units—an input layer, an output layer, and a hidden layer. A multilayer perceptron learns by means of the *generalized delta rule*, independently devised by Rumelhart, Hinton, and Williams (1986), Parker (1985), and Werbos (1974), and which is also known as the back propagation algorithm. Multilayer perceptron learning by means of the back propagation algorithm is what I will refer to as a *feed-forward network*, shown in Figure 2.1.

ALGORITHMS

ANN learning algorithms are either supervised or unsupervised. Unsupervised learning algorithms include competitive learning (devised by Rumelhart & Zipser, 1986) and the adaptive resonance theory, ART (Grossberg, 1987). Other than their mention here, unsupervised learning algorithms will not be discussed any further.

Supervised learning algorithms are of two types: *deterministic* and *statistical*. Statistical (or probabilistic) learning algorithms make psuedorandom changes in the network weight values, retaining those that result in an improvement. The Boltzman learning algorithm (Hinton & Sejnowski, 1986) and the similar Cauchy learning algorithm (Szu & Harley, 1987) are paradigmatic examples of statistical algorithms. The way that a statistical algorithm works is similar to the mechanical process of annealing a metal, and consequently, the term *simulated annealing* has been coined to describe the manner of learning in, for example, the Boltzman machine.

The main problem with statistical learning algorithms is that they often take impractically long periods of time to learn. The Cauchy learning algorithm, which makes use of the Cauchy distribution rather than the Boltzman distribution, gives better results, but convergence times are still 100 times those of back propagation. A promising way of alleviating these long training times is proposed by Wasserman (1988), where the Cauchy and the back propagation algorithms are combined.

BACK PROPAGATION

The generalized delta rule, or the back propagation algorithm, is a systematic method for training multilayer perceptrons (viz., feed-forward networks). As the name suggests, it is a generalization of the simpler delta rule for single layer networks, elaborated to accommodate one or more layers of hidden units. Training a feed-forward network with back propagation requires the following five steps:

1. Select a training pair from the input set: Apply the input vector.
2. Calculate the output of the network.
3. Calculate the difference between the actual output and the target output.
4. Adjust the weights in a way that minimizes this difference.
5. Repeat Steps 1 through 4 until error is sufficiently low.

Steps 1 and 2 constitute the "forward pass" through the network; Steps 3 and 4 represent the "reverse pass" when the weights are adjusted.

Adjusting Weights—Hidden to Output

Consider the back propagation algorithm at work for a single weight from unit p in the hidden layer i to unit q in the output layer j. The output of unit q in layer j is subtracted from its target value to produce an error signal. The value of this error is then multiplied by the first derivative of the squashing function (**out**(1 - **out**)), calculated for that layer's unit q, to yield the measure δ.

$$f(\mathbf{out}) = \mathbf{out}(1-\mathbf{out}) \quad (3)$$

$$\delta = \mathbf{out}(1-\mathbf{out})(Target-\mathbf{out}) \quad (4)$$

This measure δ is then multiplied by the value of **out** from the unit p in the layer i, and by a learning rate coefficient η, which serves to limit the effective size of a weight change to yield the value by which the weights should be adjusted, that is, Δ**w**. An identical process is performed for each weight proceeding from a unit in the hidden layer to a unit in the output layer. The two equations (6) and (7) below are illustrations:

$$\Delta\mathbf{w}_{pq,j} = \eta\delta_{qj}\mathbf{out}_{p,i} \quad (5)$$

$$\mathbf{w}_{pq,j}(n+1) = \mathbf{w}_{pq,j}(n) + \Delta\mathbf{w}_{pq,j} \quad (6)$$

where:

- $\mathbf{w}_{pq,j}(n)$ = the value of a weight from unit p in the hidden layer to unit q in the output layer at Step n before adjustment.
- $\mathbf{w}_{pq,j}(n + 1)$ = value of weight at Step n + 1, after adjustment.
- $\delta_{q,j}$ = value of δ for unit q in the output layer j.
- $\mathbf{out}_{p,j}$ = the value of **out** for unit p in the hidden layer i.

Adjusting Weights—Input to Hidden

Hidden units have no target vector, so the training procedure described above for the hidden to output weights cannot be used. Rather, back propagation trains the hidden layers by propagating the output error back through the network layer by layer, adjusting weights at each layer.

The problem is to generate a value for δ for the hidden units without benefit of a target vector. First, δ is calculated for each unit in the output layer, as in Equation 5. It is used to adjust the weights feeding into the output layer, then it is propagated back through the same weights to generate a value for δ for each unit in the hidden layer. These values of δ are used, in turn, to adjust the weights of the hidden layer.

Consider the case of a single unit in the hidden layer *i*. In the forward pass, the unit propagates its **out** value to units in the output layer *j*. During training, the reverse pass, the weights on the links between units operate in reverse, passing the value of δ from the output layer *j* back down. Each intervening weight w is multiplied by this δ value. The value of δ for the hidden unit layer *i* is produced by summing all such (δw) products from each output unit and then multiplying by the derivative of the squashing function.

$$\delta_{p,i} = \mathbf{out}_{p,i}(1 - \mathbf{out}_{p,i})(\Sigma\delta_{q,j}\mathbf{w}_{pq,j}) \qquad (7)$$

With this value of δ, the weights feeding the hidden layer can be adjusted using Equations 6 and 7. Rumelhart, Hinton, and Williams (1986) introduced a new term, called *momentum*, into the back propagation algorithm, which improves the training time and increases the stability of the process. The method involves adding a term, α, to the weight adjustment that is proportional to the amount of the previous weight change. The weight adjustment equations are modified to the following:

$$\Delta\mathbf{w}_{pq,j}(n+1) = \eta\delta_{q,j}\mathbf{out}_{p,i} + \alpha[\Delta\mathbf{w}_{pq,j}(n)] \qquad (8)$$

$$\mathbf{w}_{pq,j}(n+1) = \mathbf{w}_{pq,j}(n) + \Delta\mathbf{w}_{pq,j}(n+1) \qquad (9)$$

RECURRENT NETWORKS

The class of recurrent network architectures is a modification to the two layer feed-forward network standardly used with back propagation. One of the earliest recurrent network architectures was proposed by Jordan (1986a), shown in Figure 2.2.

In this architecture, there are links from the network's output units that propagate an output vector back down to units, the *state* units, at the same level as the input units. Because the connections on these *recurrent* links are one for one, and all have a uniform associated weight of 1.0, the state units serve to copy any given pattern of activation over the output units that are propagated back down. On the next processing cycle of the network, this copied output vector is propagated upward to the hidden units. The effect of the recurrent links and the state units is to allow the network's hidden units to "see" the previous output, so that subsequent behavior can be shaped by previous responses.

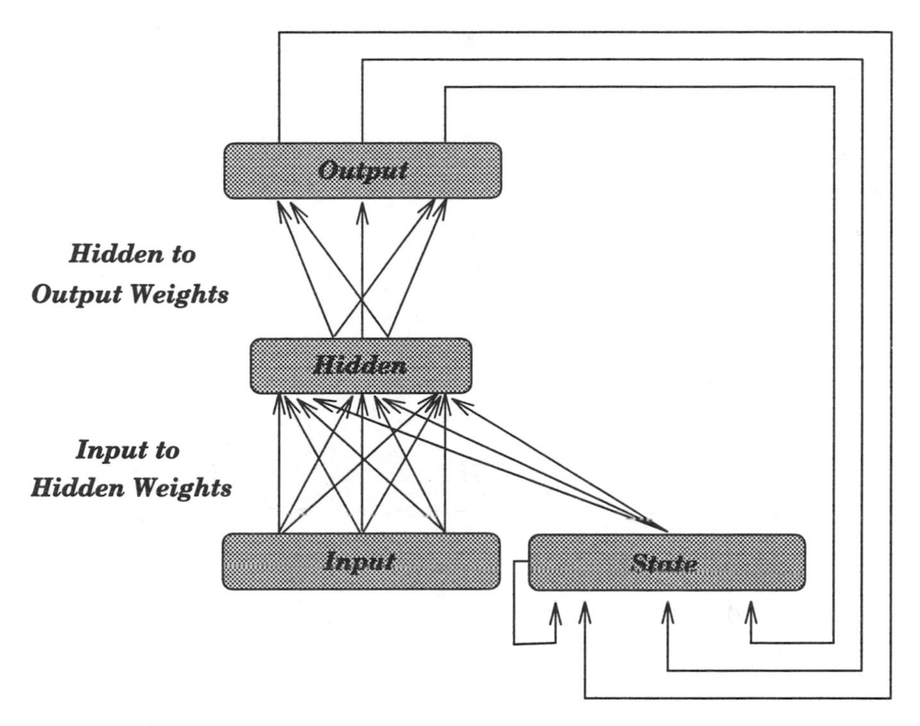

FIGURE 2.2. The architecture of the Jordan network in which recurrent connections feed back from the output units to the state units. Connections between output and state are one for one and have a weight of 1.0. The state units also feed back onto themselves with a (variable) decay. Not all connections are shown.

Elman (1990) proposed a different kind of recurrent network in which, instead of recurrent connections originating with the output units, the recurrent connections originate with the hidden units, and feed back on to a group of *context* units. The context units in this kind of *simple recurrent network* (SRN) can be thought of as a second group of hidden units in the sense that they do not receive any input from outside the network. Recurrent connections from the hidden units down to the context units are one for one, set at a value of 1.0, and are not subject to adjustment: This is so the context units will contain a faithful copy of the hidden unit vector at a given time. The connections from the context units back up to the hidden units are, however, fully distributed, with each context unit being connected to each hidden unit. The SRN architecture is illustrated in Figure 2.3.

The architecture works as follows. Given some sequential input to be processed, and some clock that regulates presentation of the

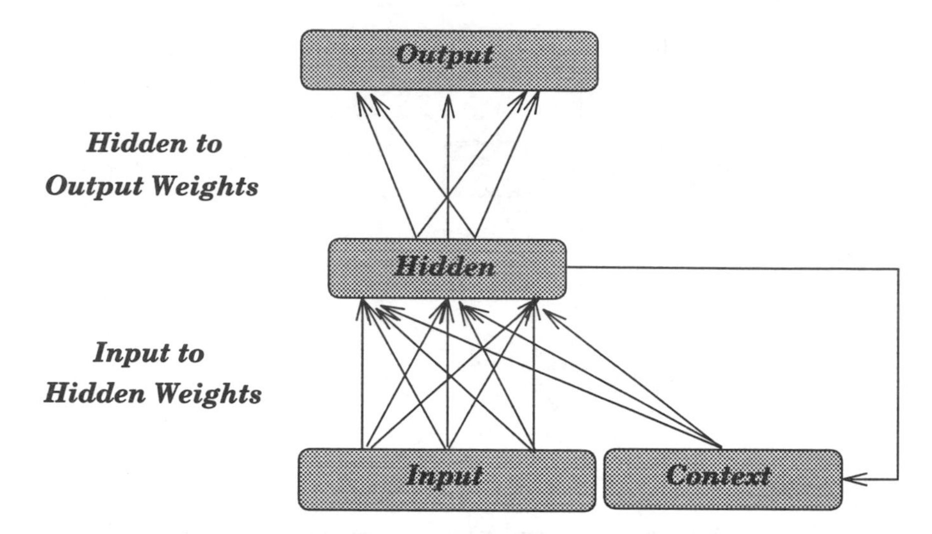

FIGURE 2.3. The architecture of the SRN in which recurrent connections feed back from the hidden units to a bank of context units.

input to the SRN, at time *t*, the input units are activated with a given training vector, and the context units are set to a *don't know* state (for example, 0, 0.5, or small random values over all units). The input and context units both propagate activation to the hidden units, which in turn propagate activation to the output units, and additionally, the hidden units also propagate activation back down to the context units: This constitutes the forward pass through the weights. Whether there is a learning phase in this time cycle *t* is dependent upon the nature of the task. If learning does take place, the standard back propagation algorithm that we have already seen is used to adjust the weights. At the next time step, $t + 1$, the above sequence of events are repeated for the next element in the sequential input, but this time the context units propagate values that are exactly those of the previous hidden units at time *t*. The context units thus provide a network with a kind of memory.

In the SRN architecture, the hidden units have the task of mapping both an external input and also a previous internal state to some desired output. Thus, the hidden unit representations that develop over the time of a sequential input are sensitive to temporal context. Cleeremans (1993) provided an excellent demonstration of the power of SRNs, using them to model the induction of finite state grammars in psychological studies of implicit learning.

RECURSIVE AUTO-ASSOCIATIVE MEMORY

The Recursive Auto-Associative Memory (RAAM) architecture, devised by Pollack (1990) is, like the architecture of the Jordan network and the SRN, essentially an elaboration of the two layer feed-forward architecture using back propagation. The RAAM architecture was motivated by the need for connectionist networks to be able to represent variable-sized, recursive data structures, such as trees or lists. Consider the kind of simple tree structure shown in Figure 2.4.

Pollack (1990) urged the theorist to conceive of two hypothetical mechanisms that could translate, in both directions, between symbolic data structures such as the binary tree in Figure 2.4 and connectionist numerical vector representations. The first of these hypothetical mechanisms, the *compressor*, should be able to encode sets of fixed width patterns into single patterns of the same size. For the binary tree in Figure 2.4, the compressor would: (*a*) take **A** and **B** and compress them to pattern R_1, (*b*) take **C** and **D** and compress them to pattern R_2, and (*c*) take R_1 and R_2 and compress them to pattern R_3.

The second hypothetical mechanism, the *reconstructor*, should be able to decode such single compressed patterns as R_3 to its constituent R_1 and R_2 patterns, and also decode R_1 and R_2 to their constituent patterns. The compressor and the reconstructor mechanisms are shown in Figure 2.5. As can be seen, the two mechanisms can be realized very easily: For a binary tree, with n-bit patterns, the compressor is a single layer network with $2n$ units in the input and n units in the output, while the reconstructor could be a single layer network with n input units and $2n$ output units.

Putting the compressor and reconstructor mechanisms together results in a $2n$-n-$2n$ network, and of course, back propagation of error can be used to incrementally adjust the weights. In order for

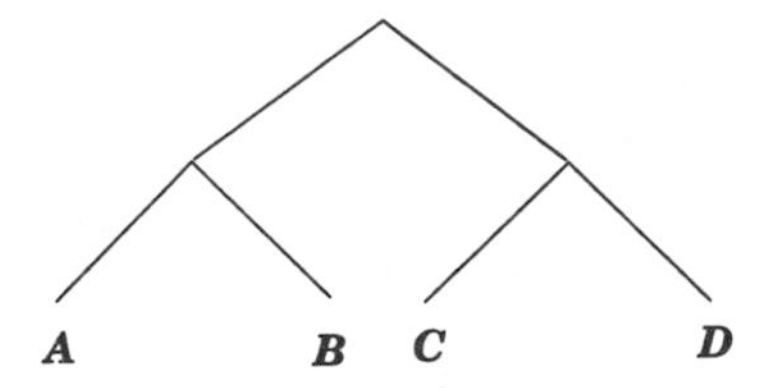

FIGURE 2.4. A binary tree structure with a valency of two.

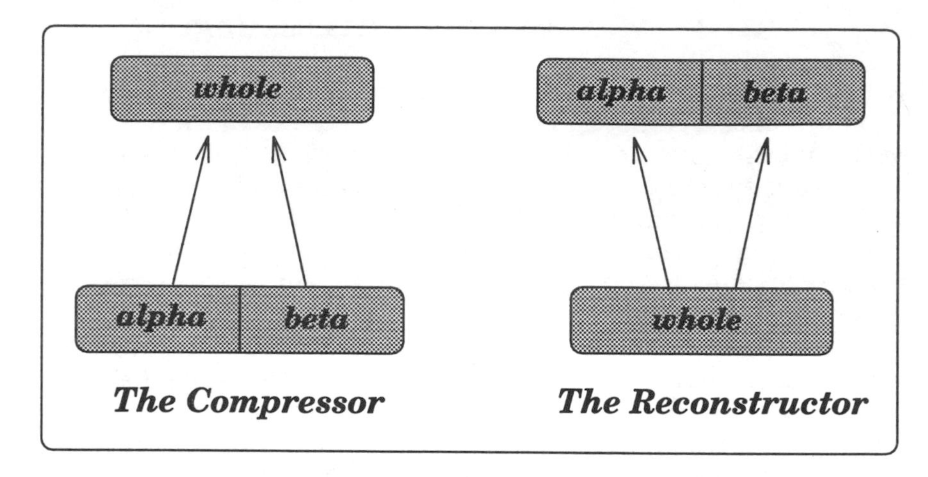

FIGURE 2.5. The compressor and reconstructor mechanisms with $2n$ units in the partitioned blocks and n units in the unpartitioned blocks.

this kind of architecture to encode tree structures, such as the one shown in Figure 2.4, its simple auto-associative nature must be augmented by the capacity for *recursively* encoding and auto-associating. In order to accomplish this, Pollack took two terminal elements of the tree, corresponding to **A** and **B**, each encoded as patterns of activation over n units, and trains the network to auto-associate them. In the course of this auto-association, the hidden units will develop a compressed representation over n units of the combination of **A** and **B**. This compressed representation is then extracted from the hidden units and used to train the network, auto-associatively, on higher order structures. This process is illustrated in Table 2.1.

The hidden unit representation R_3 at time t, constitutes a distributed encoding of the entire tree structure ((**A B**)(**C D**)).

Table 2.1.
The Recursive Construction of a Distributed Representation of the Bracketed Tree Structure ((**A B**)(**C D**)).

Input Pattern	Hidden Pattern	Output Pattern
(**A** , **B**)	$R_1(t)$	(**A***(t) , **B***(t))
(**C** , **D**)	$R_2(t)$	(**C***(t) , **D***(t))
(R_1,R_2)	$R_3(t)$	(R_1*(t) , R_2*(t))

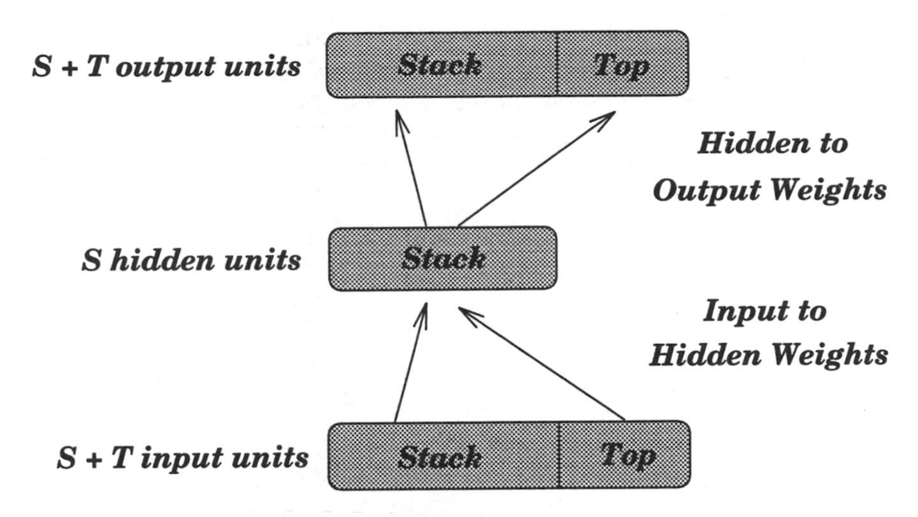

FIGURE 2.6. The architecture of the sequential RAAM network with **S** + **T** input and output units and **S** hidden units.

Sequential RAAM

The RAAM architecture permits of a novel variation, called the sequential RAAM, that is capable of encoding *sequences* of elements. The architecture of the sequential RAAM is shown in Figure 2.6, and, as can be seen, it is analogous to a simple stack.

A given sequence such as (**A**, **B**, **C**) can be represented as a left-branching binary tree, such as ((**nil A**) **B**) **C**). The first element of the sequence, **A** is presented to the sequential RAAM and coded over **S** units along with a value for **nil** coded over **T** units (usually, a value of 0.5 over all units). The hidden units develop a compressed representation that is recirculated and returned to the **S** units. The second element of the sequence, **B** is then presented to the network and coded over **T** units, and the hidden units develop a compressed representation once more, and so on. This process is illustrated in Table 2.2. The final compressed representation $R_{A,B,C}$ will be a representation for the entire sequence **A**, **B**, **C**.

NETWORK ANALYSIS

Connectionist theorists have at their disposal a variety of tools with which to work. The feed-forward architecture and the back

Table 2.2.
The Recursive Construction of a Distributed Representation of the Sequence **A**, **B**, **C**.

Input Pattern	Hidden Pattern	Output Pattern
(**nil** , **A**)	$R_A(t)$	(**nil***(t) , **A***(t))
($R_A(t)$, **B**)	$R_{A,B}(t)$	(R_A*(t) , **B***(t))
($R_{A,B}(t$, **C**)	$R_{A,B,C}(t)$	($R_{A,B}$*(t) , **C***(t))

propagation algorithm, for example, allow them to devise network simulations that can perform a large number of complex, information processing tasks. In many cases, however, merely devising a simulation to perform a given task is not enough. Additionally, the theorist would also like to be able to know *how* that simulation has performed its task. In point of fact, as Clark (1988a) has pointed out, without some means by which networks can be pulled apart, and the intricacies of their internal workings displayed, connectionist models run the risk of being *nonexplanatory* in a very real sense: After all, it does no good at all to replace a little understood mental capacity with an equally little understood network simulation emulating that capacity. What is required, in short, is some means of analyzing a network simulation once it has learned. A number of techniques for performing such analysis can be found in the literature, including *hierarchical cluster analysis*, *hyperplane analysis* (cf. Sharkey & Sharkey, 1993), *principal component analysis*, *contribution analysis* (Sanger, 1989), *weight matrix decomposition* (McMillan & Smolensky, 1988), *network rule extraction* (Mundy & Sharkey, 1992) and *decompositional analysis* (Sharkey, 1992). This book uses primarily the first of these techniques, cluster analysis, although the second, hyperplane analysis, is also employed.

HIERARCHICAL CLUSTER ANALYSIS

Any analysis of multivariate data attempts to solve the following problem: Given a number of objects, each of which is described by a set of numerical measures, devise a classification scheme for grouping the objects within classes, such that objects within classes are similar in some respect, and unlike those from other classes. In addition to the term *cluster analysis*, techniques for such multivariate analysis are variously referred to as *Q-analysis*, *typology*, *clumping*, *numerical taxonomy*, and *unsupervised pattern recognition*. This host of labels reflects the importance of multivariate analysis in such diverse fields as psychology, zoology, botany, sociology, and biology.

There are a number of different techniques of cluster analysis, including *hierarchical*, *optimization*, *density*, and *clumping* techniques. I am concerned with the first of these, the hierarchical technique, which itself has two different forms or methods: (*a*) an *agglomerative*, and (*b*) a *divisive* method. I will be concerned with the first of these methods.

Agglomerative Methods

Techniques of cluster analysis seek to separate a set of data into similarity groupings or clusters. The raw data to be analyzed consists of an (xNp) matrix of measurements, **X**, where.

$$\mathbf{X} = \begin{Bmatrix} x_{11} & x_{12} & \cdots & x_{1p} \\ x_{21} & x_{22} & \cdots & x_{2p} \\ \vdots & & & \\ x_{N1} & x_{N2} & \cdots & x_{Np} \end{Bmatrix}$$

There are generally two stages to any cluster analysis. The first involves converting the raw data matrix **X** into a matrix of similarity or *distance* measures, **D**, and the second involves the grouping of these distance measures: Variations on the theme of agglomerative hierarchical analysis arise primarily because of the different ways of defining the distance measure. The most commonly used metric of distance is the *Euclidean* metric, where the distance between any two points i and j, denoted d_{ij}, is defined as:

$$d_{ij} = \left\{ \sum_{k=1}^{p} (X_{ik} - X_{jk})^2 \right\}^{\frac{1}{2}}$$

where X_{ik} is the value of the kth variable for the ith entry. The general method of cluster analysis can be illustrated using the nearest neighbor or the *single link* method. Let us suppose that we are interested in the similarities between vectors of activation taken from the hidden units of a given network simulation. A matrix of distance measures, D_1 derived from a matrix **B** of raw data, for five such vectors is illustrated below (the numbers are multiplied by 10, for clarity):

$$\mathbf{D}_1 = \left\{ \begin{array}{lcccccc} & & \mathbf{v1} & \mathbf{v2} & \mathbf{v3} & \mathbf{v4} & \mathbf{v5} \\ & & \vdots & \vdots & \vdots & \vdots & \vdots \\ \mathbf{v1} & \cdots & 0.0 & 2.0 & 6.0 & 10.0 & 9.0 \\ \mathbf{v2} & \cdots & 2.0 & 0.0 & 5.0 & 9.0 & 8.0 \\ \mathbf{v3} & \cdots & 6.0 & 5.0 & 0.0 & 4.0 & 5.0 \\ \mathbf{v4} & \cdots & 10.0 & 9.0 & 4.0 & 0.0 & 3.0 \\ \mathbf{v5} & \cdots & 9.0 & 8.0 & 5.0 & 3.0 & 0.0 \end{array} \right\}$$

At State 1 of the procedure, Vectors **v1** and **v2** are combined to form a group or cluster, because the distance between these two vectors, denoted $d_{v1,v2}$, is the smallest entry in the matrix D_1. The distances between this group and the other three vectors are obtained from D_1 as follows:

$$d_{(v1,v2)v3} = min\left\{d_{(v1,v3)}, d_{(v2,v3)}\right\} = d_{(v2,v3)} = 5.0$$

$$d_{(v1,v2)v4} = min\left\{d_{(v1,v4)}, d_{(v2,v4)}\right\} = d_{(v2,v4)} = 9.0$$

$$d_{(v1,v2)v5} = min\left\{d_{(v1,v5)}, d_{(v2,v5)}\right\} = d_{(v2,v5)} = 8.0$$

A new matrix, D_2, can now be formed giving intervector distances and group-vector distances.

$$\mathbf{D}_2 = \left\{ \begin{array}{lccccc} & & \mathbf{v1,v2} & \mathbf{v3} & \mathbf{v4} & \mathbf{v5} \\ & & \vdots & \vdots & \vdots & \vdots \\ \mathbf{v1,v2} & \cdots & 0.0 & 5.0 & 9.0 & 8.0 \\ \mathbf{v3} & \cdots & 5.0 & 0.0 & 4.0 & 5.0 \\ \mathbf{v4} & \cdots & 9.0 & 4.0 & 0.0 & 3.0 \\ \mathbf{v5} & \cdots & 8.0 & 5.0 & 3.0 & 0.0 \end{array} \right\}$$

The smallest entry in D_2 is $d_{v4,v5}$ with a value of 3.0, and so Vectors 4 and 5 are fused to become a single group: The new distance measures, and the new matrix of distance measures, D_3 are shown below:

$$d_{(v1,v2)v3} = 5.0$$

$$d_{(v1,v2)(v4,v5)} = min\left\{d_{(v1,v4)}, d_{(v1,v5)}, d_{(v2,v4)}, d_{(v2,v5)}\right\} = d_{(v2,v5)} = 8.0$$

$$d_{(v4,v5)v3} = min\left\{d_{(v3,v4)}, d_{(v4,v5)}\right\} = d_{(v3,v4)} = 8.0$$

$$\mathbf{D}_3 = \left\{\begin{array}{lcccc} & & \mathbf{v1,v2} & \mathbf{v3} & \mathbf{v4,v5} \\ & & \vdots & \vdots & \vdots \\ \mathbf{v1,v2} & \cdots & 0.0 & 5.0 & 8.0 \\ \mathbf{v3} & \cdots & 5.0 & 0.0 & 4.0 \\ \mathbf{v4,v5} & \cdots & 8.0 & 4.0 & 0.0 \end{array}\right\}$$

The smallest entry in D_3 is now $d_{v4,v5}$, and so Vector **v3** is added to the group containing Vectors **v4** and **v5**. Finally, fusion of the two groups, **v1,v2** and **v3,v4,v5** takes place to form a single group containing all five vectors. The final result of such an analysis can be displayed in a two-dimensional diagram illustrating the fusions that have been made at each successive stage of the analysis. Such a diagram is called a *dendogram*, and the dendogram for our five vector example is shown below in Figure 2.7.

Experience with a number of different agglomerative methods of hierarchical cluster analysis revealed that, with respect to analyzing networks, some methods of analysis produced better results than other methods. Notably, the *Ward* method was found to be particularly efficacious. Ward (1963) proposed that at any stage of an analysis, the loss of information that results from the grouping of objects into clusters be measured by the total sum of squared deviations of every point from the mean of the cluster to which it belongs. At each step in the analysis, union of every possible pair of clusters is considered, and the two clusters whose fusion re-

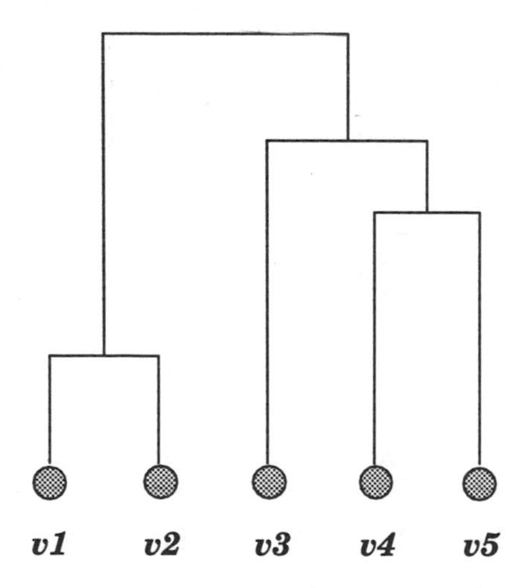

FIGURE 2.7. A Sample dendogram for the five vector example of the single link agglomerative method of hierarchical cluster analysis.

sults in the smallest increase in the error sum of squares are combined. The error sum of squares, ESS, is given by:

$$(ESS) = \sum_{i=1}^{n} X_i^2 - \frac{1}{n}(\sum x_i)^2$$

where x_i is the value associated with the *i*th object. At Stage 1, each individual object is regarded as a single member group, and so the error sum of squares is zero. The first two objects to be clustered together are those objects that result in the minimum increase in the error sum of squares.

HYPERPLANE ANALYSIS

Hyperplane analysis (cf. Sharkey & Sharkey, 1993) is a sophisticated and powerful tool for analyzing potentially all kinds of neural networks. To illustrate how the analysis works, however, I will consider only one kind of network—the feed-forward network.

In a hyperplane analysis for a feed-forward network, the input vectors to that network are viewed as *points* in an *n*-dimensional input space (where *n* refers to the number of units in the input layer). The weights from the input units to the next, hidden layer of units, the input-to-hidden or IH weights, are viewed as decision boundaries, or *hyperplanes*, which transform this *n*-dimensional space.

The hidden unit vectors, similarly, are also viewed as points occupying the *m*-dimensional hidden unit space (where *m* refers to the number of units in the hidden layer). The second layer of weights, the hidden-to-output or HO weights, are like the IH weights, also viewed as hyperplanes, which then partition this *m*-dimensional space into *decision regions*. This has the effect that every point on one side of a hyperplane will produce one kind of functionality, and all points on the other side will produce another. Typically, the difference in functionality sought is between a binary output of 1 or 0 (yes or no).

I can illustrate how a hyperplane analysis works by considering a work with two input units, two hidden units and 1 output unit (i.e. the input space and the hidden unit space are both two-dimensional). Figure 2.8 shows the values of all nine weights for such as a 2-2-1 network trained on XOR (the *bias* unit shown here is standard in all feed-forward networks: It is always "on," propa-

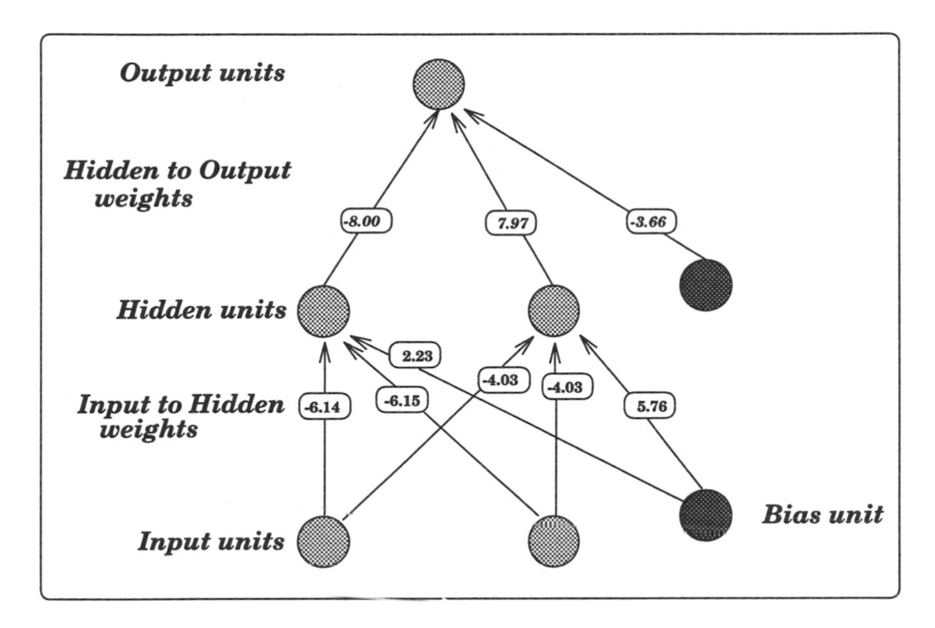

FIGURE 2.8. A learned 2-2-1 network trained on XOR.

gating a value of 1 through the network's weights). The equation used to calculate the hyperplanes associated with the receptive weights each hidden unit, and associated with the receptive weights for the output unit is shown below:

$$w_{bias} + \mathbf{x}w_1 + \mathbf{y}w_2 = 0 \tag{10}$$

where w_{bias}, w_1, and w_2 are the three receptive weights in each case, and **x** and **y** axis of the space (either input or hidden). Solving this equation for **x** and **y** for each unit, we find that, for the first hidden unit, the coordinates of this unit's hyperplane are (**0.363**, **0.362**); for the second hidden unit, its associated hyperplane has coordinates (**1.42**, **1.42**); and for the output unit, the coordinates of the hyperplane are (**–0.45**, **0.45**). Plotting the hidden unit hyperplanes in input space results in the *decision space diagram* shown in Figure 2.9.

In Figure 2.9 we can see that the two hyperplanes associated with each hidden unit have divided the input space into decision regions, with 00 and 11 being separated from 10 and 01 (because this is the difference that computation of XOR requires). This means that 00 and 11 will produce different "points" in the hidden unit space from either 01 or 10.

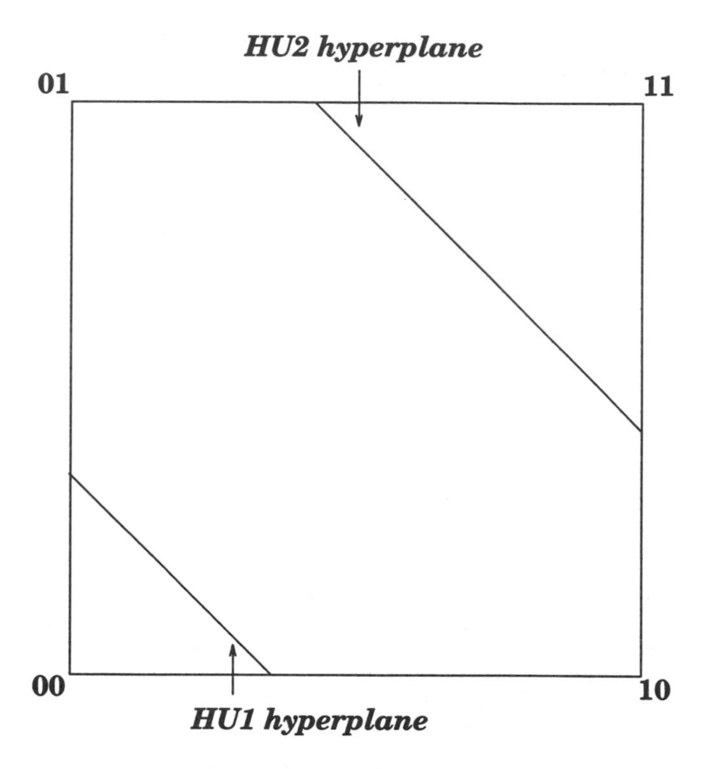

FIGURE 2.9. The decision space diagram for a 2-2-1 network trained on XOR showing two hidden unit hyperplanes in the input unit space.

Adopting a similar procedure to that for hidden unit hyperplanes in input space, we can plot the output hyperplane in the hidden unit space also; this results in the decision space diagram shown in Figure 2.10. This figure shows that the hyperplane associated with the output unit has partitioned the hidden unit space into two decision regions. The inputs 00 and 11 (or rather, the "points" that the IH weights transform them into by virtue of the IH weights) are in one region (even though they are far apart, they are still in the same decision region, on the "negative" side of the hyperplane). Whilst the other two inputs, 10 and 01 (once more, responding to the hidden unit values for these inputs) are in a different decision region. Thus, 00 and 11 will produce a different output response, namely 0, from 10 and 01, which will produce 1 on the output.

Using decision space diagrams, the theorist is able to see clearly and simply how a particular network is computing the function that it has been trained to compute (or the function that it has *actually* learned and *is* computing: Sometimes they don't coincide).

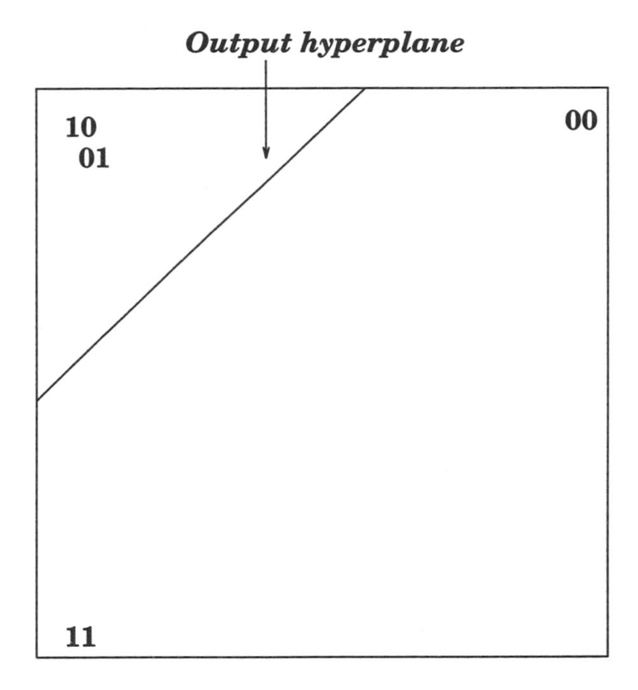

FIGURE 2.10. The decision space diagram for a 2-2-1 network trained on XOR showing an output unit hyperplane in the hidden unit space.

The main limitation of hyperplane analysis is, of course, that as soon as either the input or the hidden unit space exceeds three-dimensions (that is, if a network has more than three units in either input or hidden layers) visualization becomes much more difficult. In this event, it is necessary to take three-dimensional "slices" of the higher dimensional space using, for example, something similar to principal component analysis. This tactic, in itself, presents other problems, however, most notably of which is the potential loss of information that occurs when moving from high to lower dimensionalities.

The notion of a hyperplane analysis, and network analysis and network visualization more generally, is consequently a major, future research question for connectionists in its own right.

CONCLUSIONS

This chapter was intended as a rough guide to connectionism, as an introduction to the architectures, algorithms, and methods of network analysis of which this book avails itself. Principally, the book makes use of the back propagation algorithm, and we have seen how this error-correction procedure works. In addition, I have also documented two variants of the feed-forward architecture with back propagation: the recurrent network architecture and the RAAM architecture. The first of these architectures is initially used in Chapter 6, the recurrent network architecture is employed in Chapter 7, and the RAAM architecture is used in Chapter 8. The technique of hierarchical cluster analysis is used throughout the book in order to produce dendograms of both weights and unit activations, and hyperplane analysis is employed in Chapters 3 and 6, and is also encountered in Chapter 9.

3

Weighting for Meaning

Of all the problems that one faces when working out a comprehensive theory of mental phenomena—the natural purview of cognitive science—the study of the nature of the relationship between representations and represented is absolutely central. This, indeed, is the problem-setting context for the whole of this book, and it has quite a pedigree. Wittgenstein (1922), for example, came up against it in the *Tractatus*, at the terminus of his philosophical analysis of the indefinable names making up the constituents of atomic propositions. Taking up the Wittgensteinian cause, many philosophers and theorists at the time maintained that only via what they called *ostensive definition* (namely, some method of extralinguistic activity) could a mediation between a calculus of symbols and the reality described by its formulae be effected, and thus a solution be arrived at to the problem of representations and represented.

In modern day parlance, the nature of the relationship between representations and the represented is the stuff of *semantics*, the discipline typically concerned with constructing theories of how language and the world are related. This chapter begins by discussing one particular approach to semantics, an ad hoc synthesis of the formal and the computational, known as *procedural semantics* (PS), with its origins in an equivalence drawn between "meaning" (that is, truth conditions or intensions) and "procedure." A number of the most important criticisms leveled at PS are discussed in order to motivate the unique theoretical arguments of the rest of the chapter; namely, the *Microprocedural Proposal*, in which truth conditions (or meanings) are synonymous with a connectionist notion of procedure. Moving away from the prevalent, but needlessly restrictive, practice of viewing only units (or more precisely, their activations) as legitimate sources of representation in connectionist networks, the main theoretical force of the Microprocedural Proposal derives from its postulate of *weight representations*. Instead of a*symbolic* PS, this innovation leads to the emergence of a nonsymbolic *connectionist* PS. The salient features of this novel connectionist PS, which operates "on a symbol surface," are described and empirically fleshed out via a number of neural network simulations. Having explored the feasibility of the Microprocedural Proposal, I then extend the sophistication of the connectionist PS that it engenders by describing a PS operating *below* a symbol surface. Both inputs and outputs to this latter connectionist PS are *nonconcatenatively* compositional unit representations, and the PS is able to perform systematic, structure-sensitive operations upon these complex representations.

PROCEDURAL GRANDMOTHERS AND INTENSIONS

The arrival of Frege's *Begriffschrift*, the Platonistically conceived ancestor of modern predicate calculus, in the study of logic and philosophy in the late 19th century (1892) saw the beginning of a number of important conceptual developments in the theoretical entities known as *truth conditions*. In the application of the term to the analysis of natural language, the first such development was the definition of truth as a semantic concept by Tarski (1944), the second, Carnap's (1952) invention of intensional calculi based upon the notion of a possible world, and the subsequent equating of intensions with truth conditions. In addition, there was another conceptual innovation in truth conditions, with its origins in the famous 1936 paper of Turing, "On Computable Numbers, With an Application to the Entscheidungsproblem," which introduced the concepts of effective procedures, the Turing machine, and the Universal Turing Machine. The innovation that occurred had at its heart the fact that the term *truth condition* came to be regarded more and more as synonymous with (some appropriately abstract notion of) the term "procedure." This innovation, in turn, led to the emergence of a novel theoretical framework, couched in a computational vocabulary (rather than a formal or a mathematical one) for representing the meanings conveyed by language strings—*procedural* semantics.

Despite these clear formal origins, the precise nature of this "new theoretical framework" is not so clear. There is little consensus of opinion in the literature as to what a PS is either for, or how it might do whatever it is for. Is it an all-encompassing, explain-everything *semantic* theory, which is to say, a theory of how all of natural language relates directly to the world? Or is it, rather more modestly, a label for a variety of assumptions, tools, and techniques used in a consensual, computational investigation of meaning? Or, even more lack-luster, is it merely a collective working analogy drawn between natural language and high-level programming languages? The awkward truth is that PS has been, at one time or another, and in one guise or another, variously described as *all* of these things and a number of others besides (see, for example, the irritable and irritated exchange between Jerry Fodor and Philip Johnson-Laird in *Cognition, 6* (1978). Also see Hadley, 1989; Johnson-Laird, 1977a, 1977b; 1978; 1983; Woods, 1986).

On the one hand, some authors like PS—they are excited by the intellectual possibilities of equating "meaning" with "procedure," and are interested in exploring the implications of so doing—others, on the other hand, abhor it with a fierce passion. The following is from Jerry Fodor, taken from a wonderfully scathing paper entitled (equally wonderfully), "Tom Swift and his Procedural Grandmother": "PS suffers from: verificationism, operationalism, Empiricism, reductionism, recidivism, atomism, compound fractures of the use/mention distinction, hybris and a serious misunderstanding of how computers work" (1978, 234).

This excellent feast of "isms" is the result of a number of criticisms that Fodor and other anti-PS theorists, in general have of PS. A representative sample of these criticisms are shown below:

- PS is parasitic on a prior model theoretic semantics.
- PS interprets natural language by translating it into a machine language, possibly enriched by vocabulary items for names of input and output transducer states.
- PS is verificationist.
- PS confuses semantic theories with theories of sentence comprehension.

The first criticism is that PS is, in some sense, "parasitic" on a prior formal semantics. In fact, Fodor (1978) contrasted PS with what he called a "Classical Semantics" (CS), a labeling convention serving to isolate and bring under one rubric all the features common to a range of intensional, model theoretic semantic systems. In precisely the same way, I use the term "formal" semantics. This point about parasitism turns greatly, of course, as Hadley (1989) noted, on what one considers *parasitic* to mean. In a worst case scenario, the connotation of parasitic would imply that PS attempts to "supplant" a formal semantics, or that it touts itself as a better "alternative." In this sense of *parasitic*, as is commonly agreed, PS is worse than useless, because all that it does is add another layer of semantical complexity on top of the prior (and thus sufficient for the purposes of cognitive science) formal semantics.

Another view, however, promulgated by Hadley (1989), does not accept this worst case scenario. Rather, Hadley (1989) argued that some form of PS *ontologically presupposes* some form of formal semantics, a conclusion that is at radical odds with *any* sense of the term "parasitic." To add my own view, there would seem to be nothing parasitic about devising procedures for computing functions *similar* to those postulated in a formal model theoretic semantics. A PS need not be concerned with mechanizing precisely these mappings, with computing the *identical* functions (although some species of PS do indeed take this as their primary aim; see Hobbs & Rosenschein, 1978). Rather, the advocate of PS looks to formal semantics and sees *one* way of providing symbolic strings with an interpretation, one way to establish, for a given symbol string, what those symbol strings might refer to. In this respect, PS and formal semantics look quite similar. The point is that the relationship between PS and formal semantics need not be considered in hostile, parasitic terms: It is equally as valid to consider the relationship as benign and complementary, as one does not require the existence of the other to be useful in a given context.

The remaining three criticisms of the anti-PS position require more thought to answer. This is because they are surface features of a much deeper underlying objection to PS, which center around precisely what a theory of semantics is taken to be. Consider the last complaint on our list above, that PS confuses semantic theories with theories of sentence comprehension. About this point, advocates of the anti-PS position would seem to be adamant: *Either* PS is a semantic theory *or* it is a theory of sentence comprehension (and one shouldn't mistake one's semantic theory for one's grandmother, even if she is a procedural grandmother, as the gist of Fodor's quirkily titled paper has it). If PS claims to be a semantic theory, then it should be able to reveal everything about all of the relations between every natural language and the world. According to the anti-PS position, it cannot do this of course, because all that PS *really* is, is a mechanization of arbitrary mappings over mathematical constructs. Thus, PS must be a theory of sentence comprehension, and not a theory of semantics that is, meanings, at all.

The deeper underlying objection of the anti-PS position, to which I alluded earlier, can now be spelled out. The objection takes the form that a semantic theory must relate language directly to the world. This is solely what a semantic theory is, according to the anti-PS position, fixed and immutable. Any theoretical framework that claims to be properly semantic (in any respect) must, therefore, adhere to this convention, and any that does not, is not concerned with semantics per se. This is an admirably strict criterion for "semantichood." However, there is a bit of an odd situation here, given that the formal semantics used by the anti-PS advocate to compare PS with quite clearly does not relate language directly to the world, but rather relates (elements of a formal) language to a set theoretic model structure. Is such a classical semantics not really a *semantics* after all, then? It certainly does not adhere to the convention of a semantic theory outlined above, so what is it?

The anti-PS position, and the a priori assumption concerning what a theory of semantics is, namely, a theory of language and the world, in actual fact forces the inevitable conclusion that PS is not a semantic theory. At this point, I should say that I find nothing particularly contentious here, and I would actually agree with the spirit of the objection: PS is *not*, should not be considered to be, and must not pretend to be a semantic theory. This does not, however, by means of some kind of inescapable default, mean that PS is merely a theory of sentence comprehension, or indeed, a framework that has nothing to say about meaning. On the contrary, the most profitable way of construing a PS is, I will argue, as a theory of *meaning*.

In order to reinforce this point, and to serve as a valuable background to the subsequent discussion of a connectionist PS, let's define a *formally grounded* PS. A formally grounded PS is, or rather should be, in the business of devising a theory of *intensional grounding*, a theory of how (representations of) language relate to #worlds#. Note, not the *world*, but *#worlds#*. PS is not, or rather should not be, concerned with a theory of *extensional grounding*, the complementary theory of how the world relates to #worlds#.

What is the # symbol all about then? Well, for the reader uncertain (and justifiably so) of the distinction between the world and #worlds#, the significance of the # notation will be explained fully in Chapter 9, as part of the design of the *spatial engine*.

Briefly, however, the use of the notation is (at least in part) necessitated by the need, as I see it, to distinguish between *semantic* theory and *meaning* theory. That is, part of what a formally grounded PS is about is the rejection of a monolithic semantics, and its replacement by smaller, more specialized accounts of *meaning*, in the tradition of two-factor meaning theories (cf. Block, 1986). It may appear to the reader that the use of the # notation, the paucity of its mention here and vague talk about meaning theories (as opposed to a single semantic theory), serves only to obfuscate the important issues. This is true, but unfortunately also necessary, because the notion of a monolithic semantic theory is so pervasive that exploring other alternatives is bound to, at least initially, cause some confusion. So, for now, I ask for patience and ask that the reader wait until Chapter 9 to find out the crucial differences between extensional and intensional grounding and for clarification of the role of the # notation.

A formally grounded PS looks a bit like a formal semantics, as I have said, because both are instances of a method for relating language to an interpretation: Propositional representations (in the PS case) or well-formed formulae (in the formal case) are mapped to an interpretation (some manner of computational extension in the PS case, a set theoretic model structure in the formal case). However, both the range and the domain of analogous function mappings in the human mind are (very much likely to be) very different from their formal incarnations.

That said, a formally grounded PS, taking its cue from the formal school, is concerned with the computation of functions *akin* to intensions, the devising of procedures able to compute such functions, and the nature of the computational extensions yielded by such function mappings. The PS of Johnson-Laird (1983) is an example of a formally grounded species of meaning theory in the sense I have just outlined. In this framework, the result of executing procedures computing intensional functions, is the construction and manipulation of *mental models*. Mental models are computational constructs, they are not set theoretic constructs, and thus, calling a mental model an *extension* is likely to offend the sensibilities of a formal theorist. However, it is in the very nature of the project of transposing the formal into the computational to require that key semantic terms, notably intension and extension, have realizations radically different from their original set theoretic incarnations.

Let's review briefly here. I have considered some of the most common criticisms of PS. The first is that PS is (in some sense) parasitic on a prior formal semantics. However, the intended connotation of a useless, leechlike relationship is un-warranted. This can be seen when considering a formally grounded PS, which exists in a benign symbiosis with its formal relation. Such a PS is concerned with procedures that compute functions similar to intensions, in that the functions yield an interpretation, given a symbol string. To criticize a PS for being vague about the degree of this similarity is absurd: Part of what a formally grounded PS is *about* is actively exploring what degrees of similarity are possible and efficacious. The other worries of the anti-PS position, I have argued, revolve around a deeply engrained conception of what a semantic theory is really all about. Surface manifestations of this conception are that PS confuses semantic theories with theories of sentence comprehension, that it attempts to interpret natural language by translating it into a machine language, and so on. This is a more complicated concern, and I hinted that a resolution involves abandoning a monolithic notion of a single semantic theory in favor of smaller meaning theories (although the specifics will have to wait).

Moving on, and in contrast to the wholly *symbolic* kinds of formally grounded PS that we have been discussing up to this point, in the next part of the chapter, attention will be given over to a form of formally grounded PS that is couched in terms of non-Von Neumann-style computation. Just like the theoretical Universal Turing machine, and its descendant hardware relations that inspire the syntactic engine, connectionist networks are also universal computing devices (cf. Siegelman & Sontag, 1991). There is, therefore, no problem with lack of raw computational power associated with such machines. What is problematic, in the context of this discussion, is how to describe a formally grounded PS couched in nonsymbolic computational terms, and also how to set about demonstrating that such a proposal is feasible and workable. It is this challenge that is taken up in the next section of the chapter.

THE MICROPROCEDURAL PROPOSAL

The Microprocedural Proposal for a formally grounded connectionst PS emerged from a consideration of the problem of how to represent the meanings of spatial relational terms such as *on the left of* (cf. Jackson & Sharkey, 1991). Such terms are primitive in any natural language, which means that their meanings cannot be decomposed into more basic linguistic components. Instead of attempting a semantic decomposition of such terms and then attempting to use the putative "semantic primitives" in a computational model, the Microprocedural Proposal accepts the primitive status of spatial relational terms as a given, and views the meaning of such expression such as *on the left of* in procedural terms.

The Microprocedural Proposal draws on the terms and concepts of both formal and symbolic computational species of semantics, including the conception of meaning in terms of intension and extension, the *interpretation function* of a possible world semantics, the concept of a procedure computing the intension of an assertion, and so on, that is, the meaning theory is formally grounded. As the name suggests, and consonant with the synonym of meaning and procedure in symbolic PS, a connectionist PS requires that the term *truth condition* (viz. meaning or intension) be regarded as synonymous with a connectionist conception of the term *procedure*. So, the next question must be, What is a connectionist procedure? In terms of the Microprocedural Proposal for a connectionist PS, the procedures that compute functions yielding computational extensions are the *weights* of the network.

Microprocedures and Functions

There are two, quite different, ways of approaching the notion of a microprocedure. The first is a localist conception, intuitively illuminating but theoretically suspect and ultimately impoverished. I will consider this simpler notion first, and then reject it in favor of a second notion, which views *populations* of weights as microprocedural representations.

The notion of an individual microprocedure is essentially a very simple one, and can be illustrated by reference to its more familiar, symbolic relation. A symbolic procedure is what makes the abstract notion of a *function* concrete: A function is a mapping from one set to another, and a procedure specifies how to carry out that mapping. One can illustrate the assimilation of these two notions, to what happens in a connectionist network, by looking at just one single weight, shown in Figure 3.1.

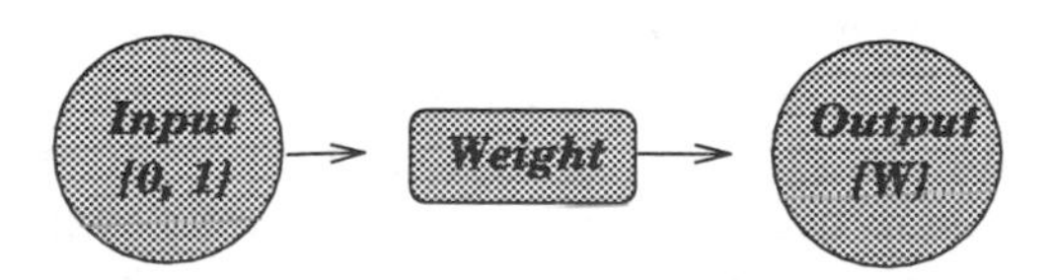

FIGURE 3.1. A single microprocedure.

Take a simple binary unit (as shown) with two states of activation, either active or inactive (for explanatory purposes, I will ignore the fact that, in a lot of connectionist work, units are more commonly continuously valued). The *domain* of the function to be computed is exhaustively specified by either **0** or **1**. We can discount the **0** case immediately, as the microprocedure would be inactive under those circumstances, so the domain of the function to be computed is simply **1**. The *range* of the function to be computed, before that set is squashed by the sigmoid, is the weight **w** times **1**, that is, **w**. A classically defined function, to reiterate, is simply the (abstract characterization of the) mapping from one set (in this example, **1**) to another set (in my example, **w**). Accordingly, an intuitively accurate characterization of the single weight is as a procedure specifying how to carry out that mapping: Couched in our favored vocabulary, the weight is thus characterized as a *microprocedure*.

Immediately, one objection that might be leveled at the notion of a microprocedure concerns how a *parameter*, or even more austere, a numerical *value*, can possibly be described as a procedure? Surely, the objection might go, symbolic procedures are *active* entities; they constitute a computational *process*. How then can the description of an inert, inactive, numerical value as a procedure be justified? And in what sense is it useful? These kinds of criticisms are part and parcel of the difficulty in assimilating the vocabulary of symbolic and connectionist explanation.

The first objection concerns how an inert numerical value can be considered a procedure. Indeed, the objection goes deeper: The implication is how it can even be considered a *representation* at all. However, this resistance is really just evidence of intellectual inertia at work: Individual *unit activations* are similarly, merely inert, numerical values also (although they do, of course, change during the course of the network computation). Take an individual hidden unit activation, for example: This can be (and often is) described as a "subsymbol," but it is more accurately and usefully considered simply as one part of the coordinates for a point in an *n*-dimensional space (where *n* is the number of units defining that space). No connectionist has a problem with this notion of representation, and so equally, no connectionist should have a problem with the notion of weight representation either, for the same reasons.

The second objection concerns the fact that procedures are active entities. Even accepting that a numerical value *could* be representational, the objection might go, this still does not mean that there is anything *procedural* about it. However, there is once more an intellectual inertia to overcome. Consider, in contrast, what happens when a *symbolic* procedure (which is active and representational) is not executing: Is it then active? Or is it inert in some fashion, such as the microprocedure? Let's turn the objection on its head: What other important function does a symbolic procedure perform (that a microprocedure doesn't), other than take an input and yield some function of that input as its output? What else does a weight do in a similar circumstance? Initially it is inert and inactive: Present it with an input, however, and the weight yields some function of that input as it output.

These kinds of objections are worth considering because, of course, conceptualizing the weights of a connectionist network in the terms given here means that microprocedures are very limited entities. They can, of course, be more complex as a function of the type of unit with which they associated—which is to say, as a function of whether the unit exhibits binary or continuous activation, whether the unit has a more sophisticated temporal profile (that is, continuous or stochastic spiking), and so on—but they are still extremely limited. Moreover, adhering to a localist conception, which a treatment of the properties of individual microprocedures does, means that one of the only advantages associated with a corresponding *unit* localism, namely, its semantic transparency is lost, and there is no other benefit in recompense. Semantic transparency refers to the fact that, because individual units can be assigned an a priori semantic significance in a localist network, their role in any computation is easily seen by virtue of this label (cf. Sharkey, 1992). Individual weights, on the other hand, cannot be usefully associated with a discrete semantic label in the same manner, and so they exhibit a semantic *opacity*.

Thus, it is important to realize that, in terms of contributing to an understanding of the representation of meaning and in terms of the design of a connectionist PS, crucial theoretical differences exist between microprocedures and *microprocedural representations*. The nature and implications of those differences are the subject of the next section.

Microprocedural Representation

Viewing populations of weights as microprocedural representations is part of a move away from the simplistic and needlessly constrained working analogy currently employed in most connectionist research in cognitive science, whereby internal representations are equated solely with patterns of unit activation. The implications of this kind of move are only slowly being realized. Haugeland (1991), for example, has expounded on the theme of weights, and representations defined over weights, as constituting a *representational genera* distinct from the genera of unit representation. Sharkey and Jackson (1994a) were more explicit, arguing that weight representations constitute a representational resource *qualitatively different* from the unit resource in (at least this) one important respect: Weight representations can be usefully understood as *context-independent* entities, in that, although plastic during learning, they remain invariant and unchanging during subsequent processing. This property of weight representations is important because it allows them to be understood as context-independent constituents of *context-dependent* unit representations.

Naturally, the way in which instances of, as I will say, the *atomic* weight resource combine to form instances of the *molecular* unit resource is not the same kind of composition that yields molecular symbols from atomic symbols in more standard "classical" explanations. More specifically, the composition is not of the concatenative species, rather it is a functionally or *nonconcatenatively* compositional species (cf. van Gelder, 1990). Moreover, this functional style of composition is context-sensitive: That is, the identity of a molecular unit representation constructed from atomic weight constituents is sensitive to the context in which the composition occurred (i.e., what particular unit representation was active on the input).

Accepting that the notion of weight representation is a legitimate one in any explanatory framework means that, in thinking about a theory of meaning, the connectionist has access to two, qualitatively different kinds of representational resource, defined over populations of units and weights, respectively. This duality of resource engenders the following axiom of a connectionist PS: *Microprocedural representations*, defined over populations of weights, encodes an intensional meaning, and *superpositional structural analogue*, defined over populations of units, encodes an extensional meaning.

I will comment on superpositional structural analogues, or SSAs as I will refer to them, before we move on. Such representations are (to use the common terminology) *fully distributed* unit representations intended to support analogical structure similarity relations. The construction and properties of these representations are discussed in a later section and in Chapter 7. Once again, I would ask for patience: Clarification will be forthcoming. For now, we return to the theme of weight representations.

Let us see what a formally grounded PS, based around the notion of weight representations, may look like. Consider a standard feed-forward network over two layers of weights after back-propagation learning has been completed. A number of input units become active. In terms of the Microprocedural Proposal, these active units *select* microprocedural weight representations to participate in a given function mapping. I will provide some more explanatory terms: We might describe the activation of the input units as engaging microprocedural representations, with the aim of yielding another unit representation over the hidden units of the network, encoding the squashed result of engaging that weight representation.

The term *engage* is used here to (deliberately) conflate the two senses of the terms *compile* and *execute* (those good old symbolic PS stalwarts). Why? Well, because there appear to be elements of both symbolic processes going on in a network. Does a network simulation (do anything remotely like) execute the microprocedural representation? In a sense it does, as it systematically yields a given result (output) when a particular configuration of binary units become active (input). No more succinct statement of a procedure executing exists to supplement this account of what happens in a network when weight representations yield an output (unit representation) when given an input (unit representation). Does a network simulation (do anything remotely like) compile the microprocedural representation? In a sense it does this also, as the weight representation is translated into a corresponding hidden unit representation. At the heart of compilation (qua translation), after all, is the moving from one representational form to another, and this is precisely what happens in the network case, where the translation is from one representational form, the context-independent weight representation, to another, the context-dependent unit representation. But one shouldn't get too attached to terms: *Compilation* and *execution* are both terms for symbolic computational processes. It is evident that *engage* is not a symbolic process in a classical sense, but there are also grounds for supposing that *engage* is not a computational process in a connectionist sense either. Why is this so?

When an information processing mechanism performs computation, it effects some series of operations upon a set of entities (the input) to yield, in entirely systematic fashion, another set of entities (the output). Such a conception of computation applies to the symbol manipulation of both von Neumann machines and the abstract syntactic engine, where syntactically structured representations are operated upon by processes sensitive to that structure. If connectionist theorists, similarly, define what their class of computing devices do as representation manipulation, then immediately, and in line with the theoretical force of the Microprocedural Proposal, they must recognize that both unit and weight representations must be admitted into this definition of computation. Let me be a little more precise, because computation (in whatever kind of computing device) is not really representation manipulation at all. Rather, more accurately, it is the manipulation of *physical instantiations* of representations, called tokens.

In a classical symbol system this is all very clear, although I should add that an atomic symbol, such as ATAHUALLPA, is both an atomic, computational token (a physical instantiation) and a referring representation, denoting or designating Atahuallpa. In a connectionist network, things are a little different, however. Computation is still, of course, defined as the manipulation of tokens, but those tokens are individual units and weights. Such tokens are not properly representations, because connectionist representation is defined over populations of units and weights.

Surely this is not entirely accurate, an objection might go. *Compositional* connectionist representations (which are nonconcatenatively compositional, to recall) result from what van Gelder (1992) called a *distributing transformation*. These latter compositional representations are defined over populations of tokens. When, however, tokens and representations coincide in connectionist systems, then nonconcatenative compositionality is lost. I will show an application of this point in a later section, in the design of the inputs for the **FF net** simulations. However, getting back to the point, whereas in the syntactic engine, in which computational tokens and referring representations are the same thing (ATAHUALLPA is both token and representation), in the connectionist case, tokens and representations are distinct entities (cf. Chalmers, 1992).

One can see an application of this point in the way that the Microprocedural Proposal is constructed. Individual weights are computational *tokens* and populations of weights are referring *representations*. Both kinds of token, individual units and individual weights, are each causally efficacious elements in any computation defined over them, in that, if the value of even just one token were to change, then the nature of the computation defined over them would change. However, each individual token contributes to representations defined over populations of tokens: It is this fact that largely legitimizes the claim that connectionist representations are (in some sense) *emergent* entities. Thus, it would seem that there are grounds for distinguishing two kinds of explanation in connectionist networks—one at a computational level, and another at a representational level. I expressed reservations earlier that the concept of engaging (a microprocedural representation) might not accurately be described as a computational process, and one can now see why. Connectionist computation is defined over tokens, not representations, and so to describe the engaging of a microprocedural representation as a computational process is to confuse the computational with the representational level of explanation.

This conception of microprocedural weight representation rapidly becomes more complex when the theorist moves away from situations where the unit resource is involved in binary-valued encodings to situations where it is involved in continuously valued encodings—that is, when tokens and representations have become dissociated and a prior distributing transformation has constructed nonconcatenatively compositional representations. In particular, and when making this move, microprocedural weight representations will be selected to participate in function mappings by, not unit tokens, but by the *superpositional* constituents of these unit representations, and we will see an application of these points in the **Combined net** simulations reported later.

Summary. The Microprocedural Proposal for a formally grounded connectionst PS emerged from a consideration of the problem of how to represent the meaning of linguistic primitives such as **on the left of**. As part of a general move away from the restrictive practice of viewing unit representations as the only representational resource sufficient for the purposes of internal representation in cognitive science, the Microprocedural Proposal argues that useful and legitimate representations are to be found in the weights of a network. Connectionist procedures are termed *microprocedural representations*, which are entities emergent from populations of individual weights. In the next section, I will see how this theoretical proposal is empirically fleshed out in working network simulations.

CONNECTIONIST PS ON A SYMBOL SURFACE

The interpretation function of a formal semantics maps from symbols to a set theoretic model structure: This is the starting point for any formally grounded PS, be it symbolic or connectionist. We can isolate two features of such theories that are particularly distinctive.

- Intensions are realized computationally as procedures yielding computational extensions.
- A form of representation is postulated to "take the place of the model structure," that is, to serve as the computational extension yielded by a computational intension.

The Microprocedural Proposal for a connectionist PS is formally grounded, and therefore exhibits these two features. Microprocedural representations are postulated as corresponding to intensions: They yield a computational extension as output given some symbolic input, and the unit representation serves as that computational extension. So, how do we turn this proposal into something concrete? One way is to investigate whether a connectionist network is able to learn an analogous mapping to that of the interpretation function of formal semantics. Give a network a set of inputs, consisting of binary vector encodings of propositional assertions about the spatial relations between objects: Give a network a set of outputs, consisting of binary vector encodings of an interpretation of those assertions. Then train a network on this task, see if it can learn it, and see if it can generalize to novel examples. The conjecture is that when the network has learned, it will have learned a set of weights, a microprocedural representation, that computes functions analogous to intensions, namely, it yields an interpretation given a symbol string.

FF Net

For the first simulation experiment, **FF net**, we use a feed-forward architecture over two layers of weights, as shown in Figure 3.2. The task of such a network is to learn, given a spatial assertion such as **A is on the left of B**, how to construct an output representation in which some instance of an **A**is on the left of some instance of a **B**. The set of input representations to the

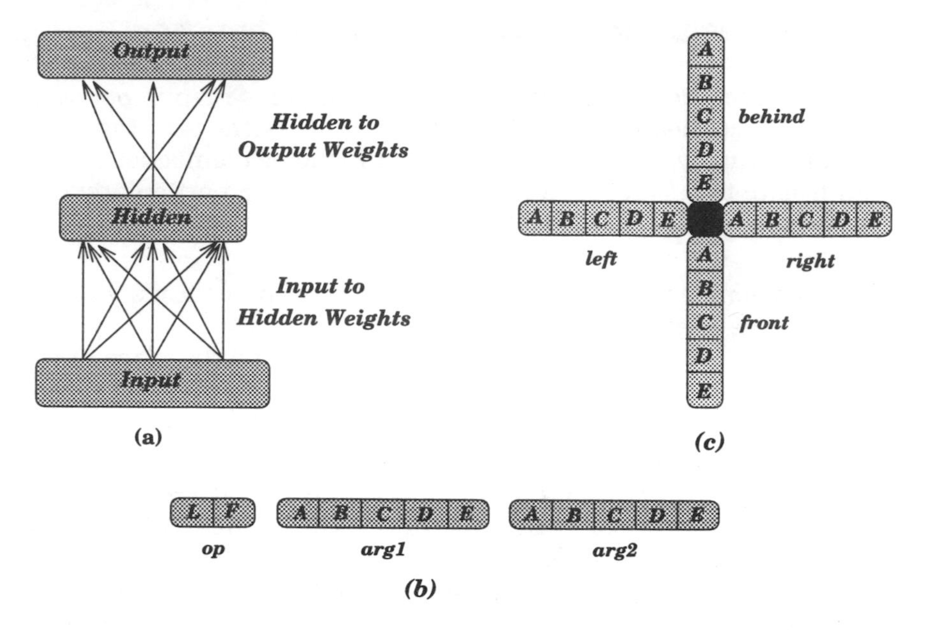

FIGURE 3.2. Shows the simplest candidate architecture for a connectionist PS.

FF net was constructed so as to adhere to the requirement that the meaning of spatial relational terms should not (and indeed, cannot adequately) be encoded, a priori, in terms of symbolic microfeatures (or in terms of subsymbolic (cf. Smolensky, 1988) microfeatures).

The input spatial assertion, **A is on the left of B**, requires three things to be encoded—the spatial term, and the two objects with which the term is associated. Design of the input representations consequently reflected this in a simple vector field method, illustrated in Figure 3.3.

This method involves a given input representation being conceptually carved up into a number of fields: In this case, we need three such conceptual fields. The first, the **op** field, contains two

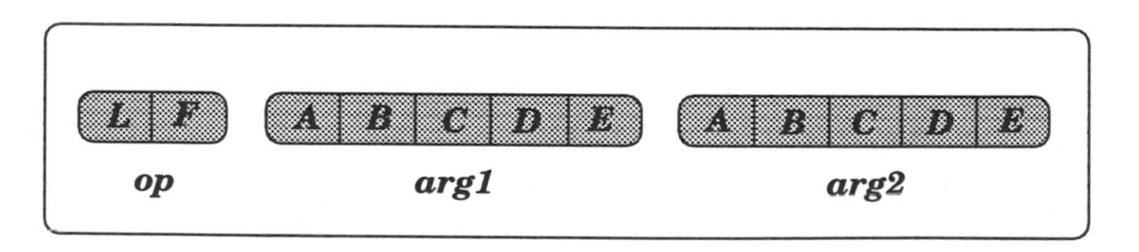

FIGURE 3.3. Shows the form of the input coding for the **FF net**.

units, which code for the two spatial terms **left** and **front**. The second and third fields, **arg1** and **arg2** respectively, each contain five units and code for the first and second symbols of a given assertion. As can be seen, the encoding contains no explicit information about what the spatial terms **left** and **front** might actually mean: LAB and FAB, for example, corresponding to the assertions **A is on the left of B** and **A is in front of B**, are only distinguishable by virtue of the fact that, in the first case, the first bit of the 12-bit vector is active in the **op** field, and in the second case, the second bit in the 12-bit vector is active in the **op** field. Similarly, the symbols that make up a given assertion are coded without explicit information. Each symbol is coded by the activation of one bit out of five in a given **arg** field and is distinguished as that symbol by its relative position in that field.

The set of output representations for the **FF net** were designed to perspicuously represent geometric relations between symbols analogous to those relations that hold between objects in the real

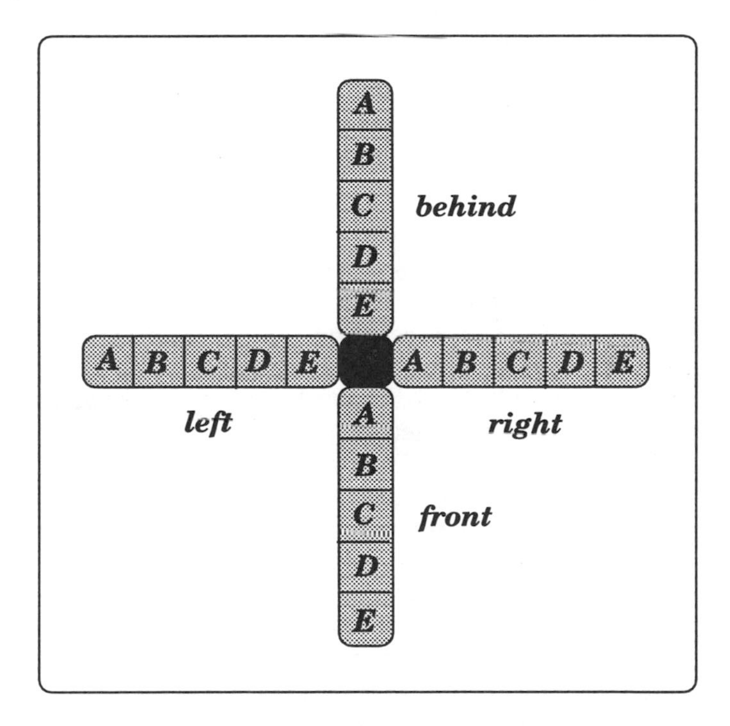

FIGURE 3.4. Shows the output coding used for the **FF net**. Over 20 units, Units 1 to 5 represent the "left" field, Units 6 to 10 represent the "front" field, Units 11 to 15 represent the "right" field, and Units 16 to 20 represent the "behind" field.

world, without an explicit (which is to say, propositional or symbolic) coding of those relations. A variant of the vector field method was chosen, illustrated in Figure 3.4. The output coding can be conceptually broken down into four fields, and visualized as a cross, with the **left** and **right** fields on the horizontal axis, and the **front** and **behind** fields on the vertical axis. Each field contains five units, and each symbol can be identified by its position within that field. For example, the vector **00100 10000 00000 00000** codes for the states of affairs in which C is on the left of A (or equally, that A is on the right of C).

Using two spatial terms and five symbols yields an exhaustive training set of 40 assertions. Ten of these assertions, removed from the corpus, were reserved to test for network generalization. Each symbol was omitted four times, and each spatial term five times. This left a training corpus of 30 assertions. Typically, a number of 12-10-20 network architectures (that is, a network with 12 input units, 10 hidden units, and 20 output units, a form of labeling convention adopted throughout the book), trained from a number of different initial random conditions will, after having learned to tolerance, exhibit a generalization to a novel test set of assertions of 90%.

Recurrent Net

A natural way of extending the Microprocedural Proposal would be to design a network simulation that was required to construct a more elaborate output representation of the extensional relations between instances of symbols. So that this more elaborate output representation might be constructed, the feed-forward architecture of **FF net** can be modified so that it involves a degree of partial recurrency, as shown in Figure 3.5

This kind of network, the SRN, is able to accept sequences of spatial descriptions, thus allowing a representation of the spatial relations between multiple instances of symbols to be constructed incrementally. The task of such a network is to learn, given a sequence of spatial assertions such as **A is on the left of B**, **C is in front of B**, and so on, how to construct an output representation in which some instance of **A** stands in the correct spatial relation with some instance of **B**, which stands in the correct spatial relation with some instance of **C**. The form of the input coding necessary to represent such relations is shown in Figure 3.6.

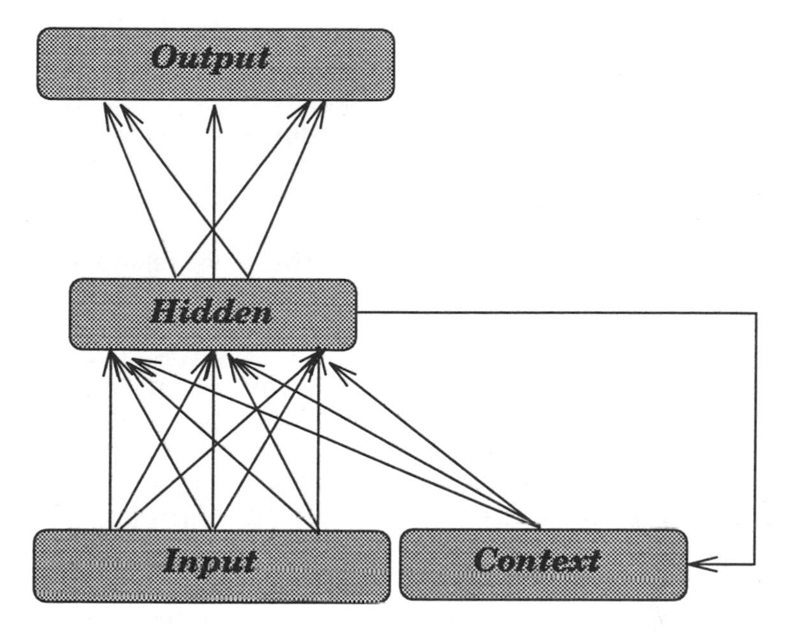

FIGURE 3.5. Shows an extension to the simplest feed-forward architecture for a connectionist PS involving a degree of partial recurrency. The recurrent links from the hidden to the context units allows the network to have a memory of its past states.

The increase in complexity for the **Recurrent net** simulation experiments over their feed-forward cousins had a number of aspects. First, four spatial terms were employed in constructing the input set of assertion sequences. This means that, instead of two units in the **op** field in the **Recurrent net** experiments, there were four. However, the number of symbol arguments from which the training corpus was constructed was reduced from five in the **FF net** experiments to three in the **Recurrent net** experiments, using

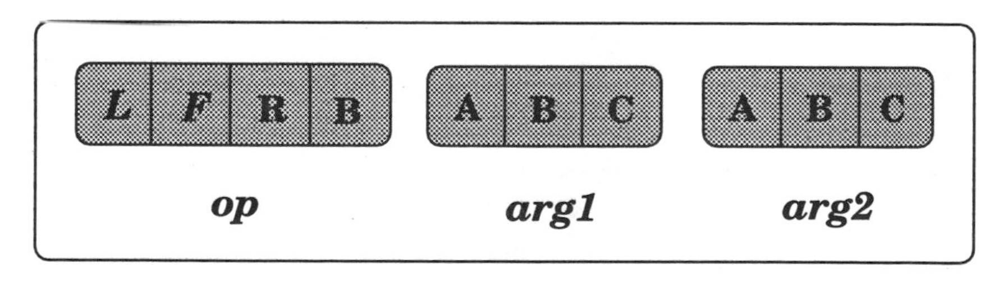

FIGURE 3.6. The input coding for the **Recurrent net**.

only **A**, **B**, and **C** (see Figure 3.6). Using three symbols also had the effect of truncating the lengths of the sequences required to fully specify the relations between tokens to only two assertions.

Secondly, the nature of the output coding was changed. Instead of four locations in which instances of symbols could appear, in the **Recurrent net** simulations there were 13 such locations, coded over 39 units, as shown in Figure 3.7. This kind of array structure is able to represent some 108 two-place relations (ie. between three symbol instances) and 24 one-place relations (between two symbol instances). Typically, a number of network simulations sharing a 10–40–39 network architecture, after having been trained to tolerance from a number of different random seeds, will exhibit a generalization to its novel test set in the range of 75 percent to 85 percent.

The Microprocedural Proposal for a formally grounded PS urges that microprocedural weight representations, which are emergent from the activations of populations of computational tokens, are a legitimate and useful representational resource, in and of themselves, qualitatively different from the unit resource. The simulation experiments reported in this section, both **FF net** and **Recurrent net**, were intended to demonstrate, clearly and simply, that a connectionist network can be understood as computing a function analogous to an intension, when it yields an interpretation given an input symbolic string. Focusing in upon one of the networks representational resources, namely the weight resource (as opposed to the unit resource), the Microprocedural Proposal urges that such representations can be understood as encoding an *intensional* meaning, a function yielding an interpretation given a symbolic description.

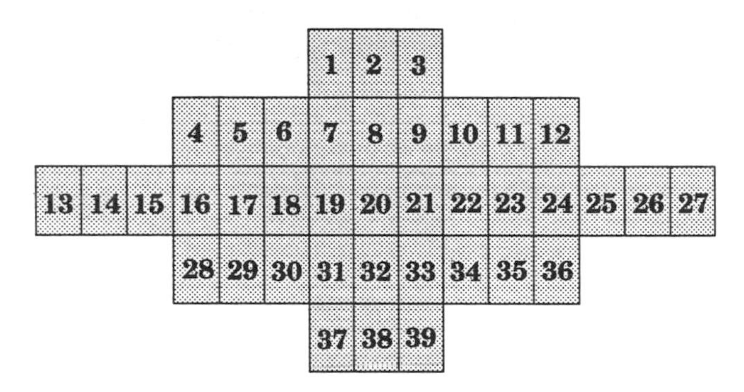

FIGURE 3.7. Shows the coding 13 locations of three symbol instances over 39 units in the **Recurrent net** simulations.

The claim is not that a network, simply because it has learned a particular kind of function mapping, forms the basis for an all-encompassing semantic theory, because a formally grounded connectionist PS, to recall, is specifically *not* concerned with semantics. Rather, it is concerned with intensional grounding. Intensional grounding, unlike semantics, is not concerned with how language relates directly to the world, but rather with how one system of representation (language) relates directly to another (#worlds#). Once again, this point will become clearer as the chapter progresses.

PS BELOW A SYMBOL SURFACE

After encountering connectionist PS and **FF net** simulation experiments, the anti-PS theorist would, no doubt, feel compelled to add at least another couple of scathing isms to the list compiled by Fodor (and quoted earlier), namely, associationism particularly and thus its corrollary, nonrepresentationalism. That is, an advocate of an anti-PS position might look at the simulation experiments—particularly the binary vector input and output unit encodings—and argue that such representations are not representational (and hence are not candidates for legitimate *mental* entities) because they are *associative* and do not exhibit a combinatorial syntactic structure (see Fodor & Pylyshyn, 1988). Of course, the anti-PS advocate would indeed be correct: The binary input and output vectors are *not* representational, at least not in terms of the kinds of representation (and manipulation thereof) commonly assumed by the classical theorist, because such binary vectors are not *concatenatively compositional* (see Chapter 7 for more details of different styles of compositionality).

This is not the whole story, however, because as well as being noncompositional in a concatenative sense, binary vector encodings are non-compositional in a functional or *nonconcatenative* sense. Because there is no prior *distributing transformation* (c.f. van Gelder, 1991b), there is no dissociation of tokens and representations, no nonconcatenative compositionality and hence no combinatorial structure sufficient for structure sensitive representation manipulation (the criticism does not apply, of course, to unit representations constructed by a distributing transformation, which *are* nonconcatentively compositional). The binary unit inputs and outputs to both the **FF net** and the **Recurrent net** simulations are efficacious *solely* to the extent that such representations are available to inspection on a putative symbol surface. The *real* power and elegance of connectionist representations only becomes apparent, however, when one dips *below* such a symbol surface. It is here that connectionist style computation comes into its own.

The purpose of the next section of the chapter, therefore, and in line with the above observation about symbol surfaces, is to detail the features and properties of a connectionist PS operating in precisely this fashion. That is, to detail a PS that, although able to support a systematic semantic interpretation on a symbol surface, is nonetheless not itself directly semantically interpretable on it. There are a number of novel features associated with this kind of PS. The first concerns the nature of the computational extensions constructed by the PS, what I have referred to as *superpositional structural analogues* (i.e., what "takes the place of the model structure"), and the second concerns the "propositional representations" that are input to the PS and on the basis of which it constructs its outputs. I will turn to both of these concerns.

Superpositional Structural Analogues

In the same way that a connectionist PS postulates that a weight resource, emergent from populations of weight tokens, is able to represent intensional meaning, it postulates that a unit resource, emergent from populations of unit tokens, is able to represent extensional meaning. Jackson (1994) detailed the method of construction, and the properties, of a form of connectionist unit representation, termed a superpositional structural analogue or SSA for short, which fulfills this latter (extensional meaning) requirement. Such a representation is nonconcatenatively compositional, is sensitive to the processing context in which it occurs and supports analogical similarity relations. SSAs are a form of representation in the tradition of discourse representation (DR) theory (cf. Spencer-Smith, 1987). DR theory is a natural companion to a formally grounded PS, comprising a mixed bag of the formal and the psychological, concerned with the nature of the representational *model* that must take the place of the set theoretic model structure during the processing of discourse (see van Dijk & Kintsch, 1983; Johnson-Laird & Garnham, 1980; Webber, 1983).

The representational efficacy of a SSA hinges on its *analogue structure*. Unfortunately, the term "analogue" is a polyseme, and recourse to a dictionary for disambiguation is likely to prove frustrating: "Analogue" is sometimes touted as a spelling variation of "analog," and hence is touted as having the same meaning, and sometimes it is not. Thus, sometimes the term *analogue* is used in the distinction between analog and digital (computers, for example), and sometimes it is used in the distinction between continuous and discrete. For purposes of clarity, here and throughout the book, only one of the senses of analogue is intended, that is, in the sense of "analogous" qua "analogical," and *not* in the sense of "analog" qua "continuous." The terms "analogue," "analogous," and "analogical" will be used interchangeably.

Unlike either the syntactic structure of representations in the symbolic theorists syntactic engine or the spatial structure of connectionist representations, which are properties of the functional architecture, analogue structure is a virtual property. SSAs are similar to the mental models constructed by the symbolic PS of Johnson-Laird (1983). Unlike the symbols postulated by the representational theory of mind (RTM, cf. Fodor, 1987, and see Chapter 9) in the Language of Thought hypothesis, mental models do not have an arbitrary syntactic structure; rather, they have one that plays a direct representational role. This feature of mental models is spelled out quite clearly by Johnson-Laird (1983), in the Principle of Structural Identity: "The structures of mental models are identical to the structures of the states of affairs, whether perceived or conceived, that the models represent." (p. 419).

The question of precisely what constitutes an *analogically structured* representation is not easily answered (for a number of different conceptions, see Johnson-Laird, 1983; Sloman, 1976). Intuitively, one can say that analogical representations have parts that correspond to parts of what they represent. That some parts of the representation indicate properties of, and relations between, parts of represented thing. In terms of SSAs, *degree* of analogue structure is defined as a function of a corresponding *lack* of syntactic structure. This view hinges on considering analogue structure as existing in a space of representational properties, with classical syntactic structure at one far corner and connectionist spatial structure at the other. Thus, the degree of analogue structure in a representation (be it connectionist or classical) is a function of a bias toward either of the syntactic or spatial corners of the space. A bias toward the syntactic corner would result in the intrusion of symbolic information (viz. syntactic structure) into the analogically structured representation.

Such an intrusion is perfectly legitimate, of course (even if unwanted on some occasions), as evidenced by the many kinds of analogical representations that have a symbolic component, for example, most simply a map. To illustrate what I mean by a deleterious intrusion of symbolic information into a structural analogue, consider Figure 3.7 once more, which shows the binary vector field encoding method that the **Recurrent net** simulation experiments employed.

The form of this coding was chosen primarily as a resource requirement: The array had to be of a certain size to accommodate all the possible permutations of symbol instances and positions relative to each other, that those instances could assume. If this were not the case then construction of an output on the basis of an input sequence would have "fallen off the edge of the world." Now consider the three example outputs shown in Figure 3.8.

Each output was constructed on the basis of three different sequences of assertions: Figure 3.8(a) is constructed on the basis of the sequence **A is on the left of B, C is behind B**, Figure 3.8(b) is constructed on the basis of the sequence **C is behind B, B is on the right of A**, and Figure 3.8(c) is constructed on the basis of the sequence **B is in front of C, A is on the left of B**. Each of the three sequences of assertions, on which construction was based, describes the same state of affairs, where an instance of **B** is both on the right of an instance of **A** and in front of an instance of **C**. We can label each output instantiation using the following notational device: **.abc** refers to construction beginning with the **A**

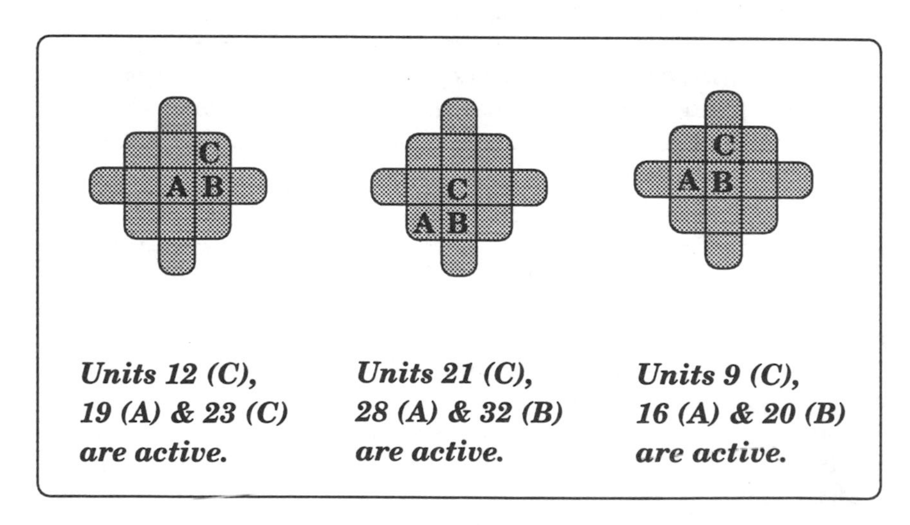

FIGURE 3.8. Three output representations of the same states of affairs resulting from three different sequences of spatial assertions.

symbol, **a.bc** refers to construction beginning with the **B** symbol, and **ab.c** refers to construction beginning with the **C** symbol. Using this notational device and looking at the three different outputs constructed by the **Recurrent net** simulation in Figure 3.8, one is suddenly struck by a very obvious flaw: What are being constructed are in fact *different* states of affairs. The three output representations in Figure 3.8, which we, an agency external to the representational system, can infer as equivalent, are described by three different 39-bit vectors: For the first output, **.abc**, Units 12, 19, and 23 are active, for the second output, **a.bc**, Units 21, 28, and 32 are active, and for the third output, **ab.c**, Units 9, 16, and 20 are active.

The question we must ask, therefore, is to what extent are these representations (viz. the 39-bit binary vectors) supporting the appropriate analogical relations? This determination can be made by an appropriate cluster analysis. By feeding the complete set of output vectors from the **Recurrent net** simulation through a cluster analysis, the lack of *significant similarity* between the structure of these representations, and the structure of the states of affairs that they are representing, is plain to see. A portion of such an analysis is shown in Figure 3.9 for the outputs constructed on the basis of multiple assertions.

The integers at the front of the labels refer to a means used for distinguishing general *classes* of outputs. Thus, the binary vector outputs labeled **4.cab** and **2.cab** in Figure 3.10(a) differ in the class of outputs to which they belong. The **4** class includes all those outputs conforming to the schema shown in Figure 3.10(a), and the **2** class includes all those outputs conforming to the schema shown in Figure 3.10(b).

Before I go on, I should add a caveat. The metric of similarity under discussion here is, of course, a Euclidean one: Specifically, vector similarity (of one raw output with any other) is determined by the location of that given vector in the 39-dimensional space occupied by the output, relative to all other vectors. This Euclidean distance measure is both extremely useful and annoyingly limited. It is useful because it provides a good way of observing similarity relations within a network's representation, but it is limited because, although Euclidean space and *computational space* (see the end of the chapter for more details) are highly correlated, they are not *identical*. This means that the measure of similarity of two vectors determined by cluster analysis is not identical to the measure of similarity that would be determined by a more fine-grained, computational analysis. This lack of granularity notwithstanding,

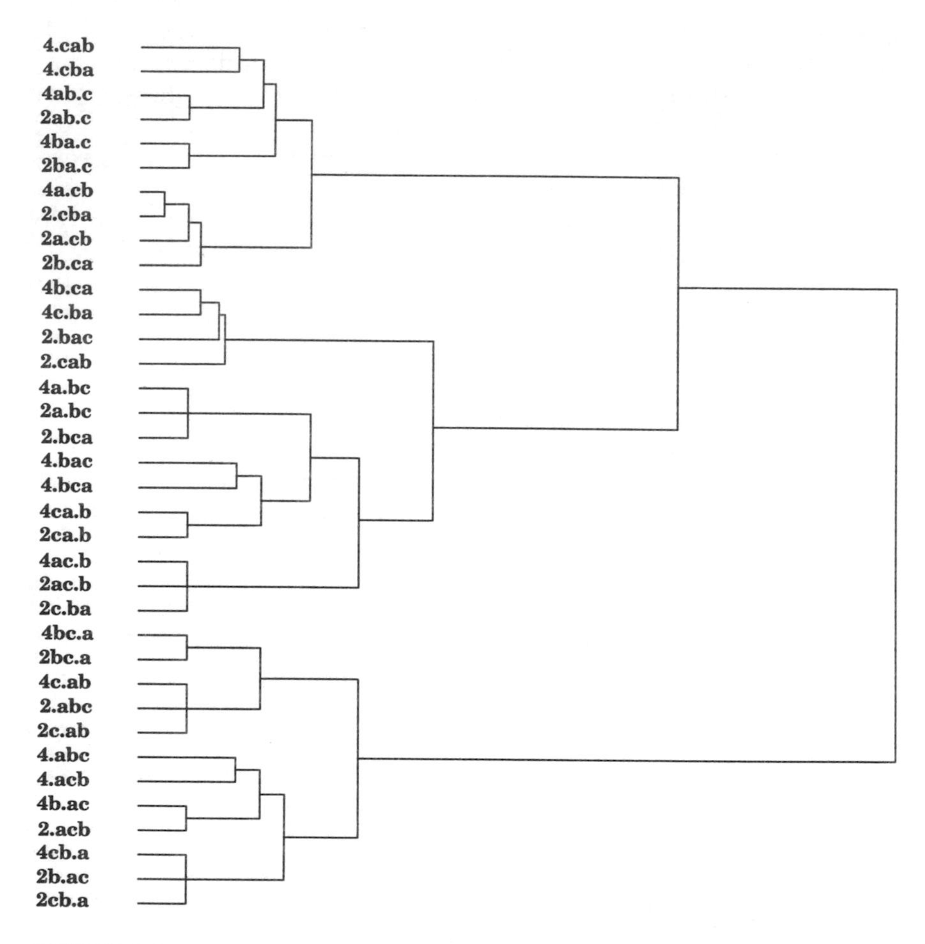

FIGURE 3.9. Dendogram showing the clustering of a portion of the raw two-place outputs used in e **Recurrent net** simulations. The labels refer to where, and with what symbol, construction began.

for most purposes a cluster analysis of a Euclidean space is a good approximation of the corresponding computational space.

Figures 3.9 shows that the output representations generated by the form of the coding employed in the **Recurrent net** simulation experiments violate a cardinal feature of a structural analogue—namely, that the representations should be non-symbolic. By necessity, the **Recurrent net** simulations had to begin construction of a given output representation at a certain point: The choice of where (and hence how) to begin construction of a given array, although important, was adjudged essentially arbitrary, and defaulted to the first symbol of the first assertion in a given

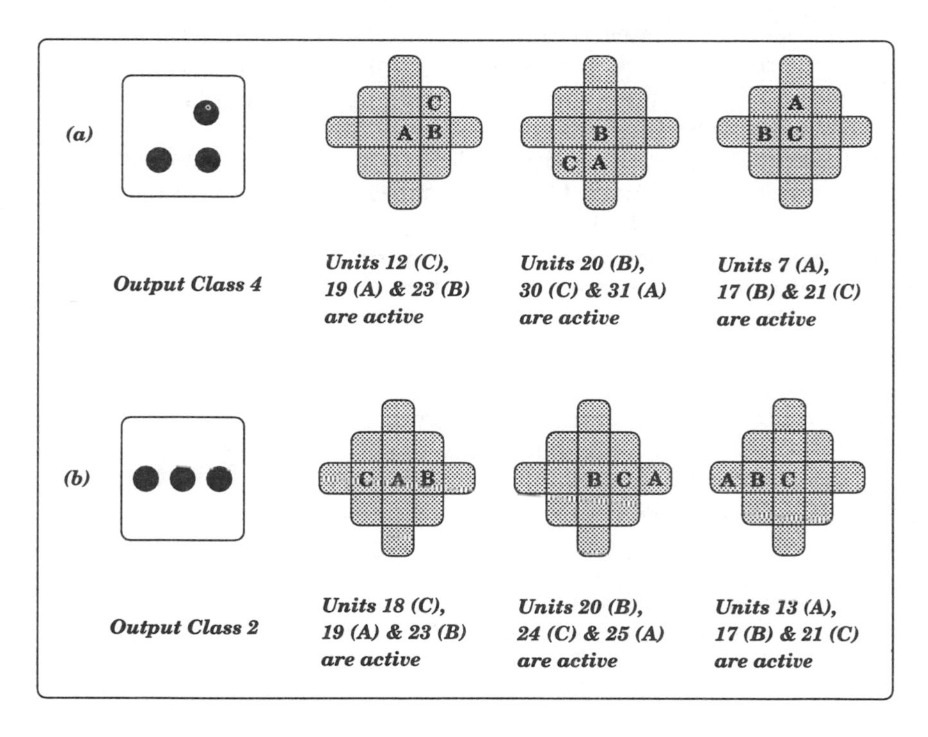

FIGURE 3.10. The grouping of raw outputs into classes.

sequence. With hindsight, this arbitrary choice can be seen more clearly as a constraint on the representational efficacy of the coding. "Anchoring" the construction of an output to Units 19, 20, and 21 in the output vector, so as to meet the resource requirement, actually results in the intrusion of *symbolic* information into the output representation. The raw 39-bit vector encodings employed in the **Recurrent net** consequently contain symbolic information about the spatial ordering of symbols within assertions and the temporal ordering of assertions within sequences.

How is the unwanted intrusion of symbolic information, via the output representation, into the structural analogue to be eliminated? What is required is for the similarities between the functional level (raw 39-bit output vector) representations to be *warped* so as to support the appropriate analogical similarities. Specifically, what this amounts to is the wholesale collapse of each binary output vector to its most primitive structural parts. This is the *collapse2 strategy*.

Using the form of the output coding used in the **Recurrent net**, with 39 units coding for the three symbol instances **A**, **B**, and **C** over 13 locations, means that 12 one-place and 36 two-place relations can be represented. For each one-place relation, there are two possible outputs of the same states of affairs, corresponding to the two different symbol instances with which construction can begin: There are, consequently, 24 one-place output vectors. For each two-place relation, there are three possible outputs, corresponding to the three different symbol instances with which construction can begin: Consequently, there are some 108 two-place output vectors. Preliminary simulations conducted to test the feasibility of the collapse2 strategy revealed that the complexity of the task might be lessened by splitting it up into two stages, so that the collapse2 strategy would involve an *incremental* collapse of each raw output, as shown in Figure 3.11.

In the first stage of such an incremental collapse, **c1**, a given output vector serves as an input and is mapped to a unique, arbitrary output. The representations developed over the hidden units of the simulations in the **c1** stage are then extracted and used to form the input set for the second stage, **c2**, where each hidden unit representation is mapped to its most primitive constituent parts. Extracting the hidden units from the final stage **c2** simulation and feeding a portion of those representations through a cluster analysis produces the dendogram shown in Figure 3.12,

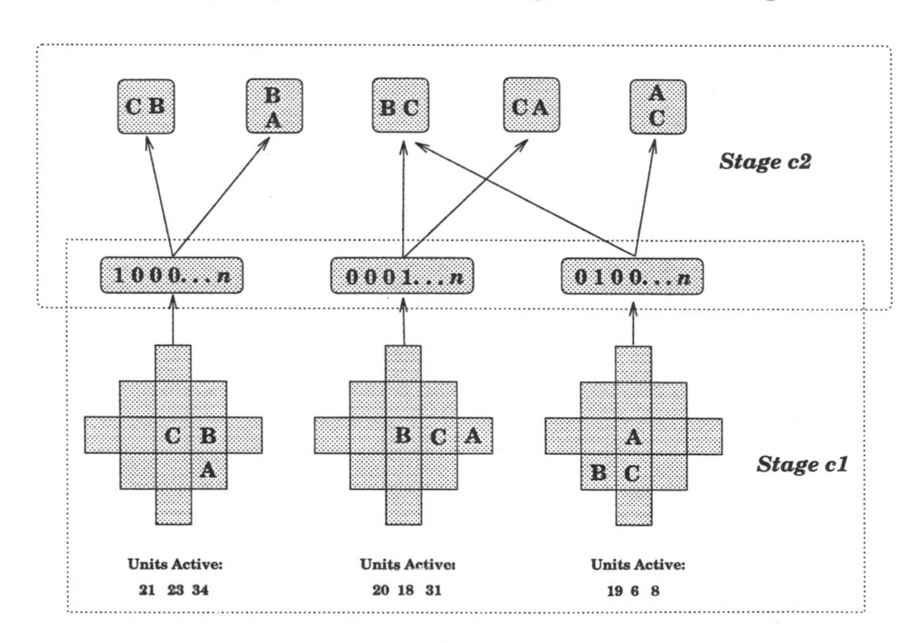

FIGURE 3.11. The two, **c1** and **c2**, stages in the collapse2 strategy.

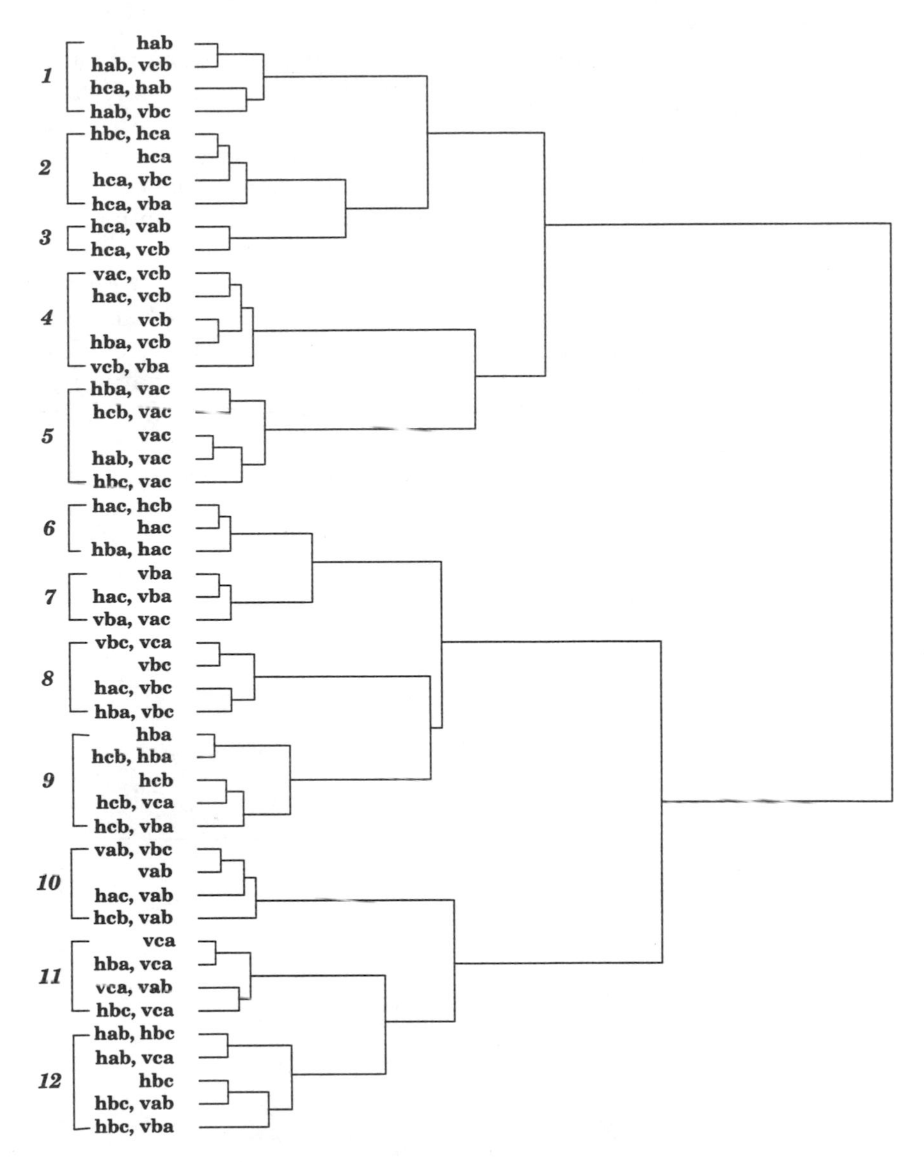

FIGURE 3.12. Cluster analysis of the hidden unit representations extracted from the final collapse2 **c2** simulation. The labels refer to what primitive relations the hidden unit representations are encoding. For example, the label **HBA,VCB** refers to a representation where **B** is on the left of **A** in the horizontal plane, and **C** is behind **B** in the vertical plane.

which shows that the cluster analysis has partitioned the hidden unit representational space into 12 locations, corresponding to the 12 primitive states of affairs.

It is worth, at this point, considering *why* the pattern of clustering in Figure 3.12 is a pleasing result. Surely, the objection might go, if the **c2** simulation is mapping the continuously valued **c1** vectors to 12 output units, then the partitioning we see is not surprising. Let us turn this objection on its head and consider how else one might expect these hidden unit representations to cluster, if not in the manner seen? Consider Figure 3.13.

Is Figure 3.13(a) more similar to Figure 3.13(b) than to Figure 3.13(c)? If it is, its clustering should reflect this, but on what basis are (a) and (b) more similar than (a) and (c)? Should Figure 3.13(a) cluster with 3.13(d) (with which it is similar in virtue of shared latters in similar relative positions), rather than 3.13(e) (with which it is not similar in virtue of shared letters in dissimilar relative positions), even though the latter is a three-place relation, like 3.13(a), rather than a one-place relation? Should representations encoding 2-place relations cluster together to the exclusion of 1-

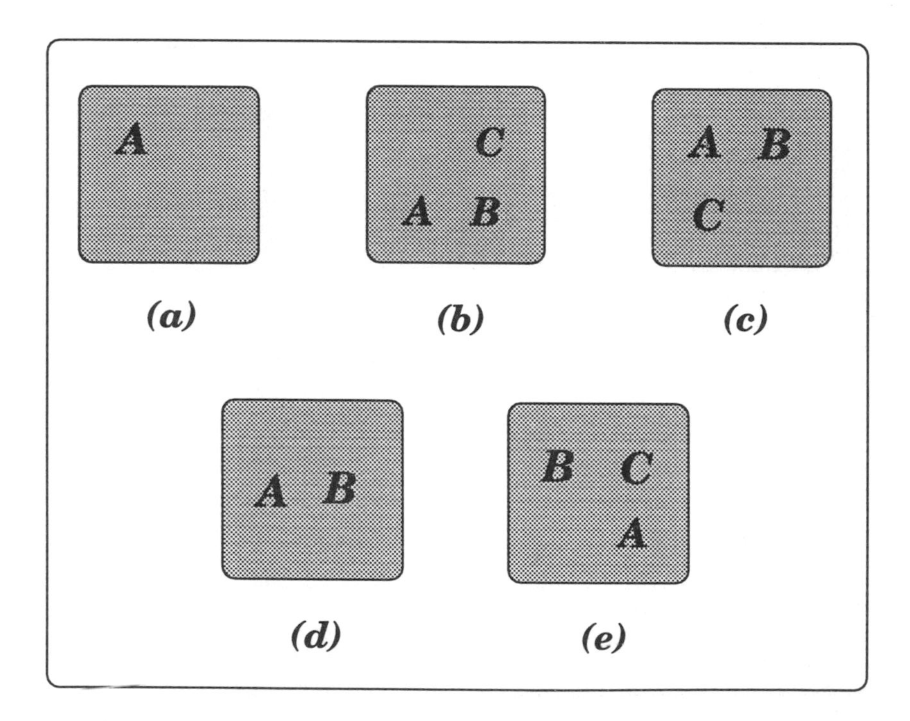

FIGURE 3.13. Is (a) more or less similar to (b) than to (c)? Or to (d)?

place relations? The latter kind of clustering would reflect the most obvious *perceptual* similarity between representations, yet this is not what we find.

On the contrary, what we find is that the cluster analysis reveals that the pair-wise Euclidean distances between output representations is not sensitive to whether that representations encodes a one-place or a two-place relation. The dendogram shows that two-place relations cluster around loci formed by one-place relations. However, any two-place relation could equally well cluster around one of two loci, corresponding to the two one-place relations that it encodes hence, one has grounds for supposing that the clustering of two-place relations around particular one-place relation loci is arbitrary. One can see this particularly clearly in Cluster 3, which does not have a one-place relation as a loci. The main contributory factor for this discrepancy involves the previously mentioned point that Euclidean space and computational space, though highly correlated, are not identical. To reiterate the point, a cluster analysis using a Euclidean distance metric is only ever a crude approximation of similarity relations in computational space.

I will refer to the unit representations from collapse2 as *superpositional* structural analogues, or SSAs. The label serves to highlight their analogical nature, but also indicates that, unlike a classical molecular symbol, the constituents of this unit representation are superposed, one on top of another, with the indentities of the constituents being destroyed in the process of composition. Being superpositional in this sense, the constituents are not available to inspection on a symbol surface. This is an unusual feature, which derives naturally from the nonconcatenative means by which the constituents were combined to form the molecular whole.

Summary. The collapse2 strategy was designed to take the binary vector encodings of spatial relations and map them to their structurally most primitive parts, thereby eliminating any information about the linguistic (or symbolic) circumstance on which their construction was based. Mapping, in two stages, to 12 primitive states of affairs, I extracted The hidden unit representations for the entire corpus of output relations from a feed-forward network at Stage **c1**, and used these nonconcatenatively compositional unit representations as input to the second stage, **c2**. Passing the representations extracted from the stage **c2** simulation through a cluster analysis, as shown in Figure 3.12, reveals that similarity relations between representations, measured in terms of a Euclidean distance metric, are not based upon any obvious perceptual metric of similarity. Rather, representations appear to cluster in terms of a functional similarity metric. The argument is that the similarity relations between these (nonconcatenatively compositional) representations can be thought of as supporting (virtual) analogical similarity relations: That is, the representations can be treated *as if they were* analogical representations.

One of the requirements of a connectionist PS, a representation able to take the place of the model structure, has been addressed. That replacement is a superpositional structural analogue (SSA). This kind of representation is nonconcatenatively compositional and has superpositional constituents. It *functions* as an analogue of states of affairs in the world, where the analogical relations are supported by spatial structure similarity relations. SSAs are a novel form of discourse representation, primarily because although they respect semantic criteria of coherency, they are not themselves semantically interpretable on a symbol surface.

Nonsymbolic Symbol Strings

This section addresses the second of the two requirements for a connectionist PS. The kind of computational extensions constructed by the PS, the first requirement, are SSAs. However, on what basis are such extensions constructed? In a connectionist PS, operating below a symbol surface, SSAs are constructed on the basis of propositional representations, on the basis of symbolic strings; yet propositions are, quintessentially, symbolic entities. So what is a connectionist proposition? Or, indeed, what is a nonsymbolic symbol string? Perhaps things would be clearer if I referred to a nonconcatenatively compositional unit representation, spatially (as opposed to syntactically) structured, but recursively constructed so as to preserve syntactic structural similarity relations between its constituents? The answer is probably not, and its more of a mouthful, so I will stick with nonsymbolic symbol strings.

The fact that connectionist representations are able to encode symbolic information, normally always available to inspection on a symbol surface, in a non-concatenatively compositional form *below* a symbol surface is likely to irritate the theorist brought up in the clear air of symbolic computation. If a representation is nonsymbolic (ie. nonconcatenatively compositional) then the gist of the complaint is how can it represent symbolic information? Here we have an instance where *descriptions of things*, and *things described* have become confused.

All modern descriptions of language, for example, are couched in terms of other languages (ie. other systems of representation) that adhere to the symbolic conception, whereby context-independent atomic symbols combine in a concatenatively compositional manner to yield more complex molecular symbols. All public natural languages, logic formalisms, and mathematical notations are "classical" in this sense, as van Gelder (1990) makes clear. A classical *description* of language, however, does not necessarily mean that the thing described—language—is a system of representation obeying the strictures of a concatenative compositionality and displaying a syntactic structure (indeed, this insight is one of the most important contributions of connectionism to the study of language and the mind). It is equally as plausible that language is a representational system comprising tokens that combine nonconcatenatively, which is to say, in a purely functional manner, and that exhibits a structure best understood in spatial terms.

The use of the term *spatial* is not meant to indicate that a connectionist representation is, internally, a spatially structured object. Rather, descriptions of language in terms of an underlying spatial structure, as opposed to an underlying syntactic structure, appeal to nonformal, nonsyntactic structural similarities usefully understood as similarities of *location* in high-dimensional computational spaces.[1] Thus, and to reiterate the point, although descriptions of language are universally classical, the thing described—language as it exists in the head—may not be classical. The claim is not that outrageous: The hypothesis that language, as it exists in the head, is a system of representation conceived of in classical terms, has just as much *empirical* evidence supporting it as the hypothesis that language, as it exists in the head, is a system of representation conceived of in connectionist terms.

A formally grounded PS assumes that its input will be propositional or symbolic strings. Fortunately, the connectionist theorist has at his/her disposal the means to construct nonconcatenatively compositional representations of symbolic strings (that are nevertheless recursively structured so as to preserve the symbolic

---

[1] Chapters 8 and 9 contain more information about this notion of spatial structure.

information in the description), in the form of the Recursive Auto-Associative Memory (RAAM) of Pollack (1990). Specifically, we will make use of a *sequential* RAAM network, the general architecture of which is shown in Figure 3.14. The RAAM network is essentially an elaboration of the two-layer, feed-forward architecture using back propagation, augmented by its ability to recursively encode and auto-associate variable-sized data structures, such as trees or lists.

Consider the single assertion, **B is on the left of C**. This would be presented to the RAAM network as a bracketed tree structure such as **C(L(B nil))**. On the first processing step, the symbol **B** is input to the network as a terminal symbol over the **branch 1** units, along with the **nil** symbol over the **branch 2** units (in this case, **nil** was represented by a uniform value of 0.5 over all **branch 2** units). The auto-associative nature of the RAAM network requires that **B** and **nil** be reproduced on the **branch 1*** and **branch 2*** output units, respectively. In performing this mapping, the network develops a compressed hidden unit representation of the composition of **B** and **nil**, which is labeled $R_{(B,nil)}$ in Figure 3.15.

On the second processing step, $R_{(B,nil)}$ has been passed down to the **branch 2** units. The **L** symbol is now presented as input on the **branch 1** units, along with $R_{(Bnil)}$ and the RAAM network is required to auto-associate **L** and $R_{(Bnil)}$ on the **branch 1*** and **branch 2*** output units, respectively. Once more, the hidden units develop a compressed representation in order to perform this mapping, which is labeled $R_{L(B,nil)}$ in Figure 3.15. On the third processing step, the input comprises **C** as a terminal symbol over

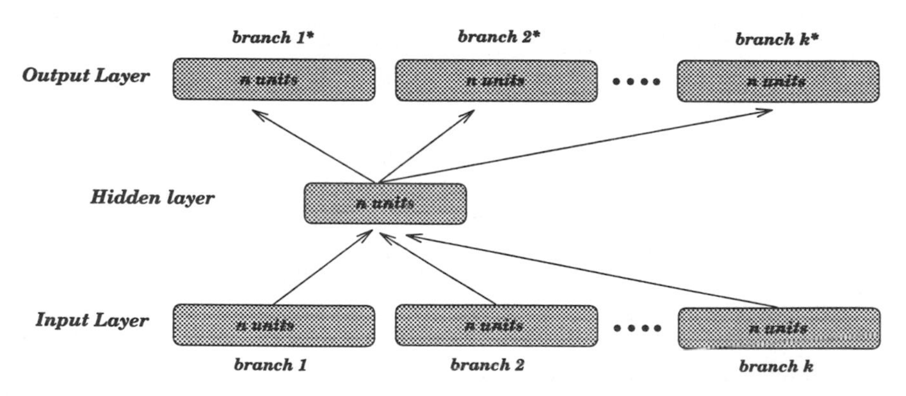

FIGURE 3.14. The general form of the RAAM architecture used for encoding and decoding arbitrary size trees. In this diagram, *k* branches, each comprising *n* units, are compressed into *n* units.

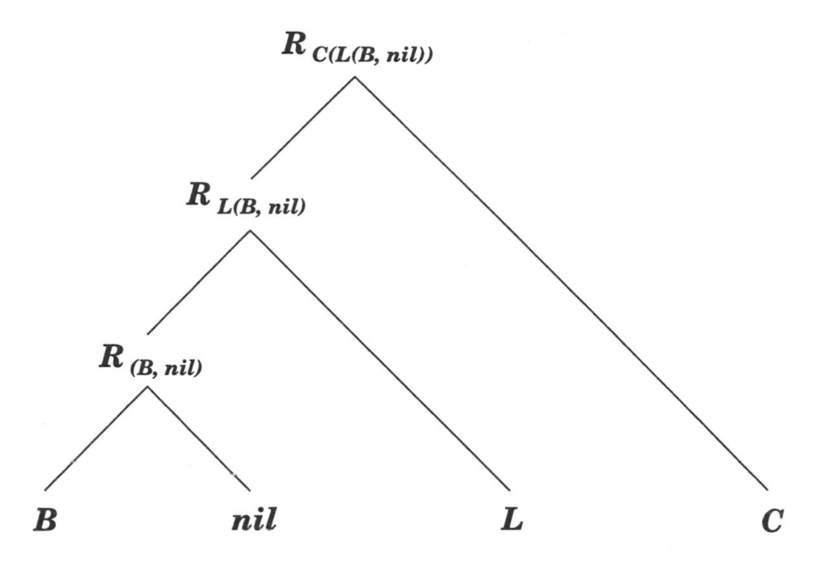

FIGURE 3.15. The sequential construction of a recursive connectionist data structure representing the assertion **B is on the left of C**.

the input **branch 1** units and $R_{L(B,nil)}$ over the **branch 2** units. Performing auto-association causes the hidden unit of the network to develop the final compressed representation, labeled $R_{C(L(B,nil))}$ in Figure 3.15, which represent the entire assertion. For sequences of assertions, this process of auto-association and recirculation is exactly the same. The sequence **B is on the left of C, A is in front of B**, would be presented to the RAAM network as a bracketed tree structure, such as **B(F(A(C(L(B,nil)))))**. Figure 3.16 shows the construction of the final complex representation from its constituents.

Using the three symbols **A, B, C**, and the four spatial terms **left, right, front**, and **behind** means that it is possible to construct 24 bracketed tree structures representing single assertions, and 384 bracketed tree structures representing sequences of assertions, giving a total of 408 possible inputs. A (7,25)–25–(7,25)[2] network configuration learned to tolerance after some 11,375 cycles, with values of learning rate and momentum of 0.075 and 0.7, respectively. For the 24 single assertions, 24 hidden unit representations

---

[2] This labeling convention reflects the architecture of the sequential RAAM: On the input, seven units code for terminal symbols, and the remaining 25 units code for the hidden unit representation recirculated back down. Because the RAAM is an auto-associative engine, the number of output units must equal the number of input units.

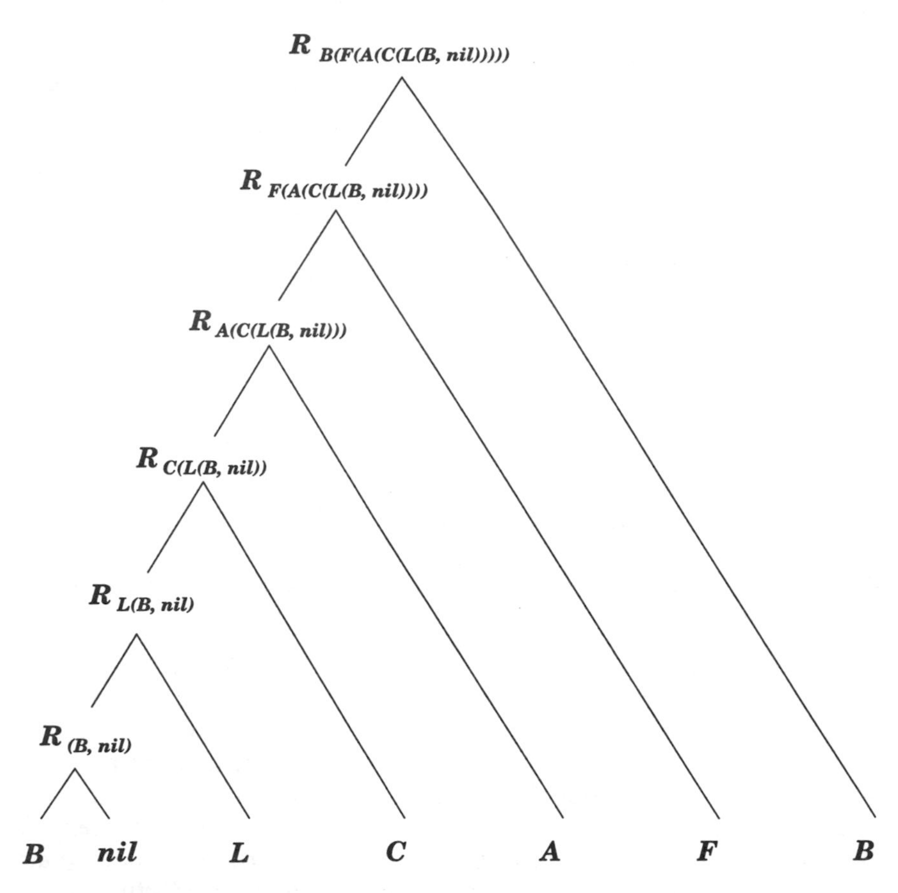

FIGURE 3.16. The sequential construction of a recursive connectionist data structure representing the sequence of assertions **B is on the left of C, A is in front of B.**

corresponding to $R_{C(L(B,nil)}$ were extracted, and for the 384 sequences of assertions, 384 hidden unit representations corresponding to $R_{B(F(A(C(L(B,nil)}$ were also extracted.

Using the RAAM network allows linguistic strings to be encoded in a manner that preserves the symbolic nature of the information that they carry, but that does not result in a representation which is itself symbolic. Quite how to characterize such representations is unclear: One intriguing possibility is as virtual symbolic representations. That is, one can treat these RAAM representations as if they were symbol strings (where "symbolic" means concatenatively combined, syntactically structured), but where virtual syntactic structure similarity relations are being supported by lower level spatial structure similarity relations.

Combined Net

Adhering to a spirit of economy of processing, the **Combined net** simulation experiments reported here revert back to a feed-forward architecture, but use training sets comprised of nonconcatenative representations extracted from the hidden units of prior simulations. The integration of the content of two successive assertions would thus not be accomplished *temporally*, as in the case of the **Recurrent net** simulation experiments, but rather, would be accomplished *structurally*. Thus, these **Combined net** simulation experiments involve structure-sensitive operations on nonconcatenatively compositional representations in the manner of, for example, Chalmers (1990), Niklassen and van Gelder (1994), and Niklasson and Sharkey (1992).

The training set input to the **Combined net** simulations comprises recursively constructed unit representations derived from the operation of a sequential RAAM network. These representations have a nonconcatenative compositional structure resulting from the composition that constructed them, but they function as linguistic strings by preserving symbolic information about the relative orderings of tokens. A useful way to characterize these RAAM representations, as I suggested, is to describe them as having a virtual syntactic structure. The training set output from the **Combined net** simulations comprise SSAs derived from the collapse2 strategy. These representations also have a nonconcatenatively compositional structure, but they differ from the input RAAM representations by functioning as structural analogues of states of affairs in the world. A useful way to describe them is as having a virtual analogue structure.

The total corpus of possible outputs of the **Combined net** coded over 18 units comprise the SSAs of both one- and two-place relations. Each one-place SSA corresponds to (ie. has associated with it) two different RAAM encodings, and thus can be constructed by the connectionist PS on the basis of those two inputs. Equally, each two-place SSA also corresponds to (ie. has associated with it) a number of different RAAM inputs, and hence can also be constructed on the basis of those different inputs. For example, the SSA illustrated in Figure 3.17(a) can be constructed from the 16 different RAAM encoded sequences of assertions shown below it, and the SSA illustrated in Figure 3.17(b) can be constructed from the eight different RAAM encoded sequences of assertions shown.

The procedure for testing the generalization performance of the various **Combined Net** simulations was not a straightforward one. Because the set of both input and output vectors were continuously valued, the outputs produced by a given network were also similarly continuously valued. This meant that merely comparing the individual output activations of the network produced via testing with a target would be quite possibly misleading. Indeed, this was found to be the case: The activation values of particular units in a given output vector were found to differ from their targets quite substantially when the activation value of each constituent unit in a given vector was considered individually.

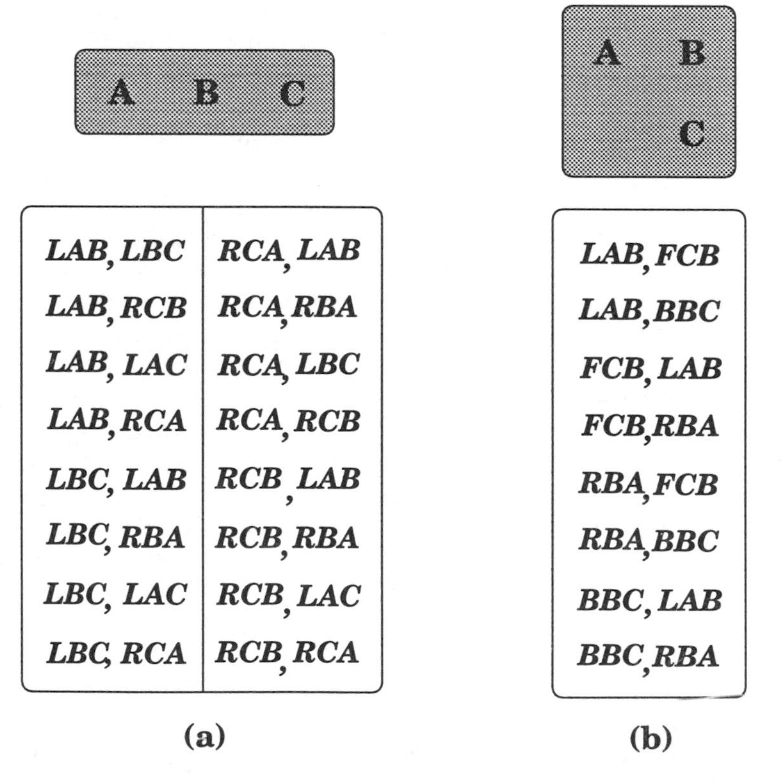

FIGURE 3.17. Illustration of two SSAs and the RAAM encoded sequences associated with them.

In order to avoid this difficulty, the continuously valued output vectors produced by a given test set of RAAM inputs was first decoded before a measure of generalization was established. The decoding was accomplished by employing the hidden to output weights of the simulation that constructed the SSA in the first place. That is, by inserting the output produced during generalization testing of the **Combined net**, into the hidden units of the appropriate SSA simulation (in fact, a stage **c2** simulation) and then allowing that pattern of activation to be processed by the hidden to output weights, means that the vector emerges on a symbol surface provided by the output units of that **c2** SSA simulation. I could then use a standard measure of generalization. The **Combined net** simulation experiments typically comprised a 25-24-18 network architecture. A number of different simulations experiments were run, with each being required to learn a different training set. Training times were typically in the order of

In light of these figures, it is worth considering that unaugmented back propagation over a standard feed-forward net is notorious for producing variable figures of generalization. There are a number of augmented architectures and algorithms that result in considerably better performance, for example, the RuleNet architecture (McMillan, Mozer, & Smolensky, 1991), which has been reported to increase generalization efficacy in the range of some 300 to 2,000% over standard back propagation. Although no augmented architectures or algorithms were used in my investigations, obviously their use would be beneficial if higher degrees of generalization were deemed necessary in order to support stated theoretical positions.

The **Combined net** simulation experiments are interesting for a number of reasons. First, the manner of representation manipulation that they exhibit involves mappings from (representations with a) virtual syntactic structure to (representations with a) virtual analogue structure, where both classes of virtual relations are supported by similarity relations defined in terms of location in a high-dimensional space, which is to say, supported by *spatial* structure similarities. Second, we have an instance of structure-sensitive operations on nonsyntactically structured representations. Third, the **Combined net** simulations show that microprocedural representations can be selected to participate in function mappings by, not individual tokens, but by the *superpositional* constituents of the nonconcatenatively compositional input RAAM representations. Let's examine each of these points in more detail.

Structure Sensitive Operations

One of the most useful things, to recall, about syntactic structure similarity relations is that representations so structured are amenable to being operated upon by processes *sensitive*to that structure. Another way to express this point is that classical processes are systematic. *Systematicity* refers to the fact that law-like, regular relations hold between processes operating on representations. In the case of the human mind, systematicity consists in the fact that, for example, the individual able to entertain the thought that *Salieri respects Mozart* must, by necessity and in order for the mind of that individual to be systematic, be able to entertain the thought that *Mozart respects Salieri*. In order for a mind to be systematic, the argument goes, a number of conditions must be satisfied, principal of which is that the content of thoughts is determined, in a uniform way, by the content of the context-independent concepts that are the constituents of such thoughts. Therefore, the thought that *Salieri respects Mozart* has the same constituents as the thought that *Mozart respects Salieri*.

Translated into the vocabulary of classical computation, systematicity depends on the fact that a molecular classical symbol, such as SALIERI RESPECTS MOZART is constituted from the atomic symbols SALIERI, MOZART, and RESPECTS. Moreover, the atomic symbol SALIERI expresses the same content in the context of RESPECTS MOZART as it does in the context of MOZART RESPECTS, picking out *Salieri*, who is an element of both the proposition *Salieri respects Mozart* and of the proposition *Mozart respects Salieri*. The classical theorist holds that systematicity can *only* be explained by appealing to the fact that mental representations have context-independent constituents, and that those constituents, when they combine, result in a molecular compound that has a syntactic structure.

Naturally, and due to the fact that connectionism is widely perceived as having access to *neither* context-independent representations nor a resultant combinatorial syntactic structure, systematicity is touted as a big problem for non von Neumann computation. In line with the theoretical force of the Microprocedural Proposal however, we can address both of these misconceptions immediately. That is, although it is correct to say that *unit* representations constitute a context-dependent resource, it is equally as true to say that *weight* representations constitute a context-independent resource. Microprocedural weight representations express the same content in any number of different contexts. They do not combine concatenatively to form compound syntactically structured representations, of course, as happens in classical symbol systems, but rather they combine nonconcatenatively, in a context-dependent functional composition, to yield a context-dependent unit representation.

By accepting the major theoretical force of the Microprocedural Proposal, connectionists can begin to answer the challenge posed by classical theory, that systematicity must be a *necessary* feature exhibited by a connectionist architecture. They are able to do this because the distinctive genus of weight representations provides precisely the resources that the classical theorist charges is missing from the connectionist approach, namely, context-independent constituents of compound representations.

Selecting Microprocedural Representations

In both of the **FF net** and the **Recurrent net** simulation experiments, the input unit representation was a binary vector. Thus, the representation was noncompositional (in both a concatenative and a functional sense). To reiterate the point, the input unit representation was efficacious solely to the extent that it was interpretable on a symbol surface. The lack of a dissociation between token and representation meant that the selection of microprocedural representations, effected by individual tokens, could be clearly seen. In the case of the **Combined net** simulations, however, inputs were nonconcatenative unit representations. Thus, the constituents of such representations were themselves distributed, nonconcatenative entities also.

It is worth mentioning at this point that a classical theorist would consider that the decomposition of a nonconcatenative representation into its superpositional constituent parts is, in principle, impossible, because such constituents are not themselves physically tokened in the compound whole: There is nothing to decompose the compound representation into. This would indeed be the case if concatenation were the only means available for constructing compound representations from more basic constituents, but it's not, of course, because there are *non*concatenative means for constructing compounds from more basic constituents also. In fact, Sharkey and Jackson (1994a) were able to show that just such a decomposition (of nonconcatenative unit representations into context-dependent, superpositional constituents) was indeed possible, and moreover, were able to show this experimentally.

So, connectionist unit representations do have context-dependent constituents, which are causally efficacious in computation. They just happen to be superpositional. What this means for the **Combined net** simulation experiments is that microprocedural representations were being selected to participate in function mappings by the constituents of the input unit RAAM representation. Note, not by individual tokens (i.e. units), because in the RAAM representation, tokens and representations have dissociated. Rather, in the **Combined net**, microprocedural representations were being selected by the context-dependent, superpositional constituents of the nonconcatenative input unit representations.

Hyperplane Analysis

Compile and *execute* are the stalwarts of a symbolic PS, and I introduced the term *engage* earlier as an intuitively illuminating (but ultimately impoverished) connectionist alternative. However, there exists another alternative way of conceiving of what happens in a connectionist PS when microprocedural representations are selected to participate in function mappings. That alternative is based on techniques for the visualization of network computation devised by Sharkey and Sharkey (1993). I remarked in an earlier section that a connectionist may find it useful to distinguish a computational level of explanation from a representational one. A computational level of explanation is concerned with the properties of individual tokens (both units and weights), how they can be manipulated, and so on, but quite what a representational level of explanation might be was not spelled out. Here it will be: Specifically, a representational level of explanation is concerned with the emergent properties of populations of tokens.

The hyperplane analysis presented in Chapter 2 is a good example of a representational level of explanation. This way of looking at network computation is extremely useful (see Sharkey & Jackson, 1994a; Sharkey & Sharkey, 1993, for more details), because it allows the theorist, as I have said, to view the emergent properties of populations of tokens (i.e., representations) clearly and simply.

The application of hyperplane analysis to a connectionist PS is very straightforward: Instead of using compile and execute as explanatory terms (or indeed the defunct engage term) in a connectionist PS, we are able to use *transform* and *partition* as explanatory terms. The microprocedural representations emergent from the populations of tokens of the IH layer of weights transform one computational space into another: This has the effect of forcing points in the input unit *n*-dimensional space that need to be treated similarly closer together in the hidden unit *m*-dimensional space, and forcing input points that require different functionalities further apart in hidden unit space. The microprocedural representations emergent from the populations of tokens in the HO layer of weights then project decision hyperplanes onto the *m*-dimensional hidden unit space, and they effect a partitioning of that space, such that the result of the function being computed can emerge on the output units.

I should stress that transform and partition do not add anything theoretically substantive to a connectionist PS, although they do, of course, constitute one more way in which it can be seen to differ from its symbolic relative. Their most important contribution is that they provide a novel means of understanding what weight representations do in a connectionist PS when they deliver an interpretation given a symbolic string. Specifically, they first transform a computational space and then partition it.

This section of the chapter has been concerned with the properties of a novel form of nonsymbolic PS, which is formally grounded, but which operates below a symbol surface. I began the section by considering why an advocate of an anti-PS position would object to both the **FF net** and the **Recurrent net** simulations, and I concluded that their principal failing lay in the noncompositional nature of the unit representations (on both the input and output).

The construction, the properties, and the novel features of a superpositional structural analogue or SSA were detailed. An SSA is a unit representation with a non-concatenatively compositional structure which supports analogical similarity relations. That is, the SSA can be treated as if it were another kind of representation altogether, namely, an analogical one. In terms of the SSAs, analogical structure was defined experimentally as the lack of a syntactic structure (ie. symbolic information, normally construed). The collapse2 strategy was designed to both banish unwanted symbolic information from a corpus of binary output unit representations, and also to effect a distributing transformation on the raw output vectors. The latter transformation had the effect of creating nonconcatenatively compositional entities in which representations have dissociated from tokens, and hence, have dipped below a symbol surface.

The nature of the "propositional representations" from which a connectionist PS learns to construct SSAs was also detailed. Such representations require that symbolic information (about ordering of tokens, and such like) be preserved, without the representation itself being symbolic. Such nonsymbolic symbol strings (or NSSSs) are nonconcatenatively compositional, spatially structured, and recursively constructed unit representations, derived from the hidden units of a sequential RAAM network. Just as SSAs were described as having a (virtual) analogue structure, so the fully distributed unit representations extracted from the RAAM network can be described as having a (virtual) syntactic structure, where both virtual analogue and virtual syntactic similarity relations are supported by spatial structure similarity relations.

The **Combined net** simulations, mapping from NSSSs to SSAs, exhibited a novel form of structure sensitive operations on spatially structured representations (as opposed to syntactically structured representations). Given that connectionist theorists are able to avail themselves of a context-independent representational resource, emergent from populations of weight-tokens, coupled with the structure sensitivity exhibited by the **Combined net** simulations, means that the connectionist is not *necessarily* impaled on what Sharkey and Jackson (1994a) have called the systematicity horn of the representational trilemma (contrary to the muse of the classical, symbolic theorist). This point becomes more obvious when one considers that, in the **Combined net** experiments, microprocedural representations were being selected to participate in function mappings by the superpositional constituents of the input RAAM representations.

CONCLUSIONS

In this chapter, an attempt has been made to show how novel conceptions of representation and structure emerging from a study of connectionist networks can contribute to the representation of meaning in cognitive science.

Initially, I discussed how an equivalence between meaning and procedure underlies the procedural approach to meaning, and considered some of the objections of an anti-PS position. The first is that PS is parasitic on a prior formal semantics. By defining a formally grounded PS, the intended connotation of the term *parasitic* becomes more benign. The most important criticism of the anti-PS position however, concerns whether PS should be considered a semantic theory—advocates of the anti-PS position arguing that it is not because it is merely a theory of sentence comprehension. Likewise, I have argued that PS is not a semantic theory either, but for entirely different reasons. Rejecting the notion of a monolithic semantic theory, I characterized PS as a theory of meaning and, specifically, as a theory of intensional grounding concerned with the relation between language and #worlds#.

I then introduced the Microprocedural Proposal for a formally grounded connectionist PS. This proposal urges that microprocedural weight representations, emergent from populations of weight-tokens, constitute a valuable representational resource in and of themselves. Such weight representations are qualitatively different from the unit resource in connectionist networks due to their context-independence. Pursuing an equivalence between meaning and procedure results in the axiom of a connectionist PS—that microprocedural representations, emergent from populations of weights-tokens, are able to encode an intensional meaning. Because a connectionist PS is formally grounded, it seeks to specify functions that yield an interpretation given a symbol string. Thus, the microprocedural representations serve as procedures that construct unique, connectionist interpretations called SSAs in the **Combined net** simulations.

The **FF net**, and **Recurrent net** simulations were meaningfully interpretable on a symbol surface due to the fact that both input and output unit representations exhibited no dissociation of tokens and representations. The value of these simulations lay in the fact that the selection of microprocedural representations by unit tokens could be clearly seen. The second, **Combined net** simulations, however, despite being able to support a meaningful interpretation, dip below a symbol surface. Both input and output unit representations are nonconcatenative compositional representations due to the dissociation of tokens and representations that resulted from the distributing transformation that initially constructed them.

The input unit representation, extracted from the hidden units of a sequential RAAM, were described as having a virtual syntactic structure, and the output unit representations, extracted from the hidden units of the final in a series of collapsing networks, were described as having a virtual analogue structure. The kind of computation exhibited by the **Combined net**simulations thus exhibited a number of novel features. First, mapping from representations with a virtual syntactic structure to representations with a virtual analogue structure involved structure-sensitive operations on nonconcatenatively compositional representations. Second, microprocedural representations were selected by the **Combined net** simulations by the superpositional, context-dependent constituents of those unit representations. Additionally, and by employing the explanatory form of a hyperplane analysis, I was able to offer a novel characterization of the behavior of a connectionist PS by replacing the symbolic *compile* and *execute* terms with the connectionist terms *transform* and *partition.*

4

Theories of Meaning: Formal

Tom Auto was out walking in the park with his **Associative Grandmother**, when he spied a crowd of people gathered around the edges of a playing field.

"Look Granny!" he said excitedly "A game!"

Indeed it was. Fourteen individuals, divided into two teams of seven via colored shirts, were running enthusiastically and at high speed about the field, throwing and catching a circular disc some 30 centimeters in diameter as they went. Strange shouts floated back to **Tom** as he watched: *"force line!," "second cut? ," "point block," "layout!,"* and other things.

"I wonder what all those strange phrases mean Granny?" asked **Tom**.

His **Associative Grandmother** looked down at him and shrugged, displacing a tartan shawl from about her shoulders. The vocabulary *was* a bit odd, and although the words by themselves made sense (**Tom** knew what *"point"* meant and he also knew what *"block"* meant), still he didn't know what their contatenation, namely *"point block,"* meant. Nevertheless, **Tom** continued to watch, intrigued, and as the game progressed, he began to see what, in the context of play, the strange phrases might mean. Presently, a question occurred to him.

"I wonder what my '*point block*{ means compared to the players '*point block*'?" "What is it that makes '*force line*' mean ' *force line*', and not, say; '*force middle*'?" another phrase he had heard wafting from the playing field), he thought. Presumably some aspect of the game, he surmised. Or *maybe* it was some aspect of what the players *thought* the game was. He asked his **Associative Grandmother**.

"Ach Tommy, you're wanting to know whether, philosophically speaking, meaning is a mental construction imposed on the game, or whether it is something determined by the game, by the world, quite separately from any other consideration," replied **Granny**, rearranging the shawls about her shoulders.

Tom Auto thought long and hard about this.

"You mean, is meaning in the *head* or in the *world*?" he paraphrased, and was pleased to see his **Associative Grandmother** nod in affirmation.

"And which is it Granny?" asked **Tom Auto**.

"Lad, don't ask such silly questions," replied **Granny**, and she cuffed him around the ear.

Modern, truth conditional, theories of meaning can be historically traced to the thinking of philosophers intoxicated by the heady atmosphere characterizing much of 16th and 17th Century physical sciences. The iconoclastic scientifica of the time, pioneered by Galileo and Kepler and culminating in Newton, seemed to show that a huge gulf existed between (what had traditionally been regarded as inviolate) appearance and reality. The reality of the objective world, if the physicists were to be believed, consisted solely of *corpuscularian matter*, whereas the appearance of the objective world (in terms of shape, color, sound, and so on) was, of course, very different. In these revolutionary times, philosophers were forced to reconstruct their conception of the world and of the mind so as to accommodate (what seemed to be) the indisputable advances of science, and accept that the sensory apparatus of a human perceiver was continually exposed to the impact of the world, as it was in itself, in the form of myriads of *corpuscles*. The result of such bombardment, so the new conception of the mind went, was a rich array of sensory data, variously called "impressions" or *ideas* that imbue the mind with any subject of experience. "Ideas" included notions of color, sound, smell, taste, and texture. Such ideas were (usually) conceived of as some manner of (pictorial) mental representation and constituted an interface between the knowing subject and the objective reality.

The prevalent conceptions of the time can be seen quite clearly in the work of John Locke. Locke accepted the existence of a (corpuscularian) objective reality (unlike Hume, for example, in whose philosophy both the knowing subject and the objective reality disintegrated into fictions generated by faculties of the imagination) but conjectured that there had to be something that lay between reality and an individual's understanding of that reality. Locke used the notion of *idea* to explicate this conjectured something. On his view, items in a language stood as proxy for ideas—words (such as "cat") standing for ideas and general words (such as "red") standing for general ideas. Hence, items in a language stood for objects in the world only derivatively, since it is ideas that stand for objects; words stand for ideas that stand for objects.

The problem-setting context for philosophers was the picture of reality that emerged in the physical sciences at this time. The materials on which the philosophers conceived themselves as required to work were the ideas in the mind, and the principles of organization whereby the mind ordered those ideas. Ideas were conceived as complex or simple, complex ideas decomposing (after suitable analysis) into simple ones. Definable words corresponded to complex ideas, indefinable words to simple ideas. This distinction gave rise to what was known as the biplanar conception of a language (or what Wittgenstein called "Augustine's picture of language"), whereby words corresponded to ideas (or more generally entities of some description) and sentences corresponded to concatenations of ideas (entities) into judgements. Ideas, on this view, were to be taken as more or less independent of language: similarly, thought, conceived as a variety of operations on ideas, was also viewed as language-independent. The prevalent conception of language was, consequently, that of a vehicle for the communication of ideas.

The biplanar conception of language prevailed in the philosophical literature—notables, such as Leibniz, excepted—up until the middle of the 19th century. It was then that the thinking of the German logician Gottlob Frege became influential. Frege held that language encoded, not ideas (conceived of as mental entities) but concepts and thoughts (conceived of as mind-independent Platonic entities). I remarked in the prolegomenon that the study of theories of meaning requires many distinctions; we have reached another—in fact. the first of two I will introduce here. The first is a very sharp divide indeed, between the schools of *Platonism* and *Psychologism*—Locke was of the school of Psychologism, and Frege was of the school of Platonism. For our specific purposes, Psychologism (also variously dubbed Intuitionism, Idealism, and more recently, Antirealism) can be understood as a commitment to viewing meaning as a mental construction: The world is only meaningful in virtue of the head. Notable advocates of the school of Psychologism include, as well as Locke, the linguist de Saussure (cf. 1960), and Wittgenstein's 1953 work, *Philosophical Investigations*. Platonism, or Realism, on the other hand, holds that meanings are not mental entities at all: The simplest realist thesis instead places meanings in the world, but this is by no means a definitive position. Among the pupils of the school of Platonism can be counted (naturally enough) Plato himself (its "founding father" perhaps), Wittgenstein (in his earlier, 1922 work, the *Tractatus*), and the philosopher Hilary Putnam (cf. 1975). Platonism and Psychologism will turn up again and again in our discussion.

The second of the distinctions that I mentioned concerns the nature of logic: Up until the mid-19th century, the study of logic had been a largely stagnant field, limited to simple syllogistic forms of reasoning, having not advanced to any great extent since the time of Aristotle. The tradition begun by the mathematician George Boole and refined by Frege, however, introduced a new kind of logic, a logical calculus, the implications of which for language, logic, and thought (and their relations) were quite staggering.

To the modern student of logic, his/her discipline is quintessentially about structural features of languages, typically concerned with studying relations of derivability and entailment between sentences. This is a distinctly 20th century conception, however. Up until the middle of the 19th century, logic was seen as primarily concerned with, not languages and their structure, but concepts and judgements. This is a representative view:

> Logic is the art of directing reasoning aright, in obtaining the knowledge of things, for the instruction both of ourselves and of others. It consists in the reflections which have been made on the four principal operations of the mind; conceiving, judging, reasoning and disposing. (*The Art of Thinking*, 1662, Introduction)

Conceiving on this view consisted of analyzing the materials of the mind and of thought (which is to say, its ideas), and sought to clarify the origins of ideas, their relations to corresponding objects (that of which they were ideas), and their classification into complex and simple ideas, and so on. From ideas, the mind forms *judgements*, by a process of concatenation of ideas, and affirms or denies one or other of the constituent ideas of a judgement (a judgement, in modern parlance, is a proposition). A judgement consequently consists of a subject and a predicate. An example judgement might be "Ultimate is fun"[1], the subject of which would be "Ultimate," and the predicate of which would be "is fun." Of the subject, the mind affirms or denies, the predicate is affirmed or denied by the mind of the subject. Logic was, then, conceived as the science of judgement and ideas, and specifically, the science of the laws of thought.

---

[1] "Ultimate" is the name of a team game (the same one that Tom Auto found so intriguing) played with a sports disc.

This conception of logic should be viewed in the context of the aforementioned distinction between Psychologism and Platonism. Psychologism, as it related to the conception of logic and its relation to thought, had a number of forms. Common to all, however, was a conception of logic as a normative science of *thought*. The laws of thought, and hence of logic, were determined subjectively by the nature and structure of the human mind. This view was promulgated by Kant and his followers, among others.

Those of the school of Platonism, on the other hand, vehemently denied that logic and its laws could be assimilated to thought and its laws. Rather, a Platonist viewed logic as studying objective relations between semipaternal, mind-independent (which is to say, Platonic) entities. A Platonic logic spurned ideas (conceived of as mental objects) and put *concepts* in their place: it spurned judgements (conceived of as mental objects) and put (contents of) judgements in their place. Both concepts and judgements were conceived to be abstract, sempiternal objects constituting part of the metaphysical structure of reality. With the invention of formal calculi, however, the whole focus of this debate changed. It is to the detail of this invention that we now turn.

THE INVENTION OF FORMAL CALCULI

Logic underwent something of a paradigm shift in the middle of the 19th century, inspired by the mathematical generalizations presented by George Boole (1815–1864). Boole wished to show the application of generalized mathematical techniques to the formal representation of patterns of logically valid reasoning. His fundamental insight was that just as there could be (and indeed was) algebra (or calculus) defined over mathematical entities (in the simplest case, numbers), so too there could be algebra defined over entities that were not mathematical. In formulating his work, he drew on an analogy between conjunction (and disjunction) of concepts, and addition (and multiplication) of numbers. Boole conceived of his new calculus as a *logical* algebra, and more specifically, as the logical algebra of thought, characterizing it as a "cross section" of rational thinking. His declared aim was: "To investigate the fundamental laws of the operation of the mind by which reasoning is performed . . . and upon this function to establish the science of Logic" (Boole, 1854, p. 1).

In his view, natural language was, at best, a clumsy tool, concealing what his new notation revealed, namely, the abstract, mathematical nature and form of human thought. It is important to remember that Boole did not see his new logical calculus as a tool for *analysis* of language, but rather he saw it as a new, logically perspicuous language in and of itself, which revealed the fundamental nature of thought: "A successful attempt to express logical propositions by symbols, the laws of whose combinations should be founded upon the laws of the mental processes which they represent, would, so far, be a step towards a philosophical language" (Boole, 1847, p. 5).

Hence, we can see that Boole's logical calculus was very much a product of the school of Psychologism, inasmuch as he explicitly referred to the laws of logic as representing corresponding mental processes of reasoning. Boole's work on logical algebra was quickly and enthusiastically taken up by a number of logicians, but formal calculi came of an age in 1884 with Gottlob Frege, who played Newton to Boole's Copernicus, in the *Foundations of Arithmetic.*

Frege was an entrenched Platonist and a relentless opponent of Psychologism in the study of language, logic, and thought. Hence the motivation for his logical calculus was different from Boole's. Whereas Boole had wished to show that logic (and hence thought) were part of mathematics, Frege aimed to show that (part of) mathematics, principally arithmetic, was deducible from logic. And logic, of course, he considered independent of language and of thought. Because Frege was a Platonist, the logical calculus he invented, called *Begriffsschrift* (concept-script), he always interpreted Platonistically. That is, the laws of logic implicit in the *Begriffsschrift* were not intended to correspond with mental processes, but intended to be absolute laws of truth, describing the metaphysical structure of reality. Like Boole, he touted the *Begriffsschrift* as a logically perfect language (he thought natural language left an awful lot to be desired) in which the ontology of the metaphysical reality, consisting of Platonically conceived concepts and judgements, could be clearly and perspicuously represented.

Frege is often accredited with virtually inventing the whole of the discipline of formal semantics. This view is not wrong, and it would be a brave (bordering on foolish) person who would argue that it was. No, the view is not wrong, so much as (adapting a quaint piece of Whitehall ideolect) extravagant with the truth. In large part this extravagance stems from playing down Frege's entrenched commitment to Platonism. Nonetheless, Frege's working out of the *Begriffsschrift* generated a spate of notions now central to formal semantics, including the now celebrated distinction between *Sinn* (sense) and *Bedeutung* (reference), and (what is now termed) the principle of compositionality. I will introduce these important notions as the details of the *Begriffsschrift* unfold.

Frege's crucial innovation was to repudiate the traditional conception of (the contents of) a judgement as composed or synthesized from subject and predicate, and to replace it with a conception of (the contents of) a judgement as decomposable into function and argument. Such function-theoretic analysis had, up until then, been the sole province of mathematics. Following the lead of Boole, Frege generalized this notion so that any entity whatsoever (and not just mathematical entities) could be thought of as the arguments (and values) of functions: Consequently, he viewed the abstract, Platonic concepts and judgements (constituting the ontology of his metaphysical reality) as viable candidates for function-theoretic analysis.

Specifically, Frege considered that a judgement *was* the value of a function for an argument. His conception of a function was very explicit: A function was a concept. That is, he considered that a (particular) concept was a (particular) function mapping an argument (or arguments) onto a judgement.

In the *Begriffsschrift*, he conceived of well-formed formulae as standing for (being proxy of) the (contents of) judgements. A function-theoretic analysis, such as Frege espoused, required, moreover, that the value of a function for an argument be an object; that is, the function-theoretic analysis forced Frege to view the (contents of) a judgement as an abstract, Platonic object. Hence, the well-formed formulae of *Begriffsschrift* were conceived of as singular referring expressions, referring or denoting, that is, to precisely one (Platonic) object. In the Fregean ideolect, singular referring expressions were termed proper names. Frege used "proper name" in a greatly extended fashion, including much more under its rubric than would be common practice in modern applications of the term. Proper names, he considered, had objects as their references: Anything that was not a proper name did not uniquely refer to an object; rather, it referred to a concept.

There is a problem of *grain* in Frege's understanding of the references of proper names, however, in that referring expressions (such as the well-formed formulae of *Begriffsschrift*) often (express meanings that) are more fine-grained than the objects for which they stand. I will use some sentences of natural language to show up this problem. Consider Frege's favorite example, the two expressions "The morning star," and "The evening star." The object to which these expressions refer is the planet Venus. Thus, they refer to the same object. Yet plainly, the two expressions are not the same, and ideally should be distinguishable via some metric. These expressions, and others like them, caused Frege to make the distinction between *Sinn* and *Bedeutung*. (The German word *Sinn* translates as sense, and *Bedeutung* translates as meaning. However, *Bedeutung* is (almost universally) translated as reference in modern, semantic ideolect, and I will follow this convention).

Frege considered the sense of a well-formed formula in *Begriffsschrift* to be concerned with the "mode of presentation" of an object (as reference) by that formula. Thus, the two expressions (in our natural language example) can be seen to differ in the mode of presentation of their reference (the planet Venus), and, are therefore assigned different senses. The metric has been established. Well-formed formulae of *Begriffsschrift* (standing for judgements) he now conceived of as distinguishable via the metrics of *Sinn* and *Bedeutung*: The reference of a well-formed formula (capitulating to the demands of a function-theoretic analysis) was either one of (the metaphysical, abstract, sempiternal objects) The True or The False. The sense of a well-formed formula (which Frege, somewhat bizarrely, called a nonpsychological "thought," a notion likely to baffle any thorough-going Representationalist) Frege conceived of as concerned with the "mode of presentation," which is to say, the manner in which a reference (either one of the The True or The False) was presented as the value of a function for an argument. The sense of a well-formed formula, put another way, was conceived of as the value of a given function for some given argument, where the function comprised the sense of the constituent concept-word of the judgement, and the argument comprised the sense of the constituent argument-expression of the judgement.

We can see from this exposition that Frege also applied the *Sinn* and *Bedeutung* distinction to the constituent elements of a well-formed formula, namely, argument-expressions and concept-words. Thus, he viewed the constituents of well-formed formula as having a reference and expressing a sense. A pertinent question at this point is *how* he applied the *Sinn* and *Bedeutung* distinction to constituents of well-formed formula.

To recall, well-formed formula in the *Begriffsschrift*, Frege conceived as singular referring expressions. In natural language, sentences take the place of well-formed formula, which are, thus, singular referring expressions. Similarly for proper names: The name **Atahuallpa**, for example, has a sense and a unique reference, picking out, as it might be, the individual who was God Emperor of the Inca people of Peru at the time of the Conquistadors.

Well-formed formula and proper names, in Frege's terminology, are *saturated* signs. A saturated sign can imply an unsaturated sign, and indeed this is the case: Consider the sentence, "Ultimate is fun." In this sentence, we can distinguish an unsaturated predicate expression, "is fun," and a proper name "Ultimate", which can be said to saturate the predicate to give a complete thought—the thought that "Ultimate is fun." Unsaturated signs correspond to what Frege thought were the constituent elements of well formed formula that were not proper names, that is, those constituent elements of a well-formed formula that did not refer to objects, but rather referred to concepts.

A concept-word was of the class of entity that was not a Fregean proper name: Thus, it did not have an object as reference, but rather, the reference of a concept-word was a concept. For argument expressions, their reference could be either an object or a concept. In the simplest case, an argument expression would refer to an argument for a function (ie. an object). However, one of the novel features of the *Begriffsschrift* was that arguments could also be concepts; That is, the argument for a (second-level) function or concept could itself be a (first-level) function. Frege used the notion of a second-level function (concept) to explicate the quantifiers of his logical calculus.

The sense of a constituent expression of a well-formed formula in the *Begriffsschrift* (the sense, that is, of either a concept-word or an argument-expression), Frege conceived of as consisting in the contribution the constituents make to the truth value (The True or The False) that was the reference of the formula.

Here we have two important things to note. The first is the contextual principle that Frege appeared to be espousing when he claimed that the sense of a constituent expression of a well-formed formula consists in its contribution to the determination of a truth value. The difficulty in interpretation here is that, when it was coined, in the 1884 *Foundations*, the *Sinn* and *Bedeutung* distinction was in the future (cf. Frege, 1892). The German in the *Foundations* read "der Bedeutung der Worter," which, in terms of his later distinction, translates as reference. However, nearly all modern interpretations (of what has become known as the Principle of Compositionality) assume that what Frege meant was "der Sinn der Worter" (following a suggestion by Dummett, cf. 1978), so that the principle comes out to read "never ask the *sense* of a word in isolation, but only in the context of a sentence."

The second important point concerns the Fregean conception of a truth condition. Frege claimed that by stipulating references for constituent elements of a well-formed formula, he also thereby fixed the conditions under which the formula referred to the value the True: Change the references of the constituent elements, and the conditions that fix the reference of the formula might change, so that (for example) the formula might refer to the the False, instead of the True. Hence, Frege thought that by fixing the references of the constituent elements, he was able to assign a sense to a well-formed formula. On this conception, one can see the beginnings of equating meaning (sense) with truth conditions: In Frege's ideolect, truth conditions would be a specification of exactly how a well-formed formula in the *Begriffsschrift* presents the value True as the value of a function for an argument.

One should be cautious, however, of Whitehallian extravagance: In modern theories of meaning, sentences are not conceived of as the names of truth values, nor are truth values conceived of as objects that are the values of functions for arguments, as they are in Frege's work. Hence, the conception of sense that Frege intended (as the mode of presentation of a reference as the value of a function for an argument) does not equate easily with the modern notion of the sense of a sentence being specified by its truth conditions.

WITTGENSTEIN'S PHILOSOPHICAL ANALYSIS

Differing markedly from that of his predecessors, Ludwig Wittgenstein conceived of logic neither as the study of fundamental timeless relations between abstract entities (as Frege had thought), nor as the study of the laws of thinking (as Boole had thought), but rather as the study of the representational principles of any sign system whatever. Thus, the *Tractatus* can be seen as quintessentially concerned with representation: The simple but powerful philosophical doctrine of logical atomism was employed to explore the nature of those principles legislating for the intelligibility of any possible system of signs.

A perspicuous philosophical analysis would, the *Tractatus* urged, reveal the abstract logical structure of any given sign system: Moreover, it was *precisely* in virtue of the fact that a given sign system had a logical structure that it was able to represent reality. Such logical structure is not, Wittgenstein argued in particular consideration of natural languages, revealed in the surface forms of natural language, but can only be discovered by an appropriate philosophical analysis.

Natural languages, for Frege and Boole, were inferior to formal calculi in terms of such things as clarity of notation, expressive power, and so on, but for Wittgenstein, natural language and formal calculi were representationally equivalent and all in good order: To be a system of representation at all, the *Tractatus* urged, a natural language must have that essential structure that makes representation possible, namely, logical structure. Thus, for Wittgenstein, devising and constructing formal calculi was not an enterprise that would terminate in a (Boolean) philosophical language, but rather a means for bringing to light what is hidden in the symbolism of language itself.

The *Tractatus* is, as has been remarked, about representation, the essence of which was taken to be the description of states of affairs by propositions: Wittgenstein discussed two types—atomic propositions and molecular propositions. An atomic proposition depicts (elementary) states of affairs and is composed of simple, unanalyzable names. Names refer to semipaternal objects in reality, such as spatiotemporal points, which were conceived by the *Tractatus* as comprising the metaphysical substance of the world. Names are arbitrary and the logicosyntactic rules by which names are combined to form atomic propositions are conventional. The logical syntax of the name must mirror the combinatorial possibilities of the objects in reality (for example, in the case of a spatiotemporal point, its combinatorial possibilities would be given by its ability to combine with other such objects to constitute a state of affairs). The relation between an atomic proposition and the (elementary) states of affairs that it depicts, in the formulation of the *Tractatus*, is essential and internal: The atomic proposition is logically isomorphic to the states of affairs depicted. Or, to put it another way, the proposition is a "logical picture" of reality of a particular states of affairs.

The doctrine of logical atomism that Wittgenstein employed uses these notions of atomic and molecular propositions, and can be couched in a logically atomic vocabulary, so as to read: Every molecular proposition is a truth function of atomic propositions. Which is to say, that every molecular proposition is equivalent to the output of truth functional "operators" applied to a (determinate) stock of atomic propositions. The "operators" mentioned here are none other than the logical constants of a formal calculi.

Differing from the logic of Frege once more, Wittgenstein did not consider that the logical constants (qua connectives, qua primitives) of a logical calculi, such as "v," "→," and so on, were themselves constituent elements of well-formed formula. Rather, he considered that they were operations rendering molecular propositions from atomic propositions. The *Tractatus* showed that truth tables can be drawn up correlating any number of atomic propositions (drawn from a determinate stock) with all possible distributions among them of truth possibilities: The concatenation of atomic propositions and connectives to form molecular propositions simply indicates a select distribution of truth possibilities among the constituent atomic propositions. There are two consequences of this strict interpretation of a logical atomism: Firstly, that any molecular proposition is true only as a function of its connective and of its constituents; and secondly, that the connectives, viewed as truth functional operators, could be given precise truth tabular explanations belonging to the syntax of logic.

It is important to realize that Wittgenstein's explanation of truth conditions in the *Tractatus* is intimately connected with the logical atomism that he espoused. Specifically, the explanation of truth conditions can only be applied to any well-formed formula that a perspicuous analysis has revealed to have been constructed by truth functional operators capable of being given a truth tabular definition. Thus, the truth conditions of the molecular proposition "$p \rightarrow q$" are those combinations of states of affairs (determined by the connective) under which it takes the value "T" in the truth table. Consequently, a truth table is the very paradigm of what it means to formulate the truth conditions of a logical formula.

Under the rubric of the logical atomism of the *Tractatus* (which revealed every molecular proposition to be a truth function of atomic propositions), the truth value of a molecular proposition will depend on how truth values are distributed among its constituents. There are, however, two limiting cases—tautologies and contradictions: a tautology is always true (irrespective of how truth values are distributed among its constituents), and a contradiction is always false (irrespective of how truth values are distributed among its constituents). The *Tractatus* maintained that all true propositions of logic are tautologies: A given formula, in other words, expressed a *logical truth* if and only if it was a tautology. Logical truth was not, moreover, held to be a property of fundamental timeless relations between abstract entities (in contradistinction to Frege), but was rather a senseless consequence of an arbitrary, conventional symbolism for combining propositions.

Wittgenstein and his philosophy heavily influenced the intellectual collective of the Vienna Circle of logicians and philosophers, and this influence can be seen quite clearly in the work of Rudolph

Carnap, who was instrumental in framing possibly the first theory of meaning, as a modern student would understand this description. As a background to his thoughts, we will consider two important theses that the Vienna Circle extracted from the *Tractatus*. The first was the taking on board of two maxims from the *Tractatus*: (a) that logical truth was a matter of logical syntax, and (b) that a formula expressed a logical truth if and only if it was a tautology. In *The Logical Syntax of Language*, Carnap urged, in support of Wittgenstein, that truth tables served as syntactic-substitution rules that replaced well-formed formula of a logical calculi (namely, Russell's *Principia*) with matrices filled with "T"s and "F"s. That a given well-formed formula is a tautology (hence that it expresses a logical truth) would then be given by the fact that, under a purely syntactic substitution, a matrix of truth values is generated in which "T" stands in each row under the main connective. Since a well-formed formula expresses a logical truth if and only if it is a tautology, logical truth is purely a matter of syntax, that is, logical truth is independent of any explicit consideration of truth and falsity (for the purposes of truth table construction, no meaning need be assigned to "T" and "F").

The corollary to this conception of logical truth as logical syntax, and the second of our Vienna Circle considerations, is the conception of logical semantics derived from the *Tractatus*, and specifically, the terminus of the philosophical analysis that Wittgenstein espoused. Wittgenstein held that formal calculi offered the means by which manipulation of signs by rules of syntax could be effected. Moreover, he considered that truth could only be defined and countenanced for the manipulation, not for the signs themselves (ie. the atomic propositions and their constituent names). Nothing much was actually said about the signs themselves in the *Tractatus*, certainly nothing about their meanings: In fact, it was in the nature of logical atomism to preclude the possibility of making any significant statement about the meanings of signs. Logical atomism precluded statements about meanings of signs, essentially, because the analysis had to terminate at a point that, for signs and their meanings, was *too early*.

Wittgenstein terminated his analysis, as we have seen, at truth functions of molecular propositions, which left the form of constituent elements of atomic propositions largely unspecified. Consequently, and crucially, the logically atomic analysis of the *Tractatus* left the relation between the indefinable names (from which an atomic proposition would be composed) and the world, that is, the *meanings* of these names, entirely underdetermined. Wittgenstein did give some thought to this problem, this terminus of logical atomism, and arguably what he had in mind was the form of explanation of name meaning that is termed *ostensive definition*: Via ostensive definition, the argument goes, the ineffable connection between symbols and symbolized can be effected.

The Vienna Circle were largely instrumental in promulgating ostensive definition as a form of explanation of meaning, on the basis of which they argued that explanations of meaning fall neatly into two types: Intralinguistic definition, belonging to logical syntax, captured all those aspects of meaning that were susceptible to being stated in a language, and extralinguistic ostension of indefinables, belonging to logical semantics, captured all those aspects of meaning not susceptible to being stated in a language. Only via ostensive definition, the Circle maintained, could a mediation between a calculus of symbols and the reality described by its formula be accomplished. The fact that ostension of indefinables was an extralinguistic activity fostered the conviction among the Circle that meaning was ineffable. They claimed, for example, using a natural language example, that the meaning of a word, such as "lemon", could not ultimately be stated, but had to be shown or given by an act of indication or pointing.

Let me review a little here and see whether I can say anything interesting about truth conditions. We have discovered part of the origins of truth conditions in the work of Frege and Wittgenstein. For Frege, truth conditions were the conditions under which a well-formed formula referred to the semipaternal object the true, conditions fixed by the stipulation of references for the constituent elements of that formula. Couched in terms of the *Sinn* and *Bedeutung* distinction, the truth conditions of a well-formed formula comprise the *sense* of that formula: Stipulating references for constituents of well-formed formula in the *Begriffsschrift* thereby fixed the sense of that well-formed formula. For Wittgenstein, truth conditions became a more expressive tool due to his use of truth tables: The conditions under which a well-formed formula (more accurately, a molecular proposition) is true could, using truth tables, be specified by drawing up a matrix of Ts and Fs.

However, the term truth condition as now used in modern theories of meaning is hardly recognizable as the same, strictly truth tabular circumscribed notion to be found in the *Tractatus*. In both meaning and use it has, if one is of a cynical bent, been perverted by the passing of novel, if empty, linguistic legislation. To those of a more optimistic nature, the term *truth condition* has evolved through a series of conceptual advancements to the point where it is of more than cursory interest to the study and analysis of natural language. Whatever one's view, there is more yet to be discussed; arguments for and against have not yet been introduced. Much was to change with the introduction of a semantic notion of truth by Alfred Tarski. It is to this that I will now turn.

IN SEARCH OF TRUTH CONDITIONS

One minor and two major developments intervened between the introduction of truth conditions into formal logic (by Wittgenstein and, to a lesser extent, Frege) and the present day application to theories of meaning. The first is the one least remarked on—namely, the direct application of truth conditions to expressions (which is to say, sentences) of natural language. When Wittgenstein talked of truth conditions in the *Tractatus*, he was referring to the truth conditions of well-formed formula in a logical calculus. However, it is now common practice to speak about the truth conditions of natural language sentences. That is, whereas in logical calculus, truth conditions had a strictly delineated role in application to well-formed formula, suddenly, their role changed so that an individual could quite freely speak about the truth conditions of a sentence such as *The matchbox is on the left of the lighter*. How did this change, this domestication of truth conditions, come about?

One of the principal causes can be shown to be the practice (which has become increasingly more prevalent) of combining symbols drawn from natural language with symbols taken from formal notations. A familiar problem posed to novice logicians, for example, is whether $\mathbf{p} \rightarrow \mathbf{q}$ means the same as **if p, then q**. Frege would have been horrified. He viewed the *Begriffsschrift* as a novel language in itself—self-contained and disjoint from every other language—and he scrupulously avoided mixing the symbols of his notation with any German. It is this practice, consisting of such innocent phraseology as **if p, then q**, that has been largely responsible for legitimizing direct comparison between formal and natural languages.

Another cause for the domestication of truth conditions can be traced to the *Tractatus* itself, in which one can discern the beginnings of a dissociation between the notion of truth condition and the logically prior notion of analysis. In the *Tractatus*, Wittgenstein argued that any significant (and molecular) proposition is a truth function of atomic propositions. That is, any molecular proposition could be exhibited as equivalent to the output of truth-functional operations applied to some members of a determinate set of atomic propositions.

Truth conditions can thus be specified for any well-formed formula constructed by truth-functional operations, because the resulting proposition will always be molecular, and molecular propositions can always be given a truth-tabular interpretation. Replacing well-formed formula with sentences is, one might think, an altogether different matter; Wittgenstein thought otherwise. A sentence such as, "Ultimate is a team game played with a sports disc," was not (or did not correspond with) an atomic proposition on his view, however. Rather, suitable philosophical analysis would reveal it to be equivalent to a truth function of atomic sentences, which could then be assigned a well-formed molecular formula of its own. Further, this latter, equivalent formula would have specifiable truth conditions simply in virtue of being molecular. The subtle equivocation of truth conditions consists of the fact that because the (hypothesized) well-formed formula corresponding to the natural language sentence has truth conditions, so too must the original sentence itself.

The point should be made clear that Wittgenstein's conception of truth conditions cannot be divorced from his conception of analysis. This is because it was only the specific practice of analysis that Wittgenstein espoused that legitimized any talk of truth conditions: If one abandons analysis, then, strictly speaking, talk of truth conditions is empty. Of course, "speaking strictly" is never something philosophers have been keen on, so the dissociation has nonetheless taken place. At the vanguard of the domestication were the logical positivists of the Vienna Circle, who counted Carnap and Schlick among their number. It was the logical positivists who, brushing equivocation aside, enthusiastically took up Wittgenstein's concept of truth conditions and made it their own in the form of Verificationism, the thesis that meaning can be identified with only those truth conditions that can in principle be established by observation or experiment.

The conception of truth conditions espoused by Wittgenstein and the Logical Positivists survives (more or less intact) to the present day: Verificationism has, of course, fallen (rightly) into disrepute, and the specific form of the philosophical analysis that Wittgenstein espoused has similarly gone out of vogue. Nonetheless, the field was set with pennants flying, and truth conditions were firmly ensconced as champions of meaning. Into this field rode Alfred Tarski and Rudolph Carnap, and, via their independently conducted but related work, they transformed the study of meaning.

TRUTH, SATISFACTION, AND ALFRED

The first major transformation in the conception of truth conditions was initiated by the Polish mathematician and logician, Alfred Tarski, in his famous (1935) paper "Der Wahrheitsbegriff in den formalisierten Sprachen" (which translates as "The Concept of Truth in Formalized Languages"; see also Tarski, 1944). The science of *formal semantics*, the study of the relations between formalized languages and their interpretations, was inaugurated on the basis of this paper and on the new conception of truth conditions that emerged from it. We have two interweaving threads to disentangle: The first concerns the nature of the formal *model theoretic* machinery that Tarski used in defining his concept of truth, which employed all the sophistication of mathematical set theory to specify the interpretations of expressions of a given language, and the second concerns Tarski's precise and narrowly circumscribed ambition to isolate a materially adequate and formally correct definition of truth for formalized languages, and, more specifically, his ambition to state a recursive definition of the expression "true sentence" (ie. true well-formed formula) in the logical calculi developed by Russell in *Principia Mathematica*. We will approach these two concerns in a logical manner.

As we have seen, the original role of the truth condition was as a syntactic entity, defined by truth tables, whose job it was to elucidate the role of the logical constants within systems of formal calculi. It is indeed, as remarked in the Prolegomenon, a curiously lumpy evolution that can transform what was once all syntax to what (on a modern reading) is now all semantics, but this is indeed what has happened: The concept of truth conditions constitutes the theoretical underpinnings of any modern theory of meaning. Hence, in Tarski's conception of truth for formalized languages as a semantic concept lies the origins of all truth conditional theories of meaning. Consequently, it is about time that this business of semantics was made a little bit more precise: We have discussed truth conditions as syntactic entities, defined by truth tables, but what is a truth condition as a semantic concept?

The fundamental notion at the heart of logical semantics is the characterization of the *validity* of an argument, where an argument comprises: (a) a set of premises, each of which is a well-formed formula; and (b) a conclusion, comprising a truth value. An argument is valid if and only if it is impossible for all of its premises to be true and its conclusion to be false. When considering an argument in the predicate calculus (which is to say, a refined, modern version of Frege's *Begriffsschrift*) things are a little different, in that an argument is valid if and only if every *interpretation* in which all of its premises together are *satisfied* is an interpretation in which its conclusion is also satisfied. Here I have introduced two italicized terms, *interpretation* and *satisfaction*: These are central to the modern discipline of semantics, and have gained much popularity. The notion of satisfaction we will encounter in the following explication of Tarski's definition of truth, but the notion of interpretation we will tackle immediately.

Logicians introduced interpretations, or "models," into the predicate calculus to help realize a formal notion of the validity of an argument: an interpretation is a complex affair consisting of a function from the syntactically well-formed formula to elements of a mathematicological *model structure*. An interpretation for a well-formed formula comprises an assignment of some nonempty set of objects, termed the domain of quantification, to serve as the values of the individual variables, an assignment to each predicate of a property or relation defined over elements of that nonempty set (viz. a property or relation is conceived of as a set of objects or a set of sets of objects), and an assignment to each well-formed formula of a truth value. The important thing to note about interpretations in formal semantics is the pervasive use of the twin mathematical notions of set and function. In determining the validity of an argument, abstract mathematical models are constructed to serve as the "semantic values" of expressions of the formal language: The sophistication of the mathematical constructs can be impressive—one can talk about functions of sets of sets of functions for example. Using mathematical set theory enables the theorist to give precise, clear statements about the relations between expressions and (the world stood proxy for by mathematical) objects. Semantics using mathematical constructs is called formal, or more specifically, *model theoretic* semantics. Table 4.1 gives a model theoretic semantics for a (very small) fragment of English, adapted from Johnson-Laird (1983).

As can be seen in the table, each proper name, such as **Jane**, is assigned an interpretation in the model consisting of a corresponding individual in the model structure. An intransitive verb is

Table 4.1.
A Simple Model Theoretic Semantics for a Small Fragment of English.

The Syntax of the Fragment

Basic Words
Nouns: **Jane**, **Alan**, **Henrietta**
Intransitive verb: **laugh**
Transitive verb: **despise**

Syntactic Rules
1. If N is a noun and V_i is an intransitive verb, then $N + V_i$ is a well-formed sentence, where "+" denotes a concatenation of words.
2. If N and M are nouns, and V_t is a transitive verb, then $N + V_t + M$ is a well-formed sentence.
3. If S_1 and S_2 are sentences, then S_1 & S_2 is a sntence, where "&" denotes a concatenation of sentences.

The Semantics of the Fragment

There is a model structure consisting of the following set of individuals: **Jane**, **Alan**, and **Henrietta**, denoted {j,a,h}.

Lexical Rules
1. **Jane** is assigned **j** is its interpretation in the model.
2. **Alan** is assigned **a** as its interpretation in the model.
3. **Henrietta** is assigned **h** as its interpretation in the model.
4. **laugh** is assigned as its interpretation in the model a set of individuals corresponding to those who laugh: {j,h}.
5. **despite** is assigned as its interpretation in the model a set of ordered pairs corresponding to who despises whom: {(j,a),(a,j),(h,a)}.

Compositional Rules
1. A sentence of the form $N + V_i$ is true with respect to the model if and only if the individual consisting of interpretation of N is a member of the set comprising the interpretation of V_i.
2. A sentence of the form $N + V_t$ is true with respect to the model if and only if the ordered pair consisting of the interpretation of N followed by the interpretation of M is a member of the set comprising the interpretation of V_t.
3. A sentence of the form S_1 & S_2 is true with respect to the model if and only if S_1 is true with respect to the model and S_2 is true with respect to the model.

likewise assigned an interpretation consisting of a set of individuals. The interpretation of **laughs**, for example consists of a set containing the individuals **j** and **h**, thereby indicating that Jane and Henrietta laugh. This interpretation of the verb can be equivalently expressed as a *characteristic* function, which returns the value *true* or *false*, depending on whether that individual is a member of the set in question.

The truth value for a sentence is built up recursively by the semantic rules, which operate in parallel with the syntactic rules: First, the lexical rules provide an interpretation for the basic words of a sentence, and second, the compositional rules build up an interpretation of complex expressions from the interpretations of their constituents. For example, the structure of the sentence, *Jane despises Alan, and Henrietta laughs*is built up compositionally in the way shown in Figure 4.1.

We can now turn to the second of the concerns indicated at the beginning of this section, that of Tarski's definition of truth itself. One of the more important considerations involved in a definition of truth, one to which Tarski was sensitive, was the nature of the language in which the definition of truth is to be formulated. Tarski argued that an appropriate definition could only be formulated for formalized languages, which is to say, languages with a fully specified, determinate syntactic structure: His definition of truth he intended to apply to the extended predicate calculus of the *Principia*, not to natural language. In modern interpretations his caution has gone out of vogue. Formal notation is freely mixed with natural language as in the following illustration of Tarski's famous paradigm:

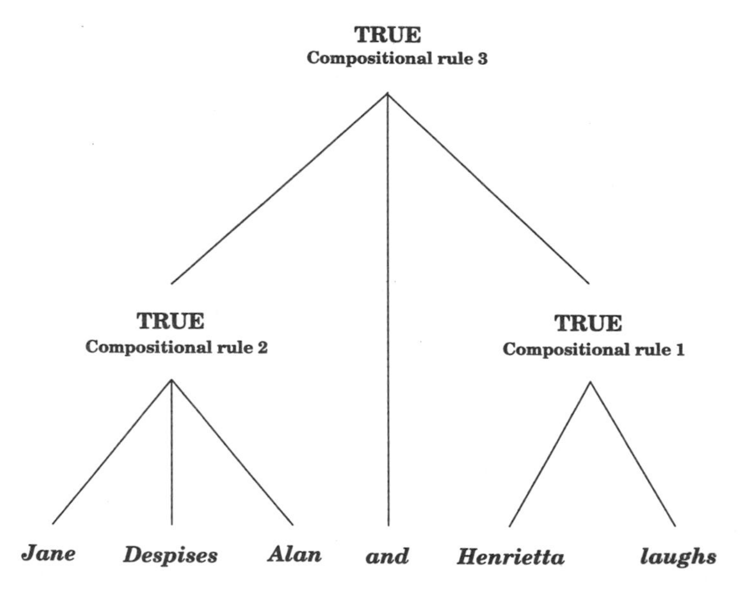

FIGURE 4.1. The structure of a sentence generated by the syntactic rules in Table 4.1.

"Ultimate is aerobic" if and only if Ultimate is aerobic.

It would be as well to remember that this mixing of natural notation with formal principles is absolutely contrary to Tarski's avowed intentions.

The intuitive sense he wished his definition to adhere to, and that formed the basis for his formal elaboration, consisted of the classical Aristotelian conception of truth.

On a modern conception, this conception of truth would be dubbed a *correspondence theory* of truth (as opposed to a coherence theory of truth), in that the truth of a sentence would be viewed in its agreement with (or correspondence to) reality: Tarski set himself the task of ablating the intuitive cancer at the heart of such conceptions of truth. He urged his reader to consider simple declarative sentences, such as "Ultimate is aerobic," and, adhering to the Aristotelian conception of truth, conjectured that such a sentence is true if Ultimate is aerobic, and that it is false if it is not aerobic. If such a statement is correct, he argued, the definition of truth (that he sought) must imply the following kind of equivalence: The sentence "Ultimate is aerobic" is true if and only if Ultimate is aerobic. It is important to note the use/mention distinction in play here. The sentence in quotation marks is the name of the sentence (its mention), but the sentence without quotation marks is the sentence itself (its use).

Tarski then generalized this procedure (which, to the layman, seems curiously *obvious* in a way) to any sentence whatsoever. That is, for any arbitrary sentence, he replaced that sentence (for reasons of clarity of notation primarily) with a letter. For example, suppose the arbitrary sentence was "Floccinaucinihilifilipication is a long word:" This would be replaced by the letter "*f*." Next, he formed a name (or mention) for this sentence, and replaced the name by another letter, say *N*. He then asked himself what the logical relation between the two sentences "*N* is true" and "*f*" might be. His conclusion was that, under the intuitive Aristotelian conception of truth, the two sentences are equivalent. Which is to say, the following equivalence holds:

(T) *N* is true if and only if, *f*.

Tarski called statements such as the one above (ie. where *f* takes the place of any sentence of the language to which the word "true" refers, and "*N*" takes the place of the name of any sentence) an equivalence of the form (**T**). Thus, the conditions under which a definition of truth (more correctly, a definition of the word "true") is materially correct are those conditions that allow the term "true" to be used in such a way that all equivalences of the form (**T**) can be asserted. Put another way, one can say that a definition of truth is materially adequate if all equivalences of the form (**T**) follow from that definition. It should be emphasized that neither the expression (**T**) itself (which is not a sentence, but rather only a schema of a sentence) nor any particular instance of the form (**T**) can be regarded as a definition of truth. Rather, every equivalence of the form (**T**) may be considered a partial definition of truth, and hence only explains in what fashion the truth of a particular sentence consists.

Tarski believed that the conception of truth just outlined belonged to the same family as *designation*, *reference*, and *satisfaction*: Which is to say, he considered truth to belong to the family of semantic concepts. Loosely speaking, semantics is a discipline that deals with certain relations between expressions of a language and the objects or *states of affairs*referred to by those expressions. Tarski called his conception of truth *semantic* because an equivalence of the form (**T**) quite clearly refers not only to a sentence (viz. the sentence of which "*N*" is the name) but also to the states of affairs to which the sentence refers. He was sensitive to the potential problems that would arise from viewing truth as a semantic concept: We will discuss two related aspects of this problem and the solutions that Tarski advocated.

The first problem we have mentioned already, and concerns the specification of the formal structure (and vocabulary) of the language in which definitions of semantic concepts are to be given: Unless this structure can be exactly specified (in terms of classes of word and expression to be considered meaningful, which primitive (or undefined) terms to admit into the language and criteria for distinguishing "sentence hood," and so on) then the problem of the definition of truth cannot be solved in a rigorous way. The second problem with viewing truth as a semantic concept concerns the more specific conditions (other than conditions of structure) that must be satisfied by any language in which (or for which) a definition of truth is stated or required. The conditions can be illustrated by considering the antinomy (or meaningful paradox) most directly concerned with truth, namely, the Antinomy of the Liar. This consists of the following kind of contradiction:

"s" is true if and only if "s" is not true.

Tarski characterized the Antinomy of the Liar as a unique construction of what he called semantically closed languages: Only in such languages was it possible for the antinomy of the liar to occur. Semantically closed languages comprise (minimally) expressions, names of expressions, and semantic terms such as "true" referring to expressions. Tarski's concern was that the language in which the definition of truth (as a semantic concept) is to be constructed should not be a language that also permits construction of paradoxes such as the antinomy of the liar; consequently, he rejected the notion of using semantically closed languages for his definition of truth. Instead, Tarski adopted the convention of using *two* languages in defining truth. The first of these languages is the one that is "talked about," and the one for which a definition of truth is required: Tarski termed this the *object language*. The second language is the one used to "talk about" the object language, and the one in which the definition of truth is to be constructed: Tarski termed this the *metalanguage*. The notions of object- and metalanguages are relative terms, in that if a theorist wished to study the metalanguage as an object language, then he would simply go to a new metalanguage, a meta-metalanguage so to speak. In this way, a whole hierarchy of languages can be envisaged.

The metalanguage can be characterized by a consideration of its properties. Firstly, the condition of material adequacy for a definition of truth, wherein all equivalences of the form *T* are implied, determines the vocabulary of the metalanguage. That is, Tarski envisaged the definition of truth itself, and all equivalences implied by it, as formulated in the metalanguage. A moment's thought will reveal that because, for example, the symbol "*f*" in an equivalence of the form (**T**) stands for an arbitrary sentence of the object language, (the vocabulary of) the metalanguage must include the (vocabulary of) the object language as part. Secondly, the symbol *N* in an equivalence of the form (**T**) represents a name of the sentence for which *f* stands. Consequently, the (vocabulary of the) metalanguage must be rich enough to provide a name for each arbitrary sentence of the object language. Thirdly, because of the lurking danger of the liar paradox, semantic terms (referring to the object language) such as "true" should only be introduced into the metalanguage by definition: If they are not (introduced into the metalanguage solely by definition), then they would have the status of undefined terms in the metalanguage, and the theorist would have to accept the (unwanted) possibility of reconstructing, for example, the antinomy of the liar in that metalanguage.

The considerations of the properties of the metalanguage discussed above are really part of a more general issue—namely, the nature of the relation between the object language and the metalanguage, and specifically, whether the metalanguage is essentially richer than the object language. An intuitive conception of essential richness can be drawn from the discussion of the antinomy of the liar, such that the essential richness of the metalanguage consists in the fact that, in the metalanguage, it is not possible to construct the antinomy of the liar. On this intuitive conception, the condition of essential richness is a necessary and sufficient condition for a definition of truth; that is, without some notion of essential richness, the antinomy of the liar can be constructed in the metalanguage, and hence a definition of truth cannot (without paradox) be stated in that metalanguage. Of what might this essential richness consist?

Tarski illuminated this question by couching his definition of truth in terms of the semantic notion of *satisfaction*. Satisfaction is a relation between arbitrary objects and *sentential functions*. A sentential function is any expression such as "x is white" or "x is greater than y." The formal structure of sentential functions is analogous to that of sentences; however, a sentential function contains free variables like x and y which cannot occur in a sentence (A sentence consequently might be viewed as a limiting case of a sentential function, where "sentence" should be read as "any class of similar formally structured constructions").

Tarski's conjecture was that the essential richness of the metalanguage consists in the type of recursive definition of satisfaction that is employed in defining truth, a type of recursive specification that is not admissible in the object language. To obtain a recursive definition of satisfaction (couched in the metalanguage), a recursive procedure is employed that indicates which objects *satisfy* the simplest sentential functions. The procedure is then employed again, to state the conditions under which given objects satisfy compound functions. The definition of satisfaction thus obtained (as a recursively specified relation between objects and sentential functions) can be used in defining not only truth for such functions, but also truth for their limiting cases, namely, sentences. Consequently, the following can be stated: A sentence is either satisfied by all objects or by no objects.

Using this definition of satisfaction as a recursively specified relation between sentential functions (and their limiting cases) and objects, formulated in a metalanguage, Tarski felt able to give the definition of a "true sentence" as any sentence satisfied by all objects; conversely, the definition of a "false sentence" is any sentence not satisfied by all objects.

TRUTH, MEANING, AND RUDOLPH

The semantic concept of truth that Tarski formulated was indeed seminal. At the time it was startling, and it provided Carnap with valuable new insight into how the problem of logical truth as logical syntax shackling meaning to the ineffable could be resolved. Carnap, to recall, was of the Vienna Circle, and heavily influenced by the *Tractatus*. Specifically, Carnap held that, firstly, natural language was in "good order," secondly that a perspicuous philosophical analysis would reveal the logical structure of that language (viz. the structure in virtue of which the language was able to represent reality), and thirdly, that an appropriate analysis must terminate in the truth functions of atomic propositions.

The most important consequence of this thinking, for our purposes, was an accompanying underdetermination of the atomic propositions themselves. The problem to which Carnap sought a solution was how to circumvent this underdetermination. From the *Tractatus*, Carnap took the conception of an atomic proposition as being composed from symbols incapable of further analysis: Because of this, the nature of the connection between a given symbol and its (metaphysical object) referent is left unclear. Hence, how a mediation between a calculus of symbols and the reality described by its formula is effected is left similarly unclear. The notion of ostensive definition was invoked, and promulgated as a viable explanation, by the Vienna Circle to accomplish such a mediation.

In Tarski's semantic concept of truth however, Carnap saw how to do away with ostensive definition as a means of explanation. He seized eagerly on Tarski's paradigmatic:

"Ultimate is aerobic" if and only if Ultimate is aerobic.

In equivalences of this form, Carnap saw a means to both rectify the under-determination of atomic propositions and also, make more precise the meanings (qua truth conditions) of sentences in both natural and formal languages. Granted, his muse went, that an atomic proposition has no *analyzable* complexity, it must still have *complexity*. Such unanalyzed complexity, Carnap conjectured, was the guise in which meanings (qua truth conditions) had been hiding all this time: Using Tarski's formal machinery, the meaning of, for example, an English sentence could be non-trivially stated by an appropriate metalinguistic statement of the form, *N* is true if and only if *f*. That is, the truth conditions for any well-formed formula of the object language could be stipulated by stating a metalinguistic equivalence.

The transformation of truth conditions from syntactic to semantic entities implicit in Carnap's treatment is worth considering. Truth conditions, as we have seen, originated in logic via truth tables. On Carnap's new interpretation, however, a Tarskian T-sentence (ie. an equivalence of the form **T**) was to be taken as a statement of truth conditions. Baker and Hacker (1983) pointed out that the only way to subsume the metalinguistic statement, *N* is true if and only if *f*, under the label of specifying the truth conditions of *f* is to pass some novel linguistic legislation: From *N* is true iff *f*, the theorist can certainly claim to have stated a condition for the truth of *f*, but he/she cannot claim to have stated the truth conditions of *f*, because *truth conditions* is a technical term, defined via truth tables, not metalinguistic equivalences.

Brushing such equivocation aside, however, we are free to concentrate on the properties of Carnap's conception of truth conditions. The applicability of the term is both extended and restricted: Extended because every well-formed formula, whether atomic or molecular, has truth conditions susceptible to being stated via a metalinguistic equivalence. Hence, Carnap envisaged a semantics, resting on the twin supports of set theory and truth conditions as metalinguistic equivalences, cut free from the demands of a terminating philosophical analysis. The applicability of the term is also restricted in that molecular propositions, on Carnap's view, could no longer be considered to have truth conditions in isolation (as it were): Because atomic propositions have truth conditions, independently of the molecular propositions they happen to be constituents of (via an appropriate operator), a specification of truth conditions for a molecular proposition is relative to the truth conditions of its constituents.

Tarski's semantic concept of truth, and its application to the philosophy of language by Carnap initiated a reorientation to the science of logical semantics and to the notion of logical truth.

Thanks to Tarski, Carnap realized that: "Logical truth in a customary sense is a semantic concept" (Carnap, 1947; p. 12.).

Upon the twin stanchions of truth conditions, conceived of as metalinguistic equivalences, coupled with mathematically sophisticated set theoretic accounts of the objects serving as the semantic values of expressions, the science of formal semantics comes fully into view. Explaining the meanings of sentences in terms of truth conditions only amounts to a clarification of meaning, however, if the concept of truth is itself clear and unproblematic: It is not entirely clear that this is the case, however. This qualification does not refer to the correctness or other of Tarski's formal definition but rather to the doctrine that informed it.

Dummett (1978) argued that the central problem of semantics should be precisely what notion of truth to admit into a theory of meaning, and his advocacy of this position brought the debate between the modern analogs of Platonism and Psychologism, Realism and Antirealism, respectively, to center stage in the philosophy of language. The debate turns to how to relate truth to verification. Realism holds that truth is verification-transcendent: Whether an individual has any means to determine the truth or falsity of a proposition has no bearing whatsoever on the question of whether that proposition is true or false. Truth, for the realist, is an objective property of propositions. Antirealism, on the other hand, eschews truth as verification-transcendent; there is no such thing as a truth that in principle cannot be shown to be true and, consequently, the distinction between being true and being thought to be true is considered irrelevant. A much easier way of describing this debate is that, for the Realist, truth (and hence meaning) is "in the world," but for the antirealist, truth (and hence meaning) is "in the head."

In the Prolegomenon, I remarked that Platonism and Psychologism would crop up again and again in our discussion: Here is an example. Most (if not all) semantic theories, in the tradition of Frege, Wittgenstein, and Carnap, can be considered antirealist theories, in that truth is considered to be a mental construction: The individual makes sense of the chaos in his/her environment by constructing representations of meanings. In contradistinction to this antirealist bias in semantics are ideas from situation semantics (cf. Barwise & Perry, 1983), which is an unashamedly realist theory. Situation semantics differs from antirealist theories on a number of important counts (see special issue on situation semantics in *Linguistics and Philosophy, 8*).

TERRA POSSIBILIA

The second major transformation in the concept of truth conditions lies in Carnap's (1947) *Meaning and Necessity*, from which arose intensional calculi based on the notion of a possible world. Carnap, to reiterate, advanced the theses that logical truth was a semantic property of any well-formed formula and that a metalinguistic statement of truth conditions was able to capture that property—capture, that is, the meaning of any well-formed formula.

One problem remained: Logical truth could be shown to be a semantic concept only if it could be defined in the metalanguage for a given object language. Tarski (1944) showed, in arguments advanced while discussing the antinomy of the liar, that semantic terms, such as "true" (referring to expressions of the object language) could only be introduced into the metalanguage by definition. That is, that any well-formed formula of the object language only expresses a logical truth when a metalinguistic statement to the effect that that formula is true can be derived from the semantic rules for that object language. The semantic rules for the object language, in Carnap's system, were formulated in the metalanguage. Consequently, the metalinguistic definition of logical truth that Carnap sought, in virtue of requiring derivations from metalinguistic rules (of the semantics of the object language), would seem to necessitate the unwanted recourse to a meta-metalanguage, in which the derivations from the metalinguistic rules could be stated.

Carnap solved this problem by judiciously relaxing the full rigors of the ascent from object language to metalanguage: Specifically, he introduced an (submetalinguistic, and supra-object-linguistic) operator, "**N**." The effect of this operator, when combined with a sentence such as "**X** is true," was to produce the metalinguistic statement "**N**(**X** is true)," which asserts that "**X**" is a logical truth. The interpretation that Carnap intended for the **N** operator was some such expresssion as "Necessarily. . . ," or "It is a necessary truth that. . . ," Hence, using the **N** operator, Carnap was able to define logical truth in the metalanguage as a semantic concept.

The **N** operator (with its interpretation "It is a necessary truth that. . .") can be described as a *modal* operator. In modern expositions, a modal logic is a calculi of necessity and possibility, dealing with statements that assert or deny necessary or possible states of affairs. The manner in which a logical calculi might deal with the two notions of necessity and possibility was one of the problems that Carnap addressed in his 1947 work. A solution, he argued, involves a modification to the model-theoretic apparatus of the predicate calculus, such that (borrowing Leibniz's notion of necessary truth as truth in all possible worlds) states of affairs can be possible as well as actual ("possible worlds" qua "possible states of affairs"). By introducing the notion of possible worlds, Carnap was able to formulate metalinguistic statements of the truth conditions of sentences such as "It is a necessary truth that. . . :" He could do this by stipulating the metalinguistic statement, "**N***p* is true if and only if '*p*' is true" in all possible states of affairs.

Carnap's use of the **N** operator to specify the semantics of modal statements was made more precise in accounts given by both Hintikka (1963) and Kripke (1963), who adopted and formalized the idea that not all possible worlds have to be entertained when considering the truth and falsity of sentences. Specifically, they defined an accessibility relation, such that a sentence is necessarily true in a particular possible world if and only if it is true in all possible worlds *accessible* from that world. Similarly, a sentence is possibly true in a particular possible world if and only if it is true in at least some possible worlds accessible from that world. Depending on the accessibility relation—depending upon, that is, the different assumptions that the theorist makes about the logical properties of the relation in a given calculus, different model-theoretic interpretations for modal statements (and more generally, modal logics) can be constructed.

According to the idiom established by the *Tractatus*, it was the truth conditions of a molecular proposition (as expressed by truth tables, and the model-theoretic explanations of the quantifiers) that determined whether that proposition expressed a logical truth. In Carnap's system, however, whether a molecular proposition expressed a logical truth was a matter of whether that proposition was true in every possible world. Carnap introduced an important term, that of *intension*, to explicate (the notion of) the truth of a molecular proposition in all possible worlds: Specifically, the intension of a well-formed formula is a determination of its truth value in every possible world.

The origin of *possible world semantics* lies in the equating of Carnap's notion of intension with (some appropriately evolved notion of) truth conditions, such that the intension of a well-formed formula specifies truth values for that formula in all possible worlds, and not just whichever world (qua states of affairs) happens to be actual. That is, the intension of a well-formed formula, in Carnap's system, is conceived of as a function from possible worlds to truth values. It is convenient at this point to return to the *Begriffsschrift* of Frege, and specifically, to two important technical contributions that he made to the modern discipline of logic. These are:

- The *Sinn* and *Bedeutung* distinction;
- The contextual principle.

Carnap's notion of intension he touted as a clarification of Frege's notion of *Sinn*. This view is very prevalent. It is not entirely correct, as we have seen that for Frege, the *Sinn* of a well-formed formula of the *Begriffsschrift* was concerned with the mode of presentation of a reference (viz. *as* the value (either of the true or the false) of a given function for a given argument), but a Carnapian intension is an altogether different beast. Nevertheless, the two terms (*Sinn* and *intension*) are now regarded as synonymous. Similarly, Frege's *Bedeteung* is often equated with its modern analogue term, *extension*. Again, the similarities are more apparent than real. The *Bedeutung* of a well-formed formula in the *Begriffsschrift* was a truth value; similarly, the extension of a sentence in modern treatments is also a truth value. Things are different, however, when we consider the constituents of well-formed formula: Specifically for an unsaturated sign (qua predicate) of the *Begriffsschrift*, its *Bedeteung* was a concept (a Platonistically conceived, abstract entity), whereas in modern parlance, the extension of a predicate is a set, or a set of sets (where "set" is meant in a very specific way, as a mathematical construct).

Frege's contextual principle urged that the *Sinn* of a constituent element of a well-formed formula was never to be asked independently of the formula in which it occurred. This translates (or rather, has been translated) into a modern reading applicable to natural language such that the intension of a sentence can be built up compositionally from the intensions of its constituents and its syntactic structure. In order to arrive at this modern interpretation, some notion of the intension for constituents of well-formed formulae had to be (as it were) invented for those constituents, as Carnap's original notion of intension was only applicable to well-formed formulae, and not to their constituents. Set-theoretic apparatus is recruited to aid in this task, where set-theoretic constructs serve as the references or *extensions* of the constituent elements. Specifically, what is required is something that (for each constant, name, predicate, or quantifier) stands to an object (for a constant or name), property or relation (for a predicate), or second-level property or relation (for a quantifier) in the same way that the intension of a sentence stands to its truth value.

Consequently, a given object in each possible world is correlated with a constant or name (to serve as its extension), and a set (or set of sets) in each possible world is correlated with each predicate or quantifier (to serve as its extension). The intensions of constituents can now be given in a clear and precise way, such that the intension of a name is conceived of as a function from possible worlds to objects, and the intension of a predicate is conceived of as a function from possible worlds to sets of objects.

Carnap's extended system of semantics may be summarized as follows: A semantic theory constitutes a set of rules formulated in a metalanguage, which lays down truth conditions for every sentence of the object language. These rules specify the intensions of well-formed formulae and their constituents using the set-theoretic apparatus of model-theoretic semantics. The intension of a sentence, as a determination of its truth value in all possible worlds, is touted as a good explication of the intuitive conception of meaning: The intension and extension distinction differentiates sharply between the meaning of a well-formed formula and that for which it stands. The system Carnap outlined was sensitive to the requirement that the meanings of compounds are built up from the meanings of their constituent parts and constructed in conformity with the principle that a meaning for a well-formed formula (qua sentence) is something that determines the conditions under which that formula is true. Carnap, it would seem, may have been the first to outline a theory of meaning for language. There are many flaws in his treatment, however, which were spotted by later logicians: Their elaborations and clarifications imbued the new discipline of semantics with hitherto unknown sophistication.

MODALITY, POSSIBILIA, AND FUNCTION

The raising of Carnap's ideas to the heights of modern formal clarity can be charted with respect to our search for truth conditions. As we have seen, Carnap's notion of *intension* was quickly equated with (a new evolved species of) truth conditions. Thus, giving intensions for well-formed formulae, and the constituents of well-formed formulae was considered an equivalent operation to specifying the truth conditions of those well-formed formulae and their constituents. Using intensions, construed as functions from possible worlds to truth values, it began to seem as if the champion of epistemological intransigence could be made to yield (in some small way) by making the meaning of well-formed formula (of given formal language), and sentences (of a given natural languages) tractable. There are degrees of sophistication immanent in this view that Carnap did not envisage, however. We can isolate two considerations that highlight such sophistication and that will serve to complete our search for truth conditions. The first concern is of a general nature (the transplanting of semantics from formal languages to natural languages by Montague and other like-minded theorists), and the second addresses the notion of intension itself (which, since its introduction by Carnap, has been clarified in nature and extended in applicability by Lewis, among others).

We saw, in the last section, how Carnap used the "**N**" operator to give semantic analyses of modal statements such as "It is a necessary truth that. . . ," by specifying the following kind of metalinguistic equivalence: "**N***p*" is true if and only if "*p*" is true in all possible worlds. Such talk of **all** possible worlds is guaranteed to offend the ontologically parsimonious on two grounds: firstly, on the grounds of the very notion of possibilia itself.

Q: When is a world *possible*?
A: When it is not *actual*.

Quine (1963) has ridiculed the notion of *possibilia*, alleging that individuation and enumeration of possible objects, possible properties, and possible states of affairs (for the purposes of semantic analysis) is at best suspect and at worst impossible and incomprehensible. Despite such equivocation, the notion of a possible world is an essential component of Carnap's explication of (the newly evolved) truth conditions as extension-determining intensions. The second of the objections to possible worlds concerns the formal realization of the notion within systems of formal logic.

Formal semantics (arguably) only came of age when (a) logicians had a good grasp of truth conditions, and (b) mathematical set theory began to be used to specify the values of expressions in a given (formal) language. Consequently, the introduction of possible worlds into the constrained, formal systems of model theory may seem (to the mathematically naive) to be inappropriate. Logicians, however, are quite happy working with possible worlds: They take the real world to be a member of the set of all possible worlds. Moreover, they consider that "things" that do not actually exist, but *might* exist in some states of affairs different from the actual states of affairs, to be perfectly legitimate components of any semantic theory. What, though, are "things?" Lewis (1972) offered the following illumination:

> I want to say, once and for all: *everything* is a thing. But I must not say that. Not all sets of things can be things; else the set of things would be larger than itself. . . . Can we choose the set of things? Not quite . . . no matter what set we choose as the set of things, the system of intensions defined over that set will not provide intensions for certain terms—"intension" for instance—of the semantic metalanguage corresponding to that choice. (Lewis, 1972, p. 177, emphasis in original)

This statement by Lewis is concerned with the generality of semantics, which is to say, the range of "things" admissible into the theory. An observation going back to Tarski is that such generality is fundamentally limited: Discussion of the antinomy of the liar, for example, reveals that no language can be its own semantic metalanguage. In the face of this intractability, Lewis adopted the arbitrary convention of assuming that an inclusive set of things exists, and that it can be modeled by set theory.

Possible worlds are realized in formal semantics by the logician defining some infinite construction(s), such as an (infinite) set or an (infinite) set of sets, to take the combinatorial load of the infinitude of possible states of affairs. Granted that such infinite constructions are realizable in set-theoretic terms, nonetheless the *scope* of most formal languages is not nearly so great as that of natural language. That is, the "size" of the possible world described by formula of a given formal calculus will usually be smaller than the "size" of the possible world described by natural language sentences.[2]

In contrast to the tidy, well-behaved formal languages derived from Frege's *Begriffsschrift* (ie. principally the predicate calculus, and more modern, esoteric languages of tensed and modal logics), natural languages seem hopelessly messy (although, as a caveat,

[2] The fact that infinity has different sizes is not an easy notion to grasp but nonetheless true: The infinity of integers, for example, is smaller than the infinity of functions from integers to integers.

one person's *messy* is another person's *flexible*). One common criticism is that the application of formal semantic machinery to natural languages is nonsensical because, following the thinking of Tarski, natural language has no determinate syntactic structure: Consequently, it is incapable of supporting a recursive definition of truth, and therefore one cannot provide a rigorous model-theoretic interpretation for it. This was a challenge taken up by the logician Richard Montague. In 1974 he wrote:

> There is in my opinion no important theoretical difference between natural languages and the artificial languages of logicians; indeed, I consider it possible to comprehend the syntax and the semantics of both kinds of languages within a single natural and mathematically precise theory. (Montague, 1974, p. 222)

Brave words indeed. This rejection, moreover, was backed up by a complex formal treatment in his 1974 work, *Formal Philosophy*. Montague Semantics (as his approach has become known) is characterized by its bewildering complexity and rebarbative notation and, more importantly, by its state-of-the-art incorporation of truth conditional and model-theoretic principles, and by its sophisticated use of possible worlds. His work is truth conditional because it adheres to the dictum that to give the meaning of a sentence is to give its truth conditions (qua intension); it makes use of possible worlds to cope with modal statements, propositional attitude statements involving *belief* and terms such as *alleged* (which cannot be treated in a simple extensional manner), and so on; and it is model theoretic because all the sophistication of mathematical set theory is employed to specify the semantic values of expressions in the object language. The depth and complexity of Montague's work requires that, for purposes of clarity, much detail must be omitted. Consideration of his treatment of noun phrases will give a flavor of the general enterprise.

In his semantics, he used a system of semantic types constructed to run parallel to the syntactic categories of the language of two basic kinds—**e**, entities, and **t**, truth values. He treated all the different kinds of noun phrases in his fragment of English, *Betty*, *a woman*, *every woman*, *no one*, and so on, in the same uniform way, as of semantic type **((e,t)t)**. What does this mean?

To recall, in Carnap's semantics, a predicate such as *plays* has a model-theoretic interpretation comprising a set of objects: Which is to say, the extension of the predicate *plays* is a set of objects (or individuals). In Montague semantics, the extension of any given predicate (such as *plays*, or *smiles*, or whatever) applied to, for example, the noun phrase *Betty* is similarly a set of individuals (included in which, of course, is *Betty* herself): only *Betty* will be a member of all of the sets that correspond to the extension of predicate. Hence, in Montague semantics, the extension of the noun phrase *Betty* is a set of sets, namely, the set of all the sets of which *Betty* is a member. Predicates are, in Montague semantics, of type **(e, t)**, becuse they are interpreted as functions from objects (qua individuals, qua entities) to truth values. Hence, a noun phrase is, accordingly, of type **((e,t)t)** because it is interpreted as a function from sets (**(e,t)**), corresponding to, in our example, the interpretation of predicates to truth values. In general, if an expression of semantic type **(a, b)** is combined with something of semantic type **(a)**, then the result will be an expression of semantic type **(b)**.

One of the idealizations of Montague Semantics (other than its use of a linguistically unsophisticated categorical grammar to specify the syntax of the fragment of English for which it was devised) is its use of *meaning postulates* to capture the synonymity of pairs of expressions such as *seek*and *try to find*.[3] Using meaning postulates in this way, the paucity of Montague's treatment of lexical semantics becomes evident. An appropriate meaning postulate for these expressions might stipulate that in a given possible world at a given time someone seeks something only if he or she tries to find it. That is, the meaning postulate provides the means by which certain interpretations can be ignored by the theorist, because it rules out certain possible worlds from the scope of the interpretation: If a theorist wished to emphasize that all women are necessarily adults, then she would simply add a meaning postulate to that effect, which would rule out any model structure in which a women is not an adult. We will return to this idealization in due course.

Turning now to the second of the considerations outlined at the start of the section, we can begin to give some precise statements about intensions (qua truth conditions) and what they look like. A state-of-the-art conception of truth conditions, in the tradition of Carnap, Hintikka, and Kripke, can be found in Lewis (1972). In this account, Lewis was initially at pains to specify the grammar of the language with which he was concerned, and thus is unusual in the context of our discussion, as linguistic concerns have

---

3 Meaning postulates will be discussed in more detail in Chapter 5.

largely passed us by. He specified four defining properties of the *categorical grammar* (cf. Ajdukiewicz, 1935) that he employed: first, a small number of basic categories—sentences (S), names (N), and common nouns (C); second, infinitely many derived categories—adjectives (C/C), verb phrases (S/N), and adverbs (S/N)(S/N); third, simple context-free phrase structure rules corresponding to each derived category; and fourth, a lexicon whose words are assigned to categories. The grammar is not a good one for natural language, but it serves for explanatory purposes.

Like Carnap, Lewis took the view that meaning is not so much something that *is*, but rather something that *does*. Also, he held that the meaning of a sentence is something that determines the conditions under which it is true or false—something, that is, that determines the truth value of the sentence in various possible worlds, subject to the accessibility relation. Like Carnap, Lewis took functions (in a very general, set-theoretic sense) to be the very paradigms of "something" that determines "something else," and consequently, considered that he had found something that does at least part of what a meaning does. Specifically, he considered that part of what a meaning does is to deliver an extension given a particular possible world as argument: Any function from possible worlds to extensions can be termed a Carnapian intension.

Unlike Carnap, however, Lewis considered that viewing possible worlds as the only kind of argument for an extension-yielding function to be too restrictive. Such a view he adopted from Montague (1974), who urged that any miscellaneous factors relevant to determining extension could be arguments for functions: Such factors might include the time and place of utterance, the preceding discourse, who else was addressed by the utterance, and so on. Lewis called such a package of miscellaneous factors an *index*, and he called the elements of an index the *coordinates* of that index: Any function from indices to extensions can be called a Lewisian intension. Lewis was sensitive to the problems that introducing superfluous factors into an index might present, and he gave some thought to what the various coordinates of that index might be. He proposed three major ones:

- A possible world coordinate;
- Contextual coordinates;
- An assignment coordinate.

In the tradition of Carnap and others, Lewis maintained that the possible world coordinate of an index corresponded to a totality of facts, determinate in all respects. Certain sentences depend for their truth value on certain facts about the world, and so are true at some possible worlds and false at others. For example, the sentence, "Boutros Boutros Ghali is Secretary General of the United Nations" is true in one possible world, but may be false in other worlds. Various contextual coordinates correspond to features of sentences that depend on context for a correct or meaningful interpretation. Thus, Lewis proposes a *time* coordinate, in virtue of tensed sentences such as "I played Ultimate yesterday;" a *speaker*coordinate, in virtue of "I play Ultimate;" an *audience* coordinate, in virtue of "Do you play Ultimate?;" an *indicated objects* coordinate, in virtue of "Those people over there play Ultimate;" and a *previous discourse* coordinate, in virtue of "The aforementioned people play Ultimate.". The assignment coordinate Lewis formulated as an infinite sequence of things, serving to give the values of any free variables in such expressions (or sentential functions) as "*x* is tall."

In discussing the intensions of derived categories (what Lewis termed compositional intensions), as opposed to the intensions of basic categories, Lewis drew on prior work from Kaplan (1968). Consider the intension of an adjective such as "alleged:"This takes a common noun such as "Libertarian," and makes a new compound noun "alleged libertarian." Adhering to a version of Frege's contextual principle, the intension of the new common noun depends on the intension of the (original) common noun in a manner determined by the meaning of the adjective. The meaning for an adjective, therefore, given meaning as something that *does* rather than something that *is*, is something that determines how one common noun intension depends on another. Consequently, the intension of an adjective can be characterized as a function from (original) common noun intensions to (new) common noun intensions. That is, the intension of "alleged" is a function that, when given "Libertarian" as argument, yields as value the intension of "alleged Libertarian."

Adjectives are one of the simplest derived categories of Lewis's categorical grammar, represented as C/C (ie. derived from the basic category C (common noun). The complexity of the intension for such a relatively simple category does not begin to approach that for other, more complex categories. Consider what the intension for an adverb, category (S/N)(S/N), might be. An adverb of this category takes a verb phrase to make a verb phrase. Consequently, an intension for this adverb is a function from verb phrase intensions to verb phrase intensions. Specifically, it is a function from functions from functions from indices to things to functions from indices to truth values to functions from functions from indices to things to functions from indices to truth values (This sentence parses unambiguously). Lewis offered no apology for the complexity of such constructions (consider what the intension for an adverb modifying adverbs, category ((S/N)(S/N)(S/N)(S/N)) might be!). Intensions (qua meanings) are complex entities, and the theorist should not expect the formal representation of such entities to be simple.

CONCLUSIONS

Like Montague Semantics, the possible world semantics of Lewis suffers from a paucity of lexical specification, and, moreover, offers no mechanism (such as meaning postulates) in recompense. Intensions (extension determining functions from indices) are not meanings, although they are close, because there are differences in meaning unaccompanied by differences in intension. It is absurd, for example, to claim that all tautologies have the same meaning, yet they all have the same intension—the function that has the value *true* at every index. Differences in intension for well-formed formula and their natural language counterparts provide, it would appear, coarse differences in meaning. Only when well-formed formula are broken down into their lexically basic constituents, and only when the intensions of those basic constituents are known and can be stated formally, can the concept of intension really be said to equate fully with the concept of meaning. Unfortunately, Lewis offers no solution to how the intensions of basic lexical items can be stated, and for one very good reason: *Formal semantics has no solution.*

This is perhaps a little uncharitable. The way that formal semantics copes with the intensions of basic lexical items is to assume that some function exists, such that that function directly provides each basic lexical item with an interpretation in the model structure. Or put another way, formal semantics *finesses* the problem of specifying the intensions of basic words by taking them as primitive, unanalyzed functions. Consider the lexical item *woman*, for example: In a formal semantics, the intension of this word would be given as a function that provides an interpretation in the model structure comprising, in Montague's treatment, a set of sets. A set of sets, however, really says nothing about the word *woman* at all, as evidenced by the fact that Montague relied on the use of meaning postulates to specify that all women are necessarily adults (a fact that a human language user would know is part of what the word *woman* means). This problem of specifying the intensions of basic lexical items is one that is taken up in the next chapter.

5

Theories of Meaning: Psychological

Conceive of a semantic continuum: At one end is formal semantics, in the tradition of Tarski and Carnap, concerned with how symbolic strings (viz. expressions in a given formal language) are mapped to model structures, and at the other end, is (a putative discipline of) psychological semantics, concerned with how symbolic strings (viz. expressions in a given natural language) are mapped to mental representations. Because of the obvious similarities between the two kinds of semantics, for the psychologist seriously interested in studying how the human mind is able to represent and use the meanings of the items of its language, formal species of semantics can offer valuable insights. Before we begin to discuss those insights, one should first be aware of the problems that such an assimilation presents. Four are particularly acute.

The first concerns the syntax or *grammar* of a language (whether formal or natural). The grammar of a language is a system of rules governing lawful concatenations of allowable constituents into more complex constituent structures: For a formal language, its grammar is relatively straightforward. The categorical grammar used by Lewis (1972) is a good example of the essentially *context free* nature of such systems of rules. Natural language, however, requires considerably more powerful grammars (or so the argument goes), the nature of which is an open conjecture. The most well-known candidate, of course, is (some species of) transformational grammar, in the tradition of Chomsky (cf. 1957; 1965), but alternative proposals, such as the generalized phrase-structure grammar of Gazdar (cf. 1981; 1982), have also been made. Although the specifics of linguistic theories of syntax are not our concern, the more specific notions of syntax and particularly *syntactic structure* are important, and will turn up again and again in later discussion.

The second of the problems concerns the mapping of a symbolic language string to elements of the model structure without reference to the human mind. Or put another way, with the problem of assimilating set-theoretic functions (intensions) with psychological functions (meanings). This gives rise to the difficulty that formal semantics has in specifying the semantics of *propositional attitudes*. We can give two examples, one easy and one hard, in illustration of this kind of problem. The easy one concerns words such as *think* and *believe*. These words cannot be treated in a simple extensional manner, because they refer to contents of the mind. Which is to say, if the meaning of a sentence (a la Frege) is built up compositionally from its constituents, then replacing one expression with another with the same extension should not affect the meaning of that sentence. However, this principle breaks down when dealing with propositional attitudes. For example, "Jane thinks that John is an awful ultimate player" may be true, but "Jane thinks that the man who made the assist is an awful ultimate player" may be false, even though the two noun phrases may be co-referential.[1]

This kind of example is not that hard, because the two coreferential phrases have different intensions: there are possible worlds in which they pick out different individuals. The real difficulty, and our second, hard example, is encountered when one expression is replaced by another with the same intension. Such examples arise when one necessary truth (which has a truth value *true* in all possible worlds) is replaced by another. For example, "Jane believes that 2 + 2 = 4" may be true, but "Jane believes that 456580504569 + 732905472304 = 1189485976873" may be false. We will not dwell on this problem long, other than to mention briefly the lengths that formal semantics is prepared to go in solving it: Hintikka (1975), for example, proposed a solution involving the notion of *impossible* worlds, such that two sentences (such as our numbers example) can be distinguished by the fact that they are true in different impossible worlds. The major difficulty with impossible worlds, other than making model theoretic interpretations ridiculously complicated is, of course, that they *are* impossible.

The third of the problems that our psychologist must address in assimilating formal semantics to a corresponding psychological semantics, concerns the nature of the mental counterpart of the model structure. It is only in virtue of objects, sets, sets of sets,

---

[1] This is the case because Jane may not know that "John" and "Ultimate player" can be taken to be the same individual.

functions, sets of functions, and so on, which serve as the semantic values of expressions of a given language, that semantics per se is possible at all. In a possible world semantics, the model structures are (potentially) infinite constructions: semanticists postulate infinite constructions to cope with the infinitude of possible states of affairs by simply indicating such in their notation. The brain, however, is a finite computational device, and infinite constructions are notoriously (and annoyingly) non-computable. There is thus very reasonable grounds for supposing that the human language user does not employ infinite constructions realized as possible worlds, or any other kind of infinite construction for that matter. The task for the psychologist, therefore, is to construct a theory that specifies how to fit the infinitude of possible worlds into the finite language user's head. Or, put another way, one of the goals of a theory of psychological semantics is to postulate a form of representation which can "take the place of the model structure" (cf. Johnson-Laird, 1983).

The fourth of the problems that all species of formal semantics face, and the problem that a theory of psychological semantics must solve if it is to account for the meaning of natural language, is the specification of the intensions of basic lexical items, part of the problem of specifying a lexical semantics in addition to a structural model theoretic semantics. I struggle to find ways of stressing how crucially important this problem is: In some theoretical treatments the problem is ignored entirely. The problem is far more pervasive than it might appear, however. The particular manifestation in semantics is the difficulty in specifying intensions for words. However, this problem is really part of a much bigger problem, namely, the problem of expressing the relation between *representations* and *represented*. This problem has many guises: Wittgenstein encountered it at the terminus of his philosophical analysis due to the fact that the logical atomism he employed precluded the possibility of making any statements about the relations between indefinable names, constituting atomic propositions and the reality described by the calculus that those indefinable names underpinned. The Vienna Circle members thought they had solved the problem by appealing to ostensive definition, and a host of theorists since Carnap (1952) have thought *meaning postulates* sufficient to address the problem (and we will discuss them later in the chapter). The essence of this *grounding* problem will turn up in a number of different forms (particularly in Chapters 8 and 9, where it is considered in detail) again and again throughout our investigations.

CONDITIONS AND CRITERIA

Are the meanings of words *decomposable*? A venerable tradition in philosophy and psychology would answer the question in the affirmative. In philosophy, the essence of the affirmation is that the meanings of words are decomposable into sets of necessary and sufficient conditions; in psychology, a corresponding notion is of a "checklist" of characteristics essential in determining that one meaning does indeed differ from another (cf. Fillmore, 1975). Such formulations of meaning make appeal to the intuitive conception that "Words have basic inalienable meanings, departure from which is either conscious metaphor or inexcusable vulgarity" (Evelyn Waugh, 1942, private letters). For example, take the meaning of the word "square," which has the following four conditions associated with any application of the term:

- A closed, flat figure;
- Having four sides;
- All sides are equal in length;
- All interior angles are equal.

Each of these conditions is by itself *necessary* in order for something to be a square and, when taken jointly, are *sufficient* to define a square and only a square. Unfortunately, it is extremely unlikely that all words can be treated in so simple a fashion. For example, Pirsig (1974) was unable to state necessary and sufficient conditions for something to be of *quality*, and in a similar vein, Wittgenstein (1953) rejected the idea that concepts (i.e. meanings) could be defined in terms of essential characteristics or necessary and sufficient conditions. In a famous passage he wrote:

> Consider for example the proceedings that we call 'games' . . . if you look at them you will not see something that is common to *all*,but similarities, relationships, and a whole series of them at that. . . . Look at board games, with their multifarious relationships. Now pass to card games; here you may find many correspondences with the first group, but many common features drop out, and others appear. When we next pass to ball games, much that is common is retained, but much is lost. Are they all 'amusing'? Compare chess with noughts and crosses. Or is there always winning and losing, or competition between players? Think of patience. In ball games there is winning and losing; but when a child throws his ball at the wall, and catches it again, this feature has disappeared. Look at the parts played by skill and luck; and at the difference between skill in chess

> and skill in tennis. Think now of games like ring-a-ring-a-roses; here is the element of amusement, but how many other features have disappeared! . . . And the result of this examination is: we see a complicated network of similarities overlapping and criss-crossing; sometimes overall similarities, sometimes similarities of detail. I can think of no better expression to characterize these similarities than 'family resemblances'; for the various resemblances between members of a family: build, features, colour of eyes, gait, temperament, etc. etc. overlap and criss cross in the same way, and I shall say: 'games' form a family. (Wittgenstein, 1953, 31–32)

Instead of necessary and sufficient conditions, Wittgenstein talked of *criteria*. Criteria are part of concepts, part of the "grammar" of things, fixed by convention (not experience or logical necessity) and they fall into no orthodox logical category. A natural way in which to conceive of criteria is as *default values*, which is to say, as characteristics of meaning that can be assumed unless there is evidence to the contrary. Consider, for example, the criteria that might be part of the concept of *dog*. Conventionally, these might include such characteristics as having four legs, fur, a tail, the ability to bark and so on. None of these criteria, however, is a necessary part of the concept of a dog: A dog may have had an accident and have only three legs, or no tail, but it is still a dog. Consequently, concepts (qua meanings) defined in terms of criteria have no clear-cut boundaries, because criteria are not clear-cut characteristics. Rather, concepts of a natural language can be seen to have fuzzy boundaries, as was shown intriguingly by Labov (1973). This has the result that in communicating a concept to another, an individual is unable to convey a precise specification of that concept, and so instead conveys a schematic model or *schema*.

Putnam (cf. 1970; 1973; 1975) reiterated this observation in his discussion of the word "lemon," the meaning of which could be decomposed into a set of characteristics such as: ovoid, yellow, waxy peel, tart taste, and so on—but are such characteristics either necessary or sufficient to identify lemons? Plainly, the answer is no, because a sweet, peel-less lemon is still a lemon, and a green lemon is still a lemon. Instead of sets of necessary and sufficient conditions, Putnam explained the meaning of "lemon" in terms of a theory, either scientific or pre-scientific, of the "essential nature" or "underlying structure" of lemons. Putnam dubbed those words whose meanings depend on theories of essential nature *natural kind terms*. According to Putnam, when an individual tells someone what the meaning of lemon is, they generally sketch a simplified theory of what it is to be a lemon: They describe a typical lemon, a normal member of the class of lemons. Such a description provides the individual's listener with what Putnam calls a *stereotype*, a notion which would seem to correspond to a schematic model defined in terms of conventional characteristics.

Notions of schemata have surfaced from time to time in the cognitive sciences: In AI there is the notion of a frame (Minsky, 1975) or script (Schank & Abelson, 1977); it finds an intriguing manifestation in the distributed representations of connectionist networks (cf. Churchland, 1990); and most notably, the notion turns up in the form of a *prototype*. The essence of prototype theory is that entries in the mental lexicon are centered around the most prototypical member of the class that the word denotes. There are two ways in which to think about prototypes: In the first instance, consider the meaning of the word *bird* (after Rosch, 1973). From a list of exemplars of the word, such as *robin*, *toucan*, *hawk*, *sparrow*, and so on, Rosch's experimental subjects reliably rated the robin as the most typical bird. On one interpretation then, the *bird* prototype is robin. A second interpretation of prototypes is, however, as an abstraction from the collection of features that serve to delineate the class of birds. That is, although Rosch's subjects rated a robin as the most typical bird, there are "possible" birds that would be even more typical, if they actually existed, and this "possible" bird (an abstraction from all the most common features of the class) would be the bird prototype.

A prototype is not a meaning or concept. Rather, it can be identified with mental representations of quadruples such as the one below, adapted from Osherson and Smith (1981):

$$< \mathbf{A}, \mathbf{d}, \mathbf{p}, \mathbf{c} >$$

In this quadruple, **A** is a conceptual domain of objects, realized formally as a mathematical set construct, **d** is a distance metric between elements of **A**, **p** is the prototypical member of **A**, and **c** is a characteristic function. We can illustrate with the concept of *bird* once more, which might be mentally represented by some such quadruple as the following:

$$< \mathbf{B}, \mathbf{d}\ (\mathit{bird}), \mathbf{p}\ (\mathit{bird}), \mathbf{c}\ (\mathit{bird}) >$$

Here, **B** is the set of birds (or rather some suitably expansive set of birds), **d(bird)** is a function from pairs of birds to real numbers:

The smaller the number assigned to a given pair (e.g., *robin, finch*), the more similar the two birds are, and the larger the number assigned to a pair (e.g., *robin, ostrich*), the more dissimilar. The third coordinate of the quadruple, **p(bird)** is the prototypical bird, the bird that has the average value on each dimension of the underlying metric space. Put another way, the prototype can be thought of as located in the center of the metric space, the dimensions of which correspond to characteristics on which exemplars of the concept can vary: For birds, such characteristics might include wingspan, clutch size, shape of feathers, and so on. The fourth coordinate, **c(bird)** is a function from elements of **B** to numbers in [0, 1]: Once again, the smaller the number assigned to a given exemplar by this function, the closer that exemplar is to the prototype. Different formulations of prototype theory differ primarily in their choice for the second and third coordinates of such quadruples. For example, Smith, Shobin & Rips (1974) used a distance metric defined in terms of number of common features shared between exemplars.

Each prototype in Rosch's framework is presumed to be bounded, and for some concepts, this boundary is not sharp. Such inherent "fuzziness" has been well demonstrated experimentally (e.g., Rosch, 1975; Labov, 1973) and explains to an extent the fact that sometimes judgements about set membership are not easy to make (e.g., Is a penguin a bird? Is an ostrich? Is a bat?). If some words have as part of their meaning a notion of "prototype with fuzzy boundaries," what is the nature of their corresponding extensions? A number of authors (cf. Putnam, 1975; Rosch & Mervis, 1975) have argued that an elucidation of extensions for schema calls for the apparatus of fuzzy sets (cf. Zadeh, 1965). Fuzzy set theory relies on a notion of "membership to a degree:" Membership in a fuzzy set consequently can be partial, ranging over some continuum of values, such as 0 to 1, 1 to 10 and so on. Truth values also can be partial in fuzzy set theory, so that a sentence such as "Ultimate is aerobic" might be assigned a truth value of 0.75, for example. Osherson & Smith (1981) argued against the efficacy of fuzzy set theory for dealing with extensions in a psychological theory of meaning.

We can draw some conclusions about conditions and criteria at this point and compare fixed versus fuzzy accounts of meaning. Plainly, there are a (small) number of words, such as "square" and "bachelor" (and a host of highly specific, invented technical terms) that appear to have a fixed meaning, which is to say, whose meanings can be exhaustively specified by sets of necessary and sufficient conditions. Equally plainly, there are words whose meanings do not behave in this way for which conditions both necessary and sufficient to fix their meaning cannot be stated and of which we can consequently say the following:

- it may be impossible to ascertain where lexical meaning ends and encyclopaedic (or world) knowledge begins.;
- often meanings may have fuzzy boundaries, in that there might be no clear point at which the meaning of one word ends and another begins.
- a single word may apply to a Wittgensteinian "family" of items, which all overlap in meaning but which do not exhaustively share any one common characteristic.

The moral? Meanings qua intensions are not uniform.

LEXICAL MEANING

Between the acoustic-phonetic analysis of an incoming language signal(spoken or written) and the mental representation of the message conveyed by that signal is the computational mediation of the *mental lexicon*. Simply stated, the mental lexicon is presumed to be the store in an individual's mind of all her knowledge related to words, containing for example, information about orthography, morphology, syntax and semantics. Since the introduction of the term, the mental lexicon has undergone several theoretical revisions: a modern conception of it is as a relay station, connecting specific sensory events and motor patterns with mentally represented knowledge structures. A psychological theory of lexical meaning, in this theoretical context, is thus a description of the properties and organization of the entries in the lexicon(corresponding to the meanings of words). We can distinguish three goals of any theory couched in such terms.

Firstly, the theory should specify the *form* of the mental representation of meaning. This goal alludes to the time-honored philosophical question: What is the meaning of a word? A theory of lexical meaning is most importantly a theory of representation and organization: It therefore needs a prior account of what exactly it is supposed to be a representational account *of*. As we have seen, a common assumption in the literature, for example, is that if a word can be *defined*, then the entry in the mental lexicon for that word can naturally be thought of as the *definition* of that word.

The definition is often couched as a set of necessary and sufficient conditions. Any explanation of meaning adhering to the efficacy of necessary and sufficient conditions is a "fixed-meaning" explanation. Many arguments and observations from linguists, psychologists and philosophers would seem to indicate, however, that(at least some) words are not capable of being defined, and that for those words, sets of necessary and sufficient conditions cannot be stated, because as Labov (1973) pointed out, the concepts from natural languages tend to have fuzzy edges and boundaries. In the literature, as we have seen, such fuzzy accounts of the meanings of words include proposals of prototypes, schema, stereotypes and frames.

Second, the theory should explain a variety of intensional phenomena. How is the meaning of "left" related to the meaning of "right," for example? Is it in terms of the distance they are apart in the lexicon? Or some other metric? In general terms, the theory should account for how intensions (i.e., meanings, namely, entries in the lexicon) are related to other intensions. Such intensional relations include those of synonomy and antonomy, entailment, polysemy, vagueness and ambiguity.

The third of the goals for a theory of lexical meaning is the most important of all for our concerns: to account for how the meanings of words are connected to the world or, in other words, to explain the extensional relations between words and the world as humans conceive it. Here we have the *real* problem of lexical semantics and another instance of the grounding problem. Consider the word "left:" What is the relation between this word and what it refers to in the world? How might you go about reconstructing this relation? As Jackendoff (1984), among others, has noted, a necessary first step is to realize that an individual does not perceive the world directly. Perception is construction of a mental representation of the world and an individual is only ever in possession of a (undoubtably impoverished and partial) *model* of the world: They are unable to compare this perceptually derived representation with the real world because it *is* their world. A solution to the problem of specifying the intensions of basic lexical items and a solution to the (more general, abstract) problem of how to solve the grounding problem must, at least in part, lie in connecting words to *models* of the world, not to the world itself.

In short, a psychological theory of lexical meaning should explain how meaning is mentally represented, how lexical entries are intensionally related to one another and how expressions are extensionally related to the world as humans conceive of it. We consider a number of such theories and use the three goals outlined here as measures of explanatory efficacy.

REPRESENTING INTENSIONAL RELATIONS

In "Connectionism and Cognitive Architecture," Fodor and Pylyshyn (1988) presented an extensive critique of the connectionist school of thought and its claim to offer a viable alternative to traditional symbolic theories of cognition. One of the principal tenets of their refutation is the assimilation of connectionism to a crude Associationism: In the context in which it is used, "Associationism" has the same distasteful connotation for Fodor and Pylyshyn as "Communism" had for McCarthy. Associationism, however, has an ancient pedigree, extending back at least to Aristotle, and it makes an important contribution to the modern notion of a *network*. A number of authors have argued that the organization of meanings in the mental lexicon can be modelled using some kind of network and have accordingly proposed theories of psychological meaning based on the notion of a *semantic* network.

One of the first such theories was developed by Quillian (1968), whose work anticipated many of the features of subsequent network models. He proposed that concepts, corresponding to the meanings of words, could be represented by the *nodes* of a network. He distinguished two sorts of node: type nodes that represent concepts and token nodes that represent instances of concepts, in virtue of links to their respective type nodes. A network is not just a collection of nodes, however: A variety of associative *links* between (type and token)nodes represent varieties of relations between concepts. In Quillian's original (1968) formulation, he proposed five kinds of link, subsuming such relations as set membership (the famous ISA link, e.g., Ultimate ISA game), set inclusion (e.g., *shittake* are fungi), part-whole relations (eg. a finger is part of a hand) and property attribution (e.g., cats are furry).

The organization of Quillian's network was designed to model aspects of the hierarchical structure of the lexicon: The classic example used to illustrate this structure is that a poodle is a dog, and a dog is an animal, but because all poodles are animals, the network should be parsimonious in that it is not necessary for the latter relation to be represented explicitly, as shown below:

Poodle → *Dog* → *Animal*

Quillian proposed his semantic network formalism in an AI context (as a part, in fact, of a "teachable language comprehender") and, because of this, sought to bolster its representational capacities with processes that set up, interrogated and drew inference's from the representations so as to create a performance model. Indeed Quillian considered that certain aspects of meaning could be dealt with quite naturally by the kinds of process he envisaged. For example, if an individual is asked to evaluate the relation between two concepts (i.e., if they are asked "Is an ostrich a bird?"), Quillian conjectured that they would perform this evaluation by searching through the network for a path between nodes: This conjectured search process he simulated in his model by activating the two nodes corresponding to (in our example) *ostrich* and *bird*, and waiting for the two (constantly and symmetrically) expanding shells of activation to intersect: At the point of intersection, the search would backtrack via a number of "tags" (on concepts) left behind by each activation shell to discover the intensional relation between the concepts. Quillian's (1968) work quickly found its way into psychology via the collaboration of Collins and Quillian (1969; 1972), and a spate of network based models followed, including those of Anderson and Bower(1973), Collins and Loftus (1975), Glass and Holyoak (1975), Rumelhart, Lindsey and Norman (1972), and Additionally, one of the earliest, and also one of the most impressive, Connectionist models to appear in the literature was the implementation of a semantic net in parallel hardware, documented in Hinton (1981). We can distinguish two important aspects of lexical meaning that are germane to such semantic network proposals.

First, the network formalism was used to investigate and explain why some assertions, such as "a poodle is an animal" take longer to evaluate (i.e., to verify as true or false) than other assertions, such as "a poodle is a dog." These kinds of results were explained in network theory by semantic distance, or path length between nodes: That is, the length of the path that had to be traversed in the first instance was greater than that which had to be traversed in the second instance, thus causing a measurable difference in verification time. Explaining such empirical data in terms of semantic distance runs into trouble, however, in dealing with instances of proto typicality, when assertions about typical instances (e.g., "A robin is a bird") are verified faster than assertions about atypical instances (e.g., "An ostrich is a bird"). In the network model of Collins and Loftus (1975),a revision to the decision process postulated in Quillian is given that accounts for such prototypicality effects. Essentially, the revision involves the introduction of a threshold function, such that the threshold must be exceeded before the intersection of two shells of spreading activation can be taken as evidence for an evaluation. Positive evidence consists of paths that establish, for example, that one concept is a super ordinate of the other, and negative evidence consists of paths that establish that one concept is not a super ordinate of another, or by establishing that two concepts are mutually inconsistent subordinates. Thus, an atypical instance is presumed to illicit more negative evidence (in favor of the correctness of the assertion) than a typical instance (which illicits less negative evidence), and so slows verification, although Sharkey (1988) illustrated some formal problems with this intersection account.

Second, many of the network models relax the constraint on the structure of the lexicon that it be strictly hierarchical. Such a relaxation is found in Glass and Holyoak (1975). In this model, the notion of semantic markers introduced by Katz and Fodor (1963) into psychology from linguistics is also taken on board, forming a hypothetical structure underlying the intensional relations between the nodes in the network. Glass and Holyoak were sensitive to the possibility that relaxing the hierarchal constraint on lexical structure would render their model empirically vacuous. They therefore tried to place constraints on their theory, although the extent to which they succeeded in this endeavor are dubious (see McCloskey & Glucksburg, 1979).

In brief summary, we can say that semantic networks constitute a general approach to the study of the organization and processing of the lexicon. Representations of words are stored in a network of nodes and links, and the meaning of a word is determined by the place of the type node representing it in the network as a whole. Semantic relations between words are represented by labelled links between the (type and token) nodes in the network. The major experimental result to which they are touted as explanations involve judgements of semantic similarity; namely, semantically related words have shorter paths between them than semantically unrelated words. Networks are primarily able to account (with varying degrees of success) for intensional relations. There is good reason to suppose, however, that as a class of theory, they are too powerful to be refuted by psychological evidence (cf. Johnson-Laird, Herrmann & Chaffin, 1984).These authors also pointed out that network theories have very little to say about extensional relations, and consequently suffer from what Johnson-Laird et al. (1984) termed a *symbolic fallacy*, an instance of the grounding problem in yet another guise. In order to confront this fallacy, some other account of the representation of word meaning seems to be required.

DECOMPOSITION OF MEANINGS

Inspired and endorsed by the school of transformational grammar, one of the earliest psychological theories of word meaning, and the second that we wish to consider in our discussion, is found in Katz and Fodor(1963). They assumed a variant of Frege's contextual principle: The semantic interpretation of a sentence is obtained by replacing the words of that sentence with their corresponding semantic representations and combining those representations according to the underlying syntactic structure of the sentence. The "semantic representation" of a word, they argued, comprised a (hierarchically structured) set of *semantic markers*, which served to decompose the meaning of the word into more primitive semantic constituents. As originally envisaged by Chomsky (1965), semantic markers were bivalent features of the form +<**feature**> and –<**feature**>.

The Katz and Fodor paper proposed a *feature* theory of meaning, which held that semantic markers were innate and linguistically universal components of meaning, comprising the end result of a process of decomposition of word meanings from all and any natural language. Underlying the meaning of "woman" for example would be the(innate and linguistically universal) semantic markers corresponding to *human*, *female* and *adult*. Knowledge of this set of semantic markers would belong to what Chomsky called Universal Grammar. The theory assumed that most, if not all, words were ambiguous. Consequently, an entry in the mental lexicon for a particular word would comprise:

- possibly several different sets of semantic markers (corresponding to each of the word's "senses") serving to capture the systematic aspects of meaning;
- a tag, or *distinguisher*, associated with each different set of semantic markers, serving to capture the non-systematic aspects of meaning.

The distinguisher, in the ideolect of the theory, functions as a *selectional restriction* that constrains the range of other meanings with which a given sense can lawfully combine. The need to account for systematic and nonsystematic aspects of meaning is easily illustrated by considering the following pairs of words, which differ systematically only in the value of the semantic feature capturing the distinction between male and female, and hence must be distinguished nonsystematically: *man/woman, boy/girl, uncle/aunt* and so on.

The purpose of the semantic markers employed in Katz and Fodor (1963) was to stipulate conditions both necessary and sufficient to fix and define the meanings of words. Moreover, the proposed markers were conceived of as innate and primitive entities, the set of which, of course, was presumed to be smaller than the vocabulary of basic lexical items they subserved. This is a fairly standard and intuitive conception of primitive:

> Semantic features cannot be different from language to language, but are rather part of the general capacity for language, forming a universal inventory used in particular ways in individual languages. (Bierwisch, 1970, 182)

As the preceding discussion of conditions and criteria showed, to be a really viable candidate theory of psychological meaning, the theory of semantic markers would have to be revised such that some notion of semantic marker (qua primitive) was devised that captured not only necessary and sufficient conditions, but also captured something like Wittgenstein's criteria, so that experimental evidence about, for example, prototypicality could be accounted for. This was exactly what happened. Schaeffer and Wallace introduced the notion of semantic markers into the vocabulary of psycholinguists in 1970, followed by the contribution of Katz (1972). However, it was in Smith, Shobin & Rips (1974) that the first serious attempt was made to integrate ideas of criteria and schema into the notion of semantic primitive. They did this by distinguishing *defining* and *characteristic* features of meaning: More accurately, they conceived of a continuum of semantic primitives, stretching from the truly defining to the merely characteristic.

Feature, decompositional, or *definitional* are all labels for theories of meaning in psychology that cleave to the idea that the meanings of words are, in principle, capable of being broken down into smaller, more specific parts. Also, they are a proliferating species. In both AI and psycholinguistics, one finds a host of models that assign complex, static semantic structures to entries in the lexicon, often concentrating on verbs (see for e.g., Gentner & Stevens, 1983; Jackendoff, 1983; Talmy, 1983). Common to all such decompositional accounts of meaning is the assumption of a certain number of primitives (often termed "conceptual" primitives) such as PLACE, EVENT, CAUSE, DIRECTION, THING, and so on. A good example is the extended family of models based on Schank's twin proposals of *primitive actions* and *conceptual dependency* (cf. Schank, 1975; Schank, 1986; Schank & Abelson, 1977).

In these models, it is proposed that approximately a dozen semantic primitives underlie all of the verbs of an individual's language: They are given labels such as MTRANS (which can be paraphrased so as to read "the transfer of mental information"), MOVE, INGEST and ATRANS ("the transfer of possessions"). Although MTRANS, INGEST and the others were touted as language-neutral meanings underlying natural language, it is unlikely that they are primitive in any real sense. First, Schank and Abelson (cf. 1977) were never able to reach a consensus on just how many primitives there were (the number has varied from 12 to 16); second, some of the touted "primitives" seem decidedly complex. INGEST, for example, could quite plausibly be broken down into the (semantic primitives of?) SALIVATE, MASTICATE and SWALLOW. Third, the 12 or so postulated primitives demonstrate a debilitating paucity of detail: What precisely is the mental faculty that MTRANS is supposed to underlie? It does little good to say "the transfer of information:" how is *information* to be viewed? Transfer how? To where?, and so on.

There is, however, an alternate account of how to interpret the notion of a semantic primitive, one of the earliest expositions of which is found in Miller and Johnson-Laird (1976). In this lengthy tome, the authors sought to show how language might plausibly be related to perception. Specifically, they investigated the idea that lexical meaning in general, and semantic primitives in particular, might be represented by computational mechanisms tightly related to the sensory transducer system. It is to the rationale of their work that we now turn.

PROCEDURAL MEANING

Recalling the previous chapter, I emphasized that two conceptual developments in the concept of truth conditions were responsible for the application of the term to natural language, the first being the definition of truth as a semantic concept by Tarski, and the second, Carnap's invention of intensional calculi based on the notion of a possible world. In actual fact, there were three such conceptual developments. The third transformation of truth conditions has its origins in Turing's famous 1936 paper, "On Computable Numbers, with an Application to the Entscheidungs problem," which introduced the concepts of effective procedures, the Turing machine and the Universal Turing Machine: von Neumann's development and construction of the first digital computer and the advent and rapid evolution of the computing sciences can all be traced to this seminal paper. The transformation that occurred had at its heart the fact that the term "truth condition" (qua meaning) came to be regarded more and more as synonymous with (some appropriately abstract notion of) the term "procedure." This in turn led to the emergence of a novel theoretical framework, couched in a computational vocabulary(rather than a formal or a mathematical one) for representing the meanings conveyed by language strings: *procedural* semantics.

The equivalence between meaning and procedure in early versions of procedural semantics was construed in quite a narrow sense. Davies and Isard (1972), for example, took the two computer science notions of *compilation* and *execution* and argued that human comprehension of natural language could be very naturally explained in terms of them:

> [In] . . . the compile and execute strategy . . . the first step that a person must perform is to compile the sentence—to translate it into a program in his or her internal language . . . once a program is compiled, the question arises as to whether the listener should run it. (Johnson-Laird 1978, p. 191)

In attempting to characterize procedural semantics, one is initially bewildered by the dearth of formulations. One reason for such diversity is that procedural semantics is not really a theory per seat all. Rather, "Procedural semantics . . . is a paradigm or a framework for developing and expressing theories of meaning, rather than being a theory in itself" (Woods 1981, p. 302).

Equally, procedural semantics can also be described as 'a label for a loose confederation of theories of meaning relying on an analogy between ordinary language and high level programming languages" (Johnson-Laird, 1978). Early formulations of procedural semantics urged that,

> Not only should we think of the production and comprehension of natural utterances as processes describable in algorithmic terms: but that *our utterances themselves should be thought of as pieces of*

> *program* whose effect is to modify one another's behavioural dispositions. (Longuet-Higgins, 1972, p. 263).

One of the difficulties in assimilating formal to procedural semantics is the computational intractability of the infinite constructions required by (specifically) possible world semantics:

> In judging that a relation holds between two or more objects, we do not consider properties or relations as sets. We do not even consider them as somehow simply intensional properties, but we have procedures that compute their value for the object in question. (Suppes, 1982, p. 29)

Consequently, some formulations of procedural semantics do away with the whole apparatus of set theory altogether, such that, "By replacing the bottom layer of [for instance] Montague's model theoretic edifice with an appropriate set of procedures, we hope to preserve computability" (Hobbs & Rosenschein 1978, p. 289).

The question of the relation between procedural semantics and formal semantics is a contentious one. Thus, for some,

> The procedural approach to the problems of semantics [is] not . . . an alternative to the more traditional Tarskian model theoretic account, but rather [is] a means to supplement that account with what, in computer terminology, would be called an "upward compatible" extension. (Woods 1981, p. 317).

Along similar lines, other authors attempt to explicate their formulations of procedural semantics by direct reference to the more formal species, using the two Carnapian notions of intension and extension: "We might speak of the intension of a program as the procedure that is executed when the program is run, and of the extension of a program as the result that the program returns when it has been executed" (Johnson-Laird 1978, p. 238). The assimilation of the formal notions of intension and extension to corresponding procedural notions is perhaps one of the most interesting contributions of procedural semantics, and we shall take a closer look.

Procedural Extensions

In a formal semantics, the extension of a declarative sentence is a truth value, either true or false. However, in a procedural semantics, the range of possible extensions for declarative sentences is much greater: Johnson-Laird (1977a) cited answers to questions, compliances with requests, additions to knowledge, modifications to plans and soon, as non-formal, uniquely procedural extensions for declarative sentences, for example. Is a procedural semantics able to specify procedural extensions for words, as well as sentences? Both Woods (1981) and Hadley (1989) argued that it is. Woods (1981), for example, considered *partial* functions—that is, functions that in particular cases assign neither true nor false to expressions. The use of such partial functions is necessitated, he argued, because many expressions of a language, such as "Is Fermat's Last Theorem true?" do not evaluate to true or false.[2]

Elaborating on the notions of procedural extensions and partial functions, Hadley (1989) also proposed a *default-oriented* procedural semantics. A formal attempt to characterize the meanings of words such as "chair." "dog", "lemon" and so on, in terms of an exhaustive list of properties or features, often falls short of delineating the class of objects to which the word applies: We encountered this problem in the section entitled "Conditions and Criteria." To reiterate, one might attempt to define the meaning of "cat" for example, as having hair, having four legs, eating meat, having sharp teeth, and so on. In almost all cases, however, one can imagine entities to which the term should apply, but that fail to have one or more of these properties. A solution, in terms of a default-oriented semantics, lies in thinking of the procedural meaning of "cat" executing a partial function, or alternatively, as involving a number of sub-procedure calls from within one major procedure returning default values. Such a default procedure would assign truth in some cases ("yes, that is a cat") and falsity in some others ("no, doesn't look like a cat to me"), but will simply never have been extended to cover all of the possible sensory stimuli that it could be given as argument.

According to Woods, the use of partial functions enables the procedural theorist to make a wholly psychologically motivated distinction between *meaning* procedures and *recognition* procedures. The rationale for it goes something like this: Consider the word "yellow." In standard expositions, this term has an entry in the mental lexicon comprising the concept YELLOW, and it has two associated procedural specifications. A meaning procedure computes the "bottom line truth conditions for the concept"

---

[2] The tide history being what it is, the sentence "Is Fermat's Last Theorem true?" now *does* evaluate to true or false, a proof having been recently found.

(Woods 1981, p. 326), a very lax statement in light of our previous discussion of formal semantics. Woods characterized the meaning procedure as an *abstract* procedure, which is to say, as a procedure that, under certain circumstances, may be unexecutable for some reason. The recognition function, on the other hand, is the procedure that is most commonly employed in judging that a concept (corresponding to a word) is applicable in a given real-world situation, although this procedure may not be foolproof; a recognition procedure associated with the concept YELLOW might be fooled (that is, executed inappropriately) in a certain kind of illumination, for example.

Procedural Intensions

In formulating a procedural semantics, Woods (1986) assumed that the natural language used by an individual publicly is translated into some manner of internal or mental language. Woods argued that such an internal language is required to cope with the pervasive ambiguity of its public cousin. Which is to say, lexical ambiguity, in respect of words such as "bachelor" (this word has a number of different interpretations, including baccalaureate degrees, certain kinds of fur seals, and an unmarried adult male); parametric ambiguity, in respect of such words as "tall," "contain," "measure" and so on (how tall is "tall?"); and more generally, under the rubric of ambiguity, a whole collection of linguistic and psycholinguistic phenomena, including anaphoric reference, ellipsis, multiple meanings, and "other locutions involving a degree of lack of specificity" (Woods 1981, p. 311). The solution for such problems of ambiguity, Woods urged, is a drastic and all encompassing disambiguation involving, quite simply, a translation of the ambiguous external language (viz. the public language that a given speaker/hearer would use in everyday discourse) into an unambiguous internal language.

One of the clearest and most thorough accounts of a procedural semantics for natural language can be found in Johnson-Laird (1983), as part of a two stage theory of comprehension. In the first stage, a natural language expression is translated into a "propositional representation": a symbolic representation couched in a language of thought (cf. Fodor, 1976). In the second stage, the propositional representation is used as a partial basis for the construction of a mental model—partial because contextual cues and implicit inferences based on general knowledge also aid the construction of the model. The second stage of Johnson-Laird's two-stage theory of comprehension is, deliberately, similar to the operation of the interpretation function in formal semantics. Unlike formal semantics, however, the functions that construct, extend, evaluate and revise mental models cannot be treated in an abstract way: The functions that map propositional representations to models must be explicitly specified. In this way, the procedural semanticist is forced to recognize the importance of specifying the intensions of words.

Johnson-Laird proposed a number of general procedures for constructing mental models and incorporated them into an illustrative program that performs spatial comprehension implemented in a simple list processing language. The general procedures are as follows: First, a procedure that begins construction of a mental model; second, a procedure that adds entities, relations or properties to a constructed model; third, a procedure that combines separately constructed models; fourth, a procedure that verifies whether asserted properties or relations hold in the model; and finally, two procedures that simulate a non-deterministic device by recursively revising the constructed model. In the course of comprehension, a propositional representation of a natural language string will elicit one of these procedures as a function of its referring expressions, the context (as represented by the model constructed to that point), and world knowledge.

For our present concern, however, the most interesting aspect of Johnson-Laird's procedural semantics is the fact that the intensions of basic lexical items are conceived of as *procedural primitives*. What, one may ask, is a procedural primitive? Let's find out.

In Johnson-Laird's procedural semantics, once a general procedure has been selected, the meaning of the assertion has to be used in running that procedure. That is, the meaning of the assertion has to be integrated into the general procedure: It is no good to insert an **A** and a **B** into a mental model arbitrarily if the assertion specifies that a particular relation, such as "in front of," holds between the two items. The integration of meaning into the general procedures is achieved by "freezing in" the values of a variable in a function. This freezing-in process creates a new, more specific function. For example, the simple function:

$$ADD\ (x,\ y);$$

simply adds together the current values of x and y. A new function can be obtained from ADD by freezing in the value of one variable.

If, for instance, the value of the first variable was frozen in to be 478685, then a new function ADD478685 would have been created via the following kind of (high level, list processing) sleight of hand:

ADD (%*478685* %) → *ADD478685*

which partially applies the function ADD so as to yield the new function ADD478685. This new function takes just a single variable and adds 478685 to whatever its value is: The result of ADD478685(3876) is 482561. This talk of "variables," "numbers," and "freezing in" becomes more pertinent when one realizes that functions, as well as numerical values, can be "frozen in" as the values of variables: Specifically, one can "freeze in" the meaning of an assertion into one of the general procedures for constructing mental models. Also of course, this notion becomes even more relevant when one recalls that intensions (qua meanings) are, on a modern conception, extension-yielding functions from indices.

Johnson-Laird (1983) illustrated the efficacy of this "freezing in" technique in a symbolic program for interpreting spatial descriptions. In this program, each general procedure has a mechanism for scanning a grid in any arbitrary direction, which works by iteratively incrementing the two x and y coordinates of any given location. Integrating the meaning of the basic spatial term *front* into these procedures involves freezing in the value of one of either the x or y coordinates—which is to say, a putative lexical entry for *front* might look something like the following:

```
function in front of;
a(%1, 0 %)
end.
```

This is the line of code from the Johnson-Laird program. The function is called *infront*, and it takes a single argument, *a*, which is always one of the general procedures described above. The second line is the procedural primitive that freezes in the values of 1 and 0, creating a more specific procedure that scans the grid by progressively incrementing the x coordinate by 1 and the y coordinate by 0. Hence, when the function *infront* is applied to the procedure for adding a new item into the grid, a new procedure is created that inserts the new item *in front of* the old item at the first available location.

The notion that procedural primitives might underly the meanings of words has been, as noted, previously investigated by Miller and Johnson-Laird (1976), who sought to show how language might be procedurally related to perception. A simplistic rejection of such an approach assumes that meaning procedures are necessarily involved in verification of facts, events or relations in the outside world. It is hardly likely, however, that the intensions of words are specified with respect to the real world, but rather are specified with respect to some model of the world—hence Johnson-Laird's use of mental models.

Whatever their status, procedural primitives are still primitives, and although decompositional theories of meaning might disagree as to the number and nature of (the putative set of) semantic primitives, at least they all agree that such a set exists: If asked the question, "Are there semantic primitives?", they will all usually answer in the affirmative. The basis for such certainty is undoubtably historical; one can see it in the way that a naive philosophy might treat the concept corresponding to a term as a simple conjunction of predicates (i.e., combination of more primitive parts), and one can see it in the parallel evolution of theories of lexical meaning and theories of concept acquisition. That is, one can say both: (*a*) that the meaning of (the word) "bachelor" is constructed from the meanings of (the primitives) "unmarried" and "man," and (*b*) that (the complex concept) BACHELOR is a construct out of the (basic) concepts UNMARRIED and MAN.

A very different view of lexical meaning in psychology, however, is promulgated in Fodor, Garrett, Walker and Parkes (1980), and it involves a rejection of:

- the idea that word meanings are decomposed into semantic primitives (or to couch the same point in their favored vocabulary, a rejection of the idea that there are *definitions*);
- a rejection of the idea that basic lexical items and morphemes have any internal meaningful structure;
- a rejection of the idea that the lexicon contains entries comprising semantic primitives (qua definitions).

How they come by such a radical conclusion will be the subject of the next section.

MEANING POSTULATES

The anti-decompositional stall is set out by the construction of what Fodor et al. (1980) called "The Standard Picture" or TSP: This they identify as a loose consensus of opinion and attitude of what a decompositional theory of meaning looks like, distilled from a variety of schools of thought, but recognizing the prominence of none, save perhaps Empiricism (and that's a pretty big school of thought). The picture that they construct accordingly views the mental lexicon (Fodor et al. called it the morphemic inventory) as partitioned into two parts: The first partition contains definable terms (viz. complex, definable lexical items such as "bachelor") and the second partition comprises the "primitive basis," which is notably smaller than the lexicon itself. Definitions of terms, such as the definition of "bachelor" as "unmarried man," relate the definable items of the first partition to items in the second partition.

Decompositional theories, according to Fodor et al., hold that definitions fix the extensions of definable terms in the first partition relative to an interpretation in the second partition. Naturally, the question arises as to what fixes the extensions of items in the primitive basis. Not another level of definitions because: (a) decompositional theories do not postulate such a level, and (b)even if they did, such a level would in turn need another level of definitions to fix the extensions of its items, and so on, downward into an infinite regress. Decompositional theories avoid such a potential regress, Fodor et al. claimed, by couching the items of the primitive basis in a sensorimotor vocabulary; in virtue of this vocabulary, the items of the primitive basis have their extensions fixed by a specifiable causal hook-up to the world via sensorimotor transducer mechanisms. Taken together, the definitions and the causal account of the hook-up serve to fix the interpretations of all the entries in the lexicon. This, according to Fodor et al., is the standard picture of decompositional theories of meaning.

The rejection of this picture that Fodor et al. undertook has several parts of interest to us, and we will consider them in due course. Before we do, however, we can say the following: All of the arguments deployed in the 1980 paper can be seen to distill down to simple reductionism of the sort that a theorist might use show that the computational power of a Turing machine is not improved by increasing the size of its vocabulary (of ones and zeros). As such, the arguments are riding on the back of the Church-Turing thesis, which is, of course, an open-ended conjecture. This objection notwithstanding, Fodor et al. deployed their arguments with respect to natural (and internal) languages and conclude that semantic decomposition is, in principle a theoretically suspect notion because is it, in principle, theoretically unnecessary.

Fodor et al. (1980) can be seen as an elaboration of the ideas put forward in Fodor, Fodor and Garrett (1975). In this earlier paper, Fodor et al. identified a precursor of TSP of decompositional theories and isolate three points that they considered serve to define what representations at a (generally agreed, loose consensual) "semantic level" might look like:

- Understanding a sentence requires recovery of the *semantic representation* of that sentence: Only in this sense are such representations psychologically "real;"
- Semantic properties and relations (such as synonomy, ambiguity and entailment) are formally definable over representations at this "semantic level"
- Lexical items of a natural language do not necessarily correspond one-for-one with syntactically simple expressions at the semantic level—that is, there are (a) *definables* at the natural language level, and (b) *definitions* at the semantic level.

According to Fodor et al. (1975), one of the principal claims of theories cleaving to a decompositional account of meaning is that understanding a sentence involves recovery of the representation of that sentence at the semantic level. To compliment what is above, by "semantic level" Fodor et al. meant a level of representation at which definable expressions are replaced by their definitions, taken from the primitive basis. Understanding an expression such as "John is a bachelor" consequently would then amount to recovery (qua computation) of the representation "John is an unmarried man." Fodor et al. (1975) pointed out, however, that if understanding a sentence of the form ". . . . bachelor . . ." involves the computation of an internal representation of the form ". . . . man . . ." then all that a decompositional theory of meaning can say (about the semantic primitives of "bachelor") is that "man" is its own internal representation. That is, to understand an utterance of the form ". . . . man . . ." is just to compute a token of a semantic representation of the form ". . . **man** . . ." About this phenomenon they say the following:

> Its part of the intuitive appeal of [such a] theory that it avoids the necessity of saying things like *'bachelor is defined as 'bachelor'* or *the internal representation of 'bachelor' is 'bachelor'*. Progress appears to be made when tokens of 'bachelor' are internally represented by tokens of some other formula (like 'unmarried man') . . . however . . . this is only progress towards the primitive basis, once one gets there, some notion of understanding a word other than re-

covering its definition will have to be involved. (Fodor et al., 1975, p. 276)

Such observation leads them to conclude that nothing changes in principle if a theorist entirely does away with the notion of decomposition and (its attendant byproduct of) semantic primitives and views the entire vocabulary of a language as primitive, so that each word functions as its own internal representation for purposes of sentence comprehension. After all, they continue, psychological theories of meaning essentially involve only functions from symbolic language strings to internal representations, and the latter are themselves merely strings of symbols in a formal (and internal)language. Nothing, to reiterate the point that Fodor et al. wished to make, hinges on the size of the primitive vocabulary that this formalism exploits—hence, the *possibility* exists that semantic primitives are not part of the psychological repertoire of the language user because they are not *necessary* for sentence comprehension.

The second of the characteristics that Fodor et al. (1975) attributed to decompositional theories, and the second of the points that we wish to discuss, is the assumption that semantic properties and relations are formally definable over representations at the semantic level. They used the relation of entailment to illustrate. To recall our discussions of formal theories of meaning in the previous chapter, logic provides the means by which the validity of arguments can be assessed, in virtue of the truth conditions of the logical connectives of some given well-formed formula, where such connectives have natural language analogues such as "all," "and," "some," "if then" and so on. To use an example that would have annoyed Frege, the expression below can be said to be valid in virtue of items in the *logical* vocabulary, "and" and "therefore."

John fell over and Jane laughed; therefore, Jane laughed.

But now consider an analogous example:

John is a bachelor; therefore, John is unmarried.

According to the picture that Fodor et al. constructed, theorists of a decompositional persuasion think like this: Just as the first expression is valid in virtue of (the meanings qua truth conditions of) items in the logical vocabulary, so too the second expression is valid in virtue of (the meanings qua truth conditions of) items in the *non*-logical vocabulary. However, the entailment relation is defined not for representations at the level of the surface form of the expression, but rather at the semantic level. Therefore, the validity of the inference from "John is a bachelor" to "John is unmarried," depends on the meaning (qua truth conditions) of "bachelor": but "bachelor" does not occur at the level of representation (i.e., the semantic level) for which the validity (of such inferences as shown above) is defined. Decompositional theories must then suppose that *informally* valid arguments (such as the inference from "John is a bachelor" to "John is unmarried") will prove to be a *formally* valid argument when couched at the semantic level, which is to say, will prove to be formally valid when definable items are recouched as their definitions at the semantic level.

In both the 1975 and 1980 papers, Fodor et al. explicitly rejected such a supposition, and instead advanced the proposal that the validity of such informal arguments is captured much better (and hence that a theory of lexical meaning can be constructed using) rules of inference culled from formal semantics termed *meaning postulates*. Meaning postulates are a device introduced into formal semantics by Carnap (cf. 1952) to limit the model-theoretic interpretations of a formal language. As conceived of by Carnap, meaning postulates were simple rules of inference that eliminated certain model structures from the interpretation of a given well-formed formula. If, for example, a theorist wished a semantics that she had constructed were sensitive to the fact that *women* were necessarily *adult*, then she would simply stipulate an appropriate meaning postulate ruling out any model structure in which a woman was not also necessarily an adult (where "women," "woman" and "adult" are, of course, referring to set theoretic constructs). Bar-Hillel (1967) showed the efficacy of meaning postulates for representing the arguments of lexical items. Such arguments are necessary in expressing the relation between the meanings of words such as *buy* and *sell*. In a quasi-formal notation, such a meaning postulate would look like the one below:

for any x, y, z (if x sells y to z, then z buys y from x)

Kintsch (1974) too, as well as Fodor et al. (1975; 1980), promulgated a psychological theory of lexical semantics using meaning postulates. On this view, the process of sentence comprehension comprises a translation of (expressions of) a natural language into (expressions of) an internal mental language (cf. Fodor, 1976 and similar to the internal language of Woods, 1981): The smallest independently meaningful parts of the vocabulary of a natural language(viz. words or morphemes) are assumed to map one-to-one with corresponding tokens in this putative language of thought. Recycling the time-worn example of adult, unmarried males (or is it species of fur seals?), the word "bachelor" would be translated one-for-one into the primitive and non-analyzable language of thought token BACHELOR.

Here's where meaning postulates come in: Because a theory of lexical meaning based on meaning postulates assumes that decomposition of word meanings does not occur, hence that there are no semantic primitives, the theorist is free to concentrate on specifying the intensional relations that hold between words, or more accurately, between the "mentalese tokens" that stand in for them in the mind. In understanding a sentence such as "A man lifts a child," a listener would (on this view) first translate it morpheme-for-morpheme into aisomorphic mentalese expression, and in determining that the sentence actually *means* that a man lifts a child, the listener would deploy a meaning postulate defined over the mentalese tokens, as shown below:

- for any x (if man (x) then human (x) & adult (x) & male (x))
- for any x (if child (x) then human (x) & not-adult (x))
- for any x, y (if x (lift) y then x (cause to move upward) y)

Summary

According to "the standard picture" of language proposed by Fodor et al. (1980), definitions fix the extensions of definable expressions to an interpretation of the primitive basis. Items in the vocabulary of this primitive basis express sensori-motor properties: Extensions of primitive terms are thus fixed by a causal account of sensori-motortransducers. Taken together, definitions and just such a causal account of sensori-motor transduction fix the extensions of the entire lexicon.

The theory of meaning postulates rejects this standard picture of language and, particularly, it rejects a sensori-motor account of the primitive basis:

> Definitions figure seriously in theories of language and the world only if: (a) all the expressions of a language are equivalent to ex-

> pressions in the primitive basis; (b) the primitive basis is notably smaller than the lexicon; (c) the extensions of expressions in the primitive basis can be fixed without further appeal to the notion of definition. The only primitive basis which has so far been seriously alleged to satisfy (a)-(c) is sensory/motor, and it is morally certain that that allegation cannot be sustained. (Fodor, Garrett, Walker & Parkes, 1980, p. 268)

Safe in their moral certitude, Fodor et al. (1980) felt free to ridicule the notion that a causal connection (such as, *prima facie*, sensori-motor transduction) mediates the relation between tokens (viz. PERSEPHONE) and tokened (viz. the individual, Persephone). In fact, they go further and state that an explication of such a relation is impossible to realize: "There is, as things now stand, no theory of language and the world, and it seems most unlikely that one will be forthcoming in the foreseeable future" (Fodor, Garrett, Walker & Parkes, 1980, p. 309).

One could be forgiven a small measure of confusion; if the theory of meaning postulates is not (after all) a theory of meaning—which is to say, a semantic theory, of the relation between language and the world—then what precisely is it? A connectionist theorist would be tempted to reply that the theory of meaning postulates is, in fact, a classical theory of the cognitive architecture in a different guise: One can almost *see* the evolution of ideas from Fodor, Fodor and Garrett (1975) through to Fodor and Pylyshyn (1988). However, the reply that is usually given is that it is a theory of sentence comprehension, concerned with the relations between tokens and tokens, and not with the relations between tokens and tokened. The meaning postulate theorist rather *must* take this view, because, after having dismissed a sensori-motor account of the relation between tokens and tokened, *offers no other account in its place*. It would appear that having characterized a semantic theory as simply too difficult to formulate, the meaning postulate theorist then feels justified in stating that such a theory is not required anyway.

However, it is precisely because of the fact that the theory of meaning postulates ignores the importance of explicating the relation between tokens and tokened that leaves it crucially impoverished and acutely vulnerable to the full force of the problem that we traced back to Wittgenstein and ostensive definition: Meaning postulates, to coin a phrase, are a *grounding failure*, and in the final section of this chapter, we can illustrate why.

A GROUNDING FAILURE

At the beginning of this section, I outlined four problems that occur when a theorist attempts to integrate formal notions of meaning, such as intension, into a psychological theory of meaning. The fourth of those problems concerned how to specify the intensions of basic lexical items. I mentioned a variety of alternate labels that this problem has attracted in a number of disciplines; in the philosophy of logic, I showed how Wittgenstein encountered it at the terminus of his philosophical analysis, and how the Vienna Circle referred to the problem by the label of ostensive definition, and in the philosophy of mind, Hofstadter (1985) has argued that recognition of the problem requires that theorists wake up from a *boolean dream*. Similarly, at the beginning of the section entitled "Lexical Meaning," I outlined three goals for a theory of lexical semantics. The third of those problems was how to specify extensional relations—which is to say, how to specify the relations between words and the world as humans conceive of it. In theories of lexical meaning, the problem has also attracted a number of labels; failure to recognize its importance has been referred to as the "symbolic fallacy" by Johnson-Laird, Hermann & Chaffin (1984), and in modern parlance, study of the problem is concerned with lexical semantic representation (Schreuder & Flores d'Arcais, 1989).

At this point, we can pose a question: How well have the theories of meaning discussed in this chapter dealt with this problem, in all its guises and with all its different labels? How well does the theory of semantic markers account for this grounding problem? Or the theory of prototypes? Or the semantic network formalism? Or meaning postulates? In short answer, all of these different theoretical accounts of meaning deal with the problem *very badly*, and we can use the theory of meaning postulates to illustrate just why this is the case.

To recall, meaning postulates are, as we have seen, rules of inference defined over mentalese tokens. An individual who infers from "John is unmarried" that "John is a bachelor" is, on the meaning postulate view, making use of an inference scheme such as the one below:

for any *x* (if UNMARRIED (*x*) & ADULT (*x*), then BACHELOR (*x*))

The illusion of the significance of such an explanation can be shown by making use of tokens taken from an unfamiliar sign system, whether formal, natural or mental, it makes no difference. A meaning postulate defined over tokens from such an unfamiliar language might look like the following:

for any x (if IJI (x) & OJU (x) then AJA (x))

where, of course, the token IJI corresponds to the word "iji." Such a postulate might well enable an individual to infer from "iji prosac" that "aja prosac," but no matter how complex the system becomes, nor how sophisticated the inferences it can make, it does not adequately address the grounding problem: A theory couched in terms of meaning postulates defined over uninterpreted tokens tells you about relations internal to a closed system of tokens (or words or symbols), but it tells you *precisely nothing* about any relations external to that system of tokens. Hence, the theory is acutely vulnerable to the grounding problem, and particularly to the problem as formulated in Chapter 8, in terms of the *symbol* grounding problem (cf. Harnad, 1990b): Meaning postulates are just tokens, and although such tokens are systematically *interpretable* as having a meaning, they just as surely have no meaning *intrinsic* to them. The meaning postulate theorists are quite happy to accept this point: *Of course* this is the case, they will argue, because meaning postulates are not intended to be part of a theory of meaning as such, but rather they are intended to be part of a theory of sentence comprehension. This qualification seems to me to be mere hand-waving, and it illustrates the widespread failure of the classical (meaning postulate) theorist to really appreciate the significance of the grounding problem.

The theory of meaning postulates was used to illustrate the pervasive disregard for the grounding problem in much of contemporary psychological theorizing, but it should be noted that this same grounding semantic network problem afflicts, and decompositional theories also. The various theories purport to account for semantic phenomena such as ambiguity and semantic similarity, but they incorporate no mechanism to reconstruct extensional relations. It is toward a solution to this problem of how to reconstruct the relation between a word and what—external to there presentational system of which that word is a vocabulary item—the word is about, to which the remainder of the book will be directed.

I have taken some care to isolate precisely what it is that this book is concerned with: beginning with the invention of formal calculi, I have charted the evolution of meaning through the philosophy of logic, the philosophy of mind, and through the psychology of meaning and have arrived at the point where I can legitimately ask "What is the nature of the relation between language and the world?" I do not share the moral certitude of Fodor et al. (1980) that this is an impossible venture; indeed, if cognitive science is to be even possible, and a theory of mental life forthcoming, it *must* address the nature of the relation between tokens and tokened must, in other words, address the grounding problem. To reiterate, and close, the question is this: How do you reconstruct the relation between the symbolic expression *on the left of* and the relation **on the left of** that holds between objects in the world?

6

Microprocedural Investigations

Consider the spatial assertion, *A is on the left of B*, and two questions that might be asked about it: (a) What does this sentence *mean*?, and (b) what is the *meaning* of the constituent lexical item **left** in the context of the sentence? Bearing in mind, as we have seen, the bewildering complexity of the term "meaning" and its applications, we should perhaps take some care in answering. The ability of an individual to answer the first question is, of course, dependent on their ability to comprehend that sentence—which is to say, their ability to construct an adequate *discourse representation* of the content of that sentence. The content of their answer will usually express something like the following: What *A is on the left of B* means is that two objects, **A** and **B** stand in a certain spatial, geometric relation to one another, a relation that can be labeled *on the left of*.

Implicit in this answer is the notion that linguistic spatial descriptions simply relate objects in space without any inequality of status of those objects. Consider the following two examples:

- The bike is near the house
- The house is near the bike

One might have expected these two sentences to be synonymous on the grounds that they simply represent the two inverse forms of a symmetrical spatial relation. In fact, they *would* be synonymous if they specified only this symmetric relation, but this is not the case, however, because the first sentence makes the non-symmetric specification that the house is to be used as a fixed point of reference by which to characterize the bike's location, and the slightly incongruous nature of the second sentence is due to the fact that a (movable) bike is not usually used to specify the location of a (static) house. A related point can be made, in a slightly different fashion, for our spatial assertion. The objects **A** and **B** do not stand in the absolute symmetrical spatial relation *on the left of.* Rather, the use of the *on the left of* is dependent on the choice of one of the objects as a point of reference, in this case, the object **A.** If a different point of reference was chosen—for example the object **B**—then, of course, the spatial relation between the objects would require a different label, namely, *on the right of.*

This simple example serves to illustrate the vagaries, exceptions and anomalies inherent in the linguistic description of spatial relational terms. It should come as no surprise, therefore, to find detailed and complex studies of such vagary in the psycholinguistic literature. Early studies, such as Fillmore (1968), Leech (1969) and Bennett (1975), tried to isolate the basic geometric and dimensional distinctions that languages mark and sought to derive adequate descriptions of such spatial categories. Talmy (1983) tried to go further by characterizing the general properties of linguistic *fine structure*, that subdivision of language that refers to categories such as space, time, perspective point, force, causation and so forth. Herskovits (1985) combined ideas and concepts from both prototype theory (cf. Rosch, 1975) and *pragmatics* to propose a two-fold theory of the meaning of spatial relational terms involving (*a*) an *ideal* meaning, which is an abstract, geometrical relation, and (*b*) a collection of *use types*, which are complex bundles of information corresponding roughly to various senses and idioms. Theories of the meanings of spatial relational terms are also found in Levelt (1984) and Garnham (1989). In the latter, a unified theory of meanings for *on the left of, in front of* and so on is presented in which a distinction is made between basic, deictic and intrinsic meaning: These three kinds of meaning are intended to mirror an important perceived distinction between observer-relative and object-relative meanings.

In this chapter, the concern is not with a global, structural semantic characterization of the meanings of spatial terms but, rather, is addressed toward the kind of concern raised in the second of the questions that we posed at the beginning of the chapter. To reiterate, that question read as follows: What, in a sentential context, is the meaning of the constituent lexical item **left**? This question is not so much concerned with a *characterization* of meaning but, rather, is concerned with how to realize an (undoubtably impoverished) *representation* of meaning in a computational system.

If asked, an individual is readily able to answer the first question posed at the beginning of the chapter, which is to say, they are readily able, via the construction of a discourse representation, to specify the states of affairs that the assertion describes. Yet that same individual will have extreme difficulty in answering the second of our questions, reiterated above. Their comprehension of what the word means cannot be divorced from the process of constructing a model of the states of affairs that the word-carrying assertion describes. That is, they find it very difficult to formulate a precise description of what the word *left* refers to in isolation from a sentential context, without using the word "right" or pointing. It would appear that underlying the vagaries of the meanings of spatial relational terms at the level of a structural semantics is a debilitating shortage of detail at the level of lexical semantics. A related point was raised, of course, in our discussion of formal semantics in Chapter 4, where the need to specify the body of the function corresponding to the intension of a basic lexical item was noted.

This phenomenon, the ability of an individual to regularly make use of knowledge (about word meanings) that bypasses their own understanding, is not restricted to language, of course. Consider, for example, the extreme difficulty that an individual encounters in trying to formulate a linguistic description of a complex motor-procedural activity, such as rolling a cigarette, learning to juggle, or throwing a side-arm in Ultimate. However, in the particular case of language, we can offer some insight into why specifying the meaning of spatial relational terms is so difficult. The phenomenon arguably occurs because spatial terms such as **left** and **front** and so forth are *natural language primitives*, which is to say, it is impossible to specify the meanings of such terms using other lexical items of a natural language vocabulary, excluding the circular definition resulting from using the words **right** or **behind**.

Spatial terms thus provide an interesting problem domain for a theorist interested in investigating the computational representation of *lexical meaning* because the straightforward strategy of using so called "semantic primitives" (introduced in the last chapter) such as the features of +ADULT, -MARRIED, and +MALE, which are sufficient to define the meaning of bachelor, are not sufficient in this case. Accepting this point might seem to require that a theorist acquiesces to the force of the meaning postulate theory detailed in the last chapter—that the notion of a semantic primitive is in itself a suspect notion. This is an erroneous assumption, and in the next section, we will see why. Briefly, the fact that spatial relational terms are *linguistic primitives*, in the sense that they are undefinable by other *words*, does not mean that such terms are also psychological qua computational primitives.

PRIMITIVES, POSTULATES AND MICROFEATURES

In connectionist research, the notion of a semantic primitive is closely related to the notion of *microfeature*. Indeed, a number of authors have pointed out that the relation is so close that "semantic feature" and "microfeature" may very well be labels for the same thing. Thus, Fodor and Pylyshyn (1988) compared the treatment of microfeatures in the connectionist literature with early proposals by Katz and Fodor (1963), arguing that microfeatures: "are perfectly ordinary semantic features, much like those that lexicographers have traditionally used to represent the meanings of words" (Fodor & Pylyshyn, 1988, p. 19).

Consequently, both microfeatural accounts of meaning and semantic primitive accounts of meaning would be prey to the anti-decompositional stance of the meaning postulate theory. Connectionists, however, tend to conceive of the notion of microfeature as different, in some significant degree, from its simply semantic relative; in part this view is as a result of the debate between *localist* and *distributed* connectionism.

Sharkey (1991) discussed the status of microfeatures, and made a distinction between two types, both distinctive of distributed representations. The first he referred to as semi-localist microfeatures or *symbolic* primitives, used, for example, by McClelland and Kawamoto (1986). In their model, a word such as "ball" would be represented by a number of microfeatures, such as non-human, soft, small, rounded and so on: These microfeatures are termed symbolic because they refer to properties in the world and are themselves independently meaningful or semantically interpretable. Other authors, however, for example Hinton (1981) and Smolensky (1988), discuss microfeatures in terms of semantically *un*interpretable entities. In this case, no one individual microfeature refers to a property of the world. Instead, reference to such properties emerges from a pattern of activation across many microfeatures. Sharkey (1991) termed such semantically uninterpretable microfeatures *non-symbolic* primitives.

In considering the design of a connectionist network that makes use of representations of word meanings, the semantics of most natural language words can be adequately specified using symbolic primitives, the objections of the meaning postulate theorist notwithstanding. The meaning of a spatial term such as *left*, however, to reiterate the point, cannot be adequately captured using such symbolic primitives because such terms are themselves natural language primitives—it is not possible to define the word *left* in terms of other words where those words are serving as putative symbolic primitives. This problem of representing and encoding meaning can be circumvented if the connectionist employs non-symbolic primitives to specify the meanings of spatial terms. Adopting this approach would usually amount to coding a semantically uninterpretable feature set over input activations (cf. Harris, 1990).

How would the meaning postulate theorist view the status of non-symbolic primitives? To begin with, it is important to realize that meaning postulate theorists—being, as we have noted, *classical* theorists—have no interest in explicitly rejecting a connectionist notion of primitive, any more than a linguistic notion of primitive. Rather, they are in the business of rejecting *any* notion of primitive. The fact that the connectionists have at their disposal the notion of non-symbolic primitives, which is to say, primitives that are individually uninterpretable with respect to the world, would not, therefore, substantially alter the meaning postulate theorist's views. However, I should add, rejecting the notion of a primitive basis requires that the meaning postulate theorists propose some other way in which meaning, in a general sense, can be introduced into their purely syntactic engine. Thus, the symbol grounding problem rears its head again. The fact that the theory of meaning postulates does not specify any mechanism by which such an introduction can be achieved is, as we saw at the end of Chapter 5, one of the principal reasons to reject it.

However, these considerations aside and the meaning postulate theory notwithstanding, it is possible to entertain a different conception of primitive, quite apart from either the linguistic notion or, indeed, the symbolic and nonsymbolic notions. That conception comprises viewing a primitive as a *procedural* entity, and not merely as a procedural entity couched in the vocabulary of the von Neumann machine, which is to say, a procedure characterized purely in classical terms. Rather, the possibility exists for considering a kind of primitive, which is not a symbolic nor indeed a non-symbolic microfeature encoded over input activations, and which is neither a classical procedural primitive. Instead, it is an entity contained in the weights of a connectionist network and which has *evolved* after a given period of learning to meet specific task requirements. Such a conception of a primitive draws its inspiration from two sources: first, from the notion introduced in the last chapter, of a procedural primitive (cf. Johnson-Laird, 1983), and second, from the notion of a uninterpretable non-symbolic primitive. It is to the details of such a conception that we now turn.

THE MICROPROCEDURAL PROPOSAL

The proposal to view primitives as procedural entities, or more accurately, the proposal to view the representations of meaning as distinctly connectionist procedural entities was first mooted in Jackson and Sharkey (1991). This paper sought to describe a novel computational framework for representing the meanings of spatial relational terms, such as *in front of*, *on the right of* and so on. The framework drew on the terms and concepts of both formal and symbolic computational species of semantics, including the conception of meaning in terms of intension and extension, the *interpretation function* of a possible world semantics mapping lexical items into mathematical model structures, the concept of a procedure computing the intension of an assertion and so on. Recognizing the influence of both these species of semantics led to the framework being dubbed a *connectionist* procedural semantics (hereafter connectionist PS). As the name suggests, a connectionist PS requires that the term "truth condition" (qua meaning qua intension) be regarded as synonymous with a connectionist conception of the term "procedure." As has been noted, an early motivation for a connectionist PS was a concern to counter a prevalent view in much of the literature that useful and legitimate connectionist representations should *only* be vectors of unit activation. The connectionist PS framework counters this bias by viewing weights as representations and specifically as non-symbolic, *microprocedural* representations.

The notion of a microprocedure is essentially very simple, and it can be illustrated by reference to its more familiar symbolic relation. A symbolic procedure is what makes the abstract notion of a *function* concrete: A function is a mapping from one set to another, and a procedure specifies how to carry out that mapping. One can illustrate the assimilation of these two notions to the computation performed by a connectionist network by looking at just one single weight, shown in Figure 6.1.

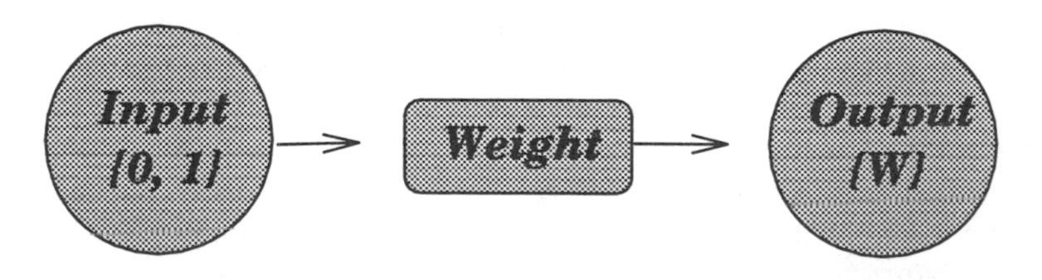

FIGURE 6.1. A single microprocedure.

Take a simple binary unit with two states of activation, either active or inactive (for explanatory purposes, we will ignore the fact that units are more commonly continuously valued). The input set, consequently, is exhaustively specified by either **0** or **1**. We can discount the **0** case immediately as the microprocedure would be inactive under those circumstances, so the input set is simply **1**. The second set, which the function is mapping the first set into before it is squashed by the sigmoid, is given by the weight **w** times **1** (i.e., **w**). A classically defined function, to reiterate, is simply the (abstract characterization of the) mapping from one set (in this example, **1**) to another set (in our example, **w**). Accordingly, an accurate characterization of the weight is as a procedure specifying how to carry out that mapping: Couched in our favored vocabulary, the weight is accurately characterized as a *microprocedure.*

Of course, conceptualizing the weights of a connectionist network in the terms given here means that microprocedures are very limited entities. They can, of course, be more complex as a function of the type of unit with which they associated—which is to say, as a function of whether the unit exhibits binary or continuous activation, whether the unit has a more sophisticated temporal profile (i.e., continuous or stochastic spiking) and so on, but they are still limited. It is consequently important to realize that, in terms of contributing to an understanding of the representation of meaning, crucial theoretical differences exist between (single) microprocedures and what a connectionist PS calls *microprocedural representations.* The term "microprocedure" refers to one single weight, and the term "microprocedural representation" refers to a population of weights.

This is all very well. Naturally, one must ask *how* the behavior of the kinds of simulations to be reported in this chapter are explained using the notions of microprocedures and microprocedural representations. Specifically, how do microprocedural representations explain what happens, after back propagation learning is completed, when a particular input vector is activated? Although we will go into considerably more detail in answering this question in later sections, we can give a flavor of a microprocedural explanation as follows.

Using the general form of the coding used in all of the simulations in this chapter, the input vector **10 10000 01000** codes for the spatial assertion *A is on the left of B.* Each active unit in this binary vector serves to select which weights will participate in a given mapping. In the case of the example assertion, some 30 weights will be selected to participate in the mapping. Couched in the vocabulary of a connectionist PS, a given input is *engaging* a microprocedural representation. The word "engage" is employed here because neither of the symbolic processes of "execution" nor "compilation" (terms that would have to be deployed in any explanation couched in terms of a symbolic procedural semantics) are entirely applicable in this case. There would appear to be elements of both symbolic processes going on, however. The microprocedural representation seems to, in a sense, execute as it returns a result, given a mapping. However, there is also a sense in which the microprocedural representation can be described as participating in a compilation (qua translation) of the input assertion string onto the hidden unit vector of activation.

The main aim of this chapter is to report the results from connectionist simulations inspired by the notion of procedural semantics. The simulations reported in this chapter were trained on a number of variations on a simple spatial task. Employing symbolic or subsymbolic primitives to specify the semantics of the terms *left* and *front* would usually amount to coding the semantics of these words as vectors of activations over input units. In this work, however, no such encoding was used. Instead, the simulations were designed with the idea in mind that the meaning of a word like *left* would become encoded in terms of how the network actually processed it: In terms, that is, of microprocedural weight representations.

LANGUAGE AND STATES OF AFFAIRS

Investigating the meaning of a spatial relational term such as **left**, without recourse to a prior conception of what that meaning comprises—which is to say, eschewing the use of symbolic or nonsymbolic primitives—means that a theorist is able to begin her inquiry from as small a number of assumptions as possible. In this section, I intend to spell out precisely what assumptions I am working from and the way those assumptions are implemented.

All of the simulations detailed in this chapter share a feed-forward architecture, and perform gradient descent in a space of error by means of the standard back propagation algorithm (see Chapter 2). All of the networks perform variations on a simple task: Given an input set comprising representations of spatial assertions, the network must learn to map these inputs to an output set comprising representations of states of affairs that the input representations describe. Drawing on our prior discussion of formal semantics in Chapter 4, the nature of this task can be likened to the corresponding model-theoretic case, where well-formed formulae are mapped to an interpretation in the model structure. This characterization most accurately captures the general form of the task that the simulations were required to learn.

Input Coding

The sets of representations input to all of the simulated networks in this chapter were constructed so as to adhere to the requirement that the meaning of spatial relational terms should not (and indeed, cannot adequately) be encoded, a priori, in terms of microfeatures. Consider the spatial assertion *A is on the left of B*, and the representation of it that is input to a network simulation.

There are three things to encode: the spatial term and the two objects with which the term is associated. Design of the input representations, consequently, reflected this in a simple vector field method. This method involves a given input representation being conceptually carved up into a number of fields; in this case, we need three such conceptual fields. The first, the **op** field (depending on the simulation) contains either 2 or 4 units. Let's stick with the simplest case for now and assume that the **op** field contains 2 units, which code for the two spatial terms **left** and **front**. The second and third fields, **arg1** and **arg2** respectively, each contain five units and code for the first and second symbols of a given assertion. This method of encoding is illustrated in Figure 6.2.

As can be seen, the encoding contains no explicit information about what the spatial terms **left** and **front** actually *mean*: LAB and FAB, for example, corresponding to the assertions *A is on the left of B* and *A is in front of B*, are only distinguishable by virtue of the fact that, in the first case, the first bit of the 12-bit vector is active in the **op** field, and in the second case, the second bit in the 12-bit vector is active in the **op** field. Similarly, the symbols that go to make up a given assertion are coded without explicit information: Each symbol is coded by the activation of one bit out of five in a given **arg** field and is distinguished as that symbol by its relative position in that field.

Output Coding

The output sets for all of the simulated networks in this chapter were designed to perspicuously represent states of affairs and,

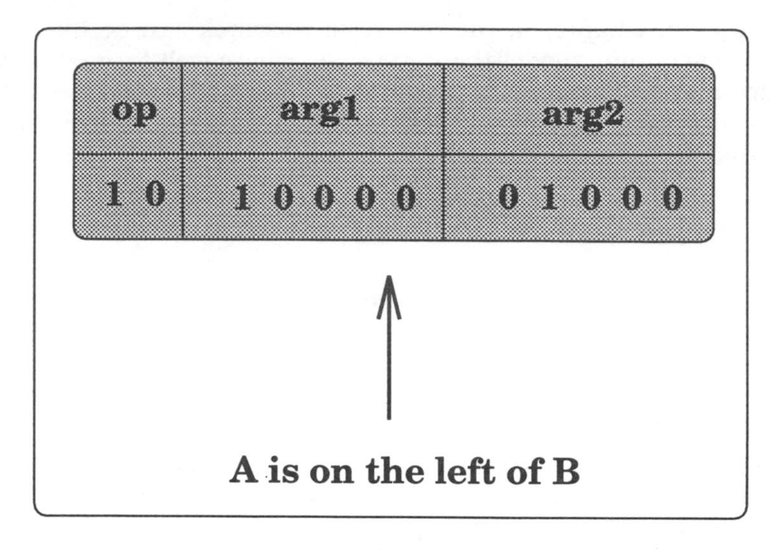

FIGURE 6.2. The rationale for the construction of input representations of spatial assertions.

specifically, geometric relations between objects. The form of output coding should encode relations between symbols analogous to those relations that hold between objects in the real world without an explicit coding of those relations. A variant of the vector field method was chosen, illustrated in Figure 6.3.

The output coding can be conceptually broken down into four fields and visualized as a cross, with the **left** and **right** fields on the horizontal axis and the **front** and **behind** fields on the vertical axis. Each field contains five units and each symbol can be identified by its position within that field. For example, the vector **00100 10000 00000 00000** codes for the states of affairs in which **C** is on the left of **A** (or equally, that **A** is on the right of **C**). This conceptual cross is also shown in Figure 6.3.

LW SIMULATION 1

Given an input set, coding over 12 units representations of spatial assertions, the simulated network was required to perform a mapping to an output, coding over 20 units representations of states of affairs. The set of input vectors used in the simulation code for spatial assertions of the form LAB, LCD, FDE, FAD, and

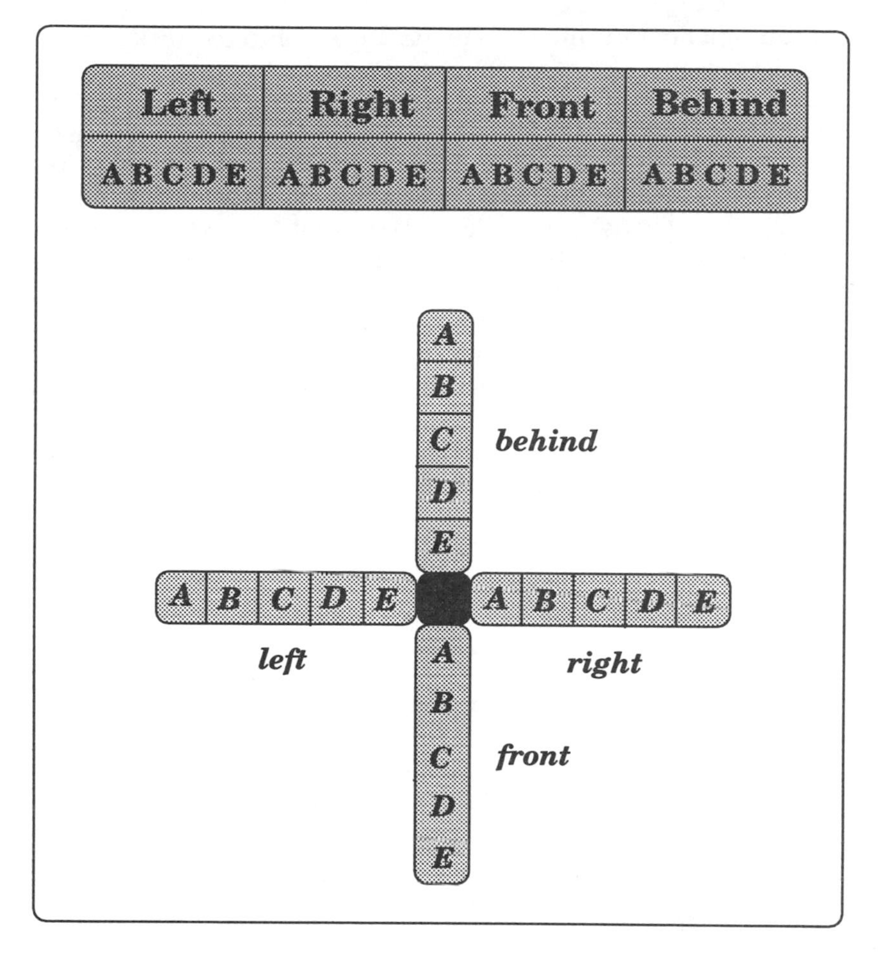

FIGURE 6.3. A conceptual illustration of the form of vector field coding used to construct output representations.

so on, and they were constructed from a corpus of lexical items including the two spatial terms, **left** and **front**, and the five symbols, **A**, **B**, **C**, **D** and **E**. The output vectors code for the states of affairs to which the input assertions refer.

Using two spatial terms and five symbols yields an exhaustive training set of 40 assertions. Ten of these assertions were removed from the corpus and reserved to test for network generalization. Each symbol was omitted four times, and each spatial term five times. The network configuration (12–10–20) learned to a tolerance of 0.05 after 967 cycles with the momentum term at 0.6 and the teaching constant at 0.1. The network exhibited 90% generalization to the test set of novel assertions, using an activation of greater than 0.4 on each output unit as criteria.

The Microprocedural Proposal argues that legitimate and useful representations of meaning can be encoded in the weights of a network, in response to the requirements of a mapping similar to that found in a formal semantics, where well-formed formulae are mapped to an interpretation in the model structure. Support for the proposal that meaning might be encoded microprocedurally can, consequently, be obtained by pulling a learned network apart and looking at precisely *how* the network has performed the task required of it. Accordingly, a hierarchical cluster analysis was performed on the weights and hidden unit activations of this learned network. The analysis was broken down into three components corresponding to the networks two layers of weights and its hidden unit activations.

Input to Hidden Weights

Figure 6.4 shows that the input to hidden weights of LW Simulation 1 have been divided into two major clusters. The first includes the weights associated with the input unit coding for the *front* term; the second includes the weights associated with the input unit coding for the *left* term. The two clusters are thus determined by the two spatial terms of the input assertions. The two sets of symbols coded by the **arg1** and **arg2** fields are distributed uni-

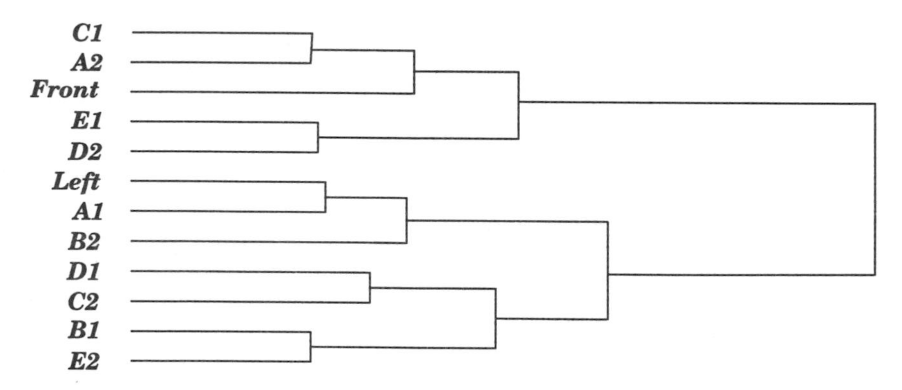

FIGURE 6.4. Projective weights of input for LW simulation 1. The numbers after the symbols refer to which field the symbol occupied in a given input, either **arg1** or **arg2** fields.

formly in the n-dimensional space around the vector of weights for the two terms. This distribution occurs because, essentially, there is no principled reason why it should not. All of the symbols were combined an equal number of times with each spatial term during learning, and any clustering exhibited by the projective weights and associated with each of the units in both the two **arg** fields we can assume is arbitrary. The fact that the division of the weights uses the projective weights associated with the two units coding for **left** and **front** as the determining factor is borne out by an examination of the hidden unit activations, shown in Figure 6.5.

Hidden Unit Activations

Figure 6.5 shows that the hidden units take on two clearly distinguishable classes of activation: One representing the generic class of *left* patterns and one representing the generic class of *front* patterns. Once again, the symbols going to make up the spatial assertions appear to be distributed uniformly within the two clusters determined by the spatial terms.

Hidden to Output Weights

Figure 6.6 shows two major clusters. One includes the vectors of weights associated with output units coding for symbols in the **left** and **right** fields and corresponding to the horizontal axis of the cross. The second includes the vectors of weights associated with the output units coding for symbols in the **front** and **behind** positions of the array, corresponding to the vertical axis of the cross.

What do these three cluster analyses tell us about how the network has performed its task? Additionally, what might they tell us about the meanings of **left** and **front**? The first cluster analysis shows that the network has encoded the regularity that **left** is a qualitatively different thing from **front** and that the network has encoded this regularity in the lower weights. The same sharp separation is reflected in the partitioning of the hidden unit representations, shown in Figure 6.5. Moreover, Figure 6.6 shows that the network has also encoded the regularity that **left** and **right** should be considered similarly—and differently—from **front** and **behind**, and has encoded this regularity in the upper weights.

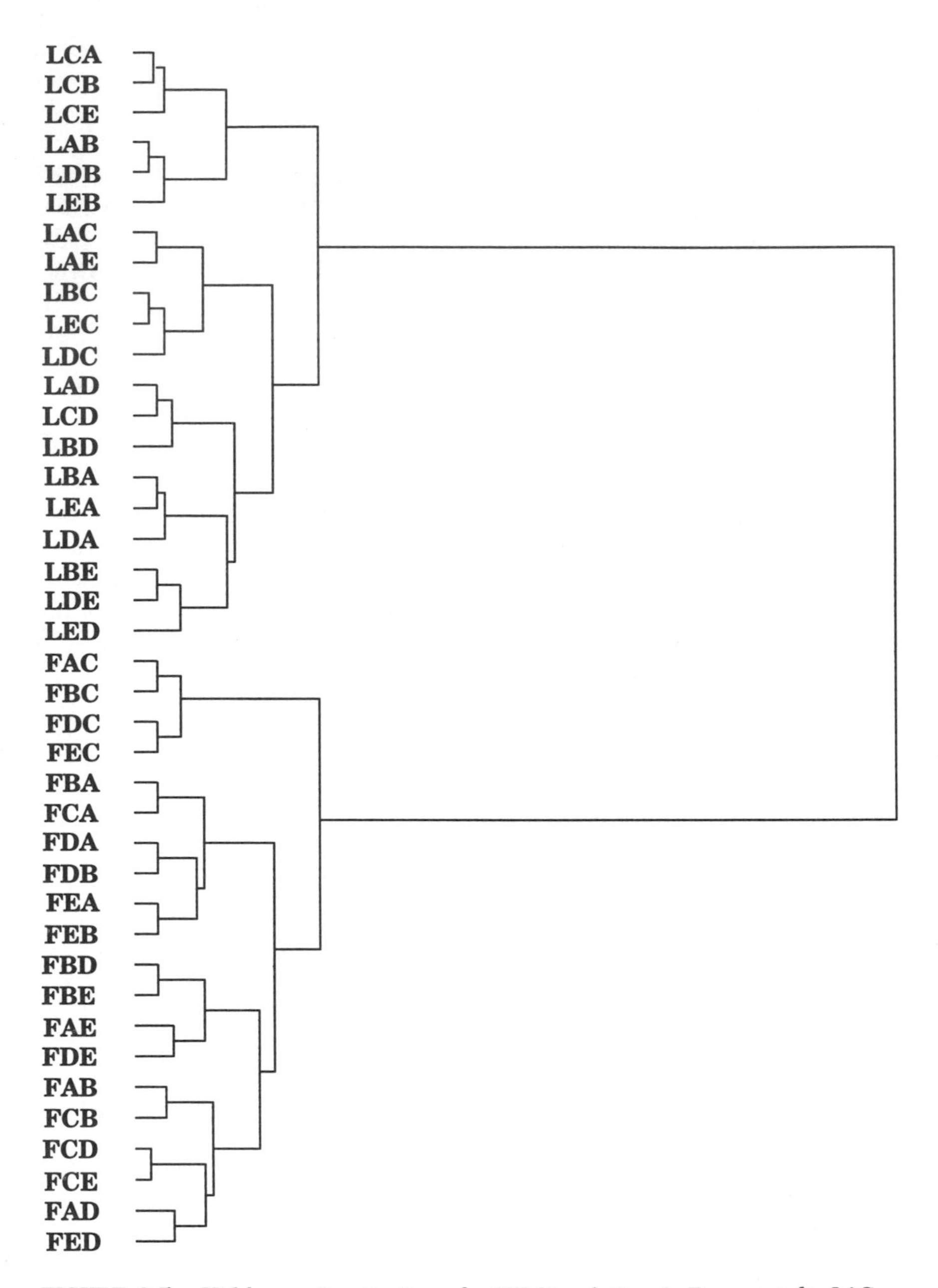

FIGURE 6.5. Hidden unit activations for LW Simulation 1. For example, LAC means "A is on the left of C" and refers to the particular input assertion that gave rise to the hidden unit representation.

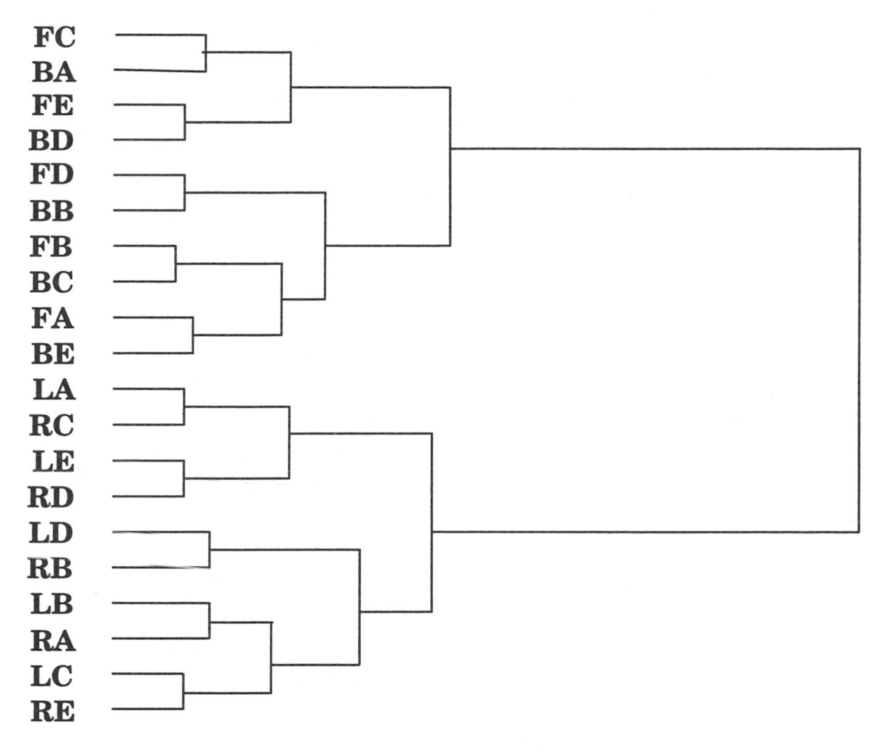

FIGURE 6.6. Receptive weights of output for LW Simulation 1. For example, **BD** refers to the output unit coding for D in the behind field of the output vector.

However, what the cluster analysis does *not* reveal is any detail about the meanings of the four spatial terms; each of the three analyses shows that **left** and **right** are treated similarly by the network, as are **front** and **behind**, but it does not show how the network *distinguishes* between these two similar spatial terms.

Accordingly, the next simulation was designed to explore the extent to which a network is able to distinguish **left** and **right** (or equally, **front** and **behind**) by appropriately encoding the meanings of these terms. Specifically, LW Simulation 2 was run using an input set of assertions constructed using **left** and **right**, rather than **left** and **front**. The conjecture was that the representations the network must develop in order to accurately map an input set of assertions constructed from only these two complimentary terms would result in a clean partitioning of the weights and activations of the network that would be more amenable to analysis and interpretation.

LW SIMULATION 2

The input set for this second simulation codes over 12 unit representations of spatial assertions constructed using **left** and **front**, and the simulated network was asked to perform a mapping from these inputs to an output set of representations of states of affairs, coded over 10 units. The coding of the inputs for this second, LW simulation was essentially no different than that used in LW Simulation 1. However, for example, whereas **01 01000 00001** in the first simulation can be taken to read as "B is in front of E," the same vector in simulation 2 now reads as "B is on the right of E." That is, the two units in the **op** field of the input vector in LW Simulation 2 should be taken as encoding the two spatial terms **left** and **right** in contra-distinction to the **op** field in the previous simulation, which was taken to encode **left** and **front.**

The output vectors code for the states of affairs to which the input assertions refer. Instead of there being 20 units in the output vector, however, in this simulation there are only 10. That is, the output vector can be conceptually broken down into two fields, and visualized as a linear array, with the **left** and **right** fields adjacent on a horizontal axis. For example, the vector **00100 01000** codes for the states of affairs in which **C** is on the left of **B**. In other words, the conceptual cross illustrated in Figure 6.3 has been truncated, with the vertical axis being lost. As in the previous simulation, using two spatial terms and five symbols means that it is possible to construct an exhaustive set of 40 assertions. Ten of these were removed from the corpus of 40 and were reserved to test the generalization performance of the learned network to novel inputs.

Results and Discussion

The 12–7–10 network configuration learned to a tolerance of 0.05 after 13,120 cycles, with the momentum term at 0.7 and the teaching constant at 0.075. The generalization of the network to the test set of 10 novel assertions was tested by noting activation greater than 0.35 of individual units in each of the two fields. The measure turned out to be very poor—at 20% for the 12–7–10 configuration—considerably worse than the measure of generalization for LW Simulation 1. Why this should be so was not immediately apparent, and variations on the configuration of the network,

Table 6.1.
Generalization of LW simulation 2. The I symbol in each of the left and right columns refer to the network *inverting* its target (i.e., given LBC as an input test assertion, the networks' output was consistent with an input test assertion of RBC).

	7		8		9		10		12	
	L	**R**	**L**	**R**	**L**	**R**	**L**	**R**	**L**	**R**
LAB	√	×	√	×	I	I	I	I	√	×
LBC	I	I	I	I	I	I	I	I	I	I
LCD	√	√	√	√	I	I	I	I	√	×
LDE	I	I	×	√	I	I	√	×	√	×
LEA	√	×	√	×	√	×	√	×	√	×
REA	I	I	I	I	I	I	×	×	I	I
RBD	×	×	×	√	√	√	I	I	×	×
RCB	×	√	×	×	√	×	×	×	I	I
RDA	×	√	×	√	√	√	×	√	×	×
REC	×	×	×	×	√	×	×	×	I	I

using 7, 8, 9 10 and 12 hidden units respectively, produced little or no improvement in performance as shown in Table 6.1.

Comparing the cluster analyses obtained from this simulation with those obtained from the first simulation, we can offer some insight into the poor generalization performance of the network to its novel test set. Figure 6.7 shows that the constituent weights have not been divided into major clusters, as happened in the first simulation. Instead, the clustering appears to be far more fragmented and is not determined by the two spatial terms of the set of input assertions. In fact, five minor clusters can be distinguished.

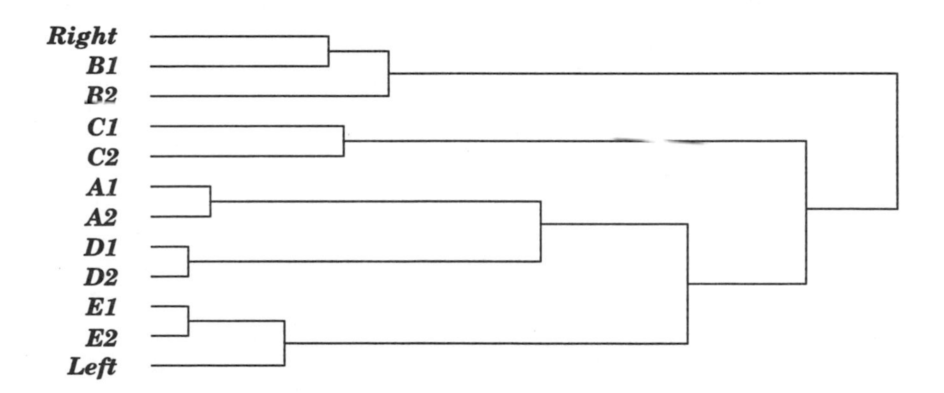

FIGURE 6.7. Projective weights of input for LW simulation 2.

The first includes the vector of weights associated with the unit coding for **right** and the symbol **B** in both fields; the second includes the vector of weights associated with the unit coding for **left** and the symbol **E** in both fields; and the other three minor clusters are all associated with the remaining symbols found in both fields of the input. This dendogram provides a graphical illustration of the poor generalization performance of the network, in that the constituent weights show little systematic clustering related to the performance of the mapping. This lack of systematic clustering is reflected in the hidden unit activations.

The dendogram in Figure 6.8 shows that the hidden units do *not* take on two clearly distinguishable classes of activation as seen in Figure 6.6 and, consequently, have not successfully partitioned a computational space into *left* and *right* patterns. Instead, there appear to be many minor clusters, with the hidden unit partitioning reflecting an encoding of the weighted inputs having more to do with the symbols making up the assertions than the spatial terms themselves.

The poor generalization performance of LW Simulation 2 is a cause for concern for a number of reasons. First, the efficacy of all connectionist networks are normally judged on their behavior toward (a set of) novel inputs. A low generalization performance for a particular simulation to its test set means that the network has not learned *systematically*, and systematicity, as every connectionist knows, is a crucially important notion. A second and related cause for concern, is that a low measure of generalization means that the postulated microprocedural meanings encoded in the weights of the networks are not as robust as they might be. The result of this lack of robustness is that the network easily mistakes spatial assertions constructed using **left** with those constructed using **right**.

This pattern of results can be clearly seen in Table 6.1, where the **I** symbol refers to the situation where, given REA as test for instance, the network produces an output representation consistent with LEA. Such "I" errors were at least as prevalent as correct outputs. It is also interesting to note that the LW Simulation 2 *never* made the mistake of confusing, for example, REA with RAE; it would only confuse REA with LEA. One can see why this should be so if one considers the raw binary data comprising the input set for LW simulation 2.

Figure 6.9 shows that the raw inputs cluster in terms of *symbol*: That is, LDA and RDA, for example, cluster together because

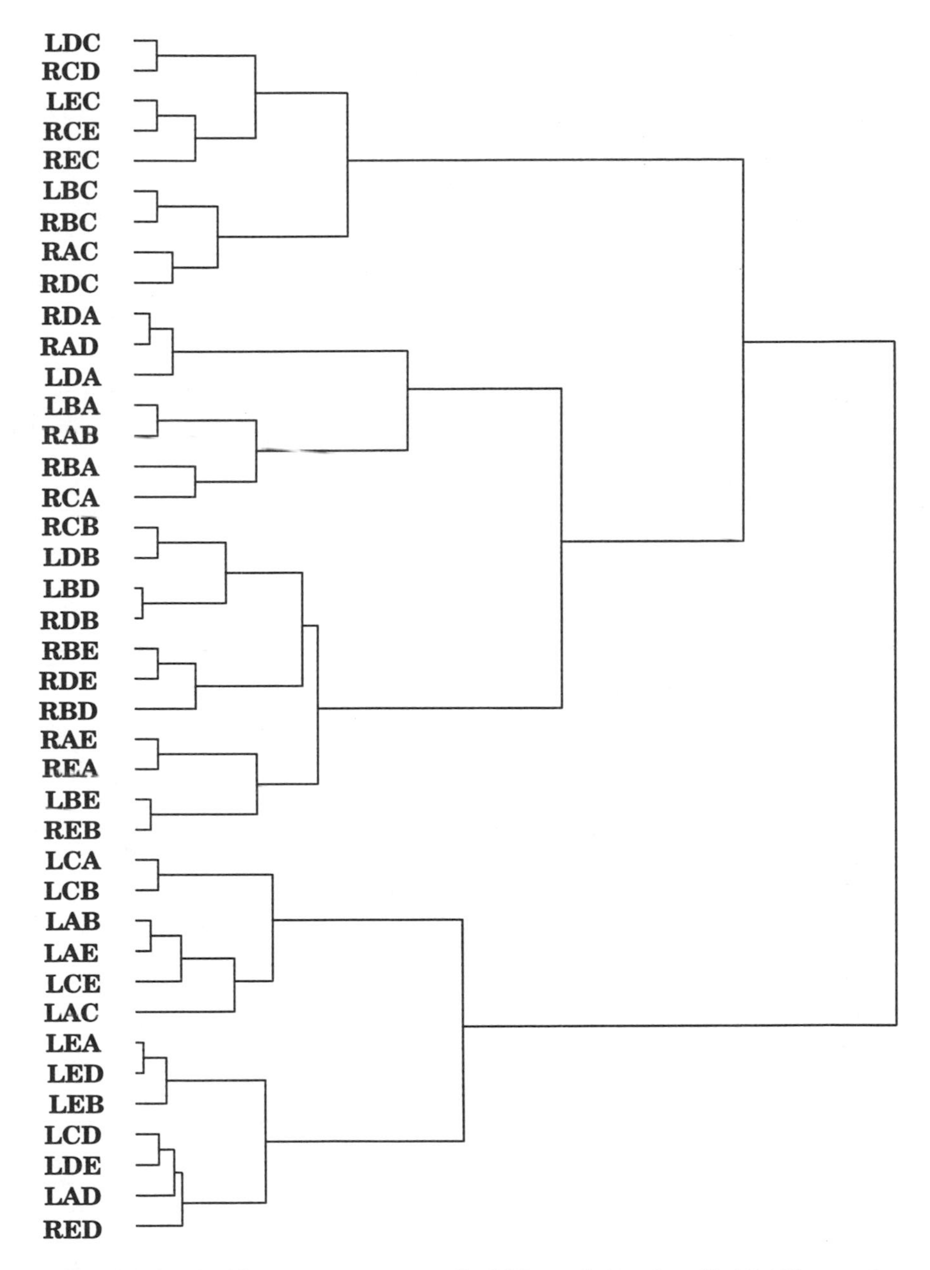

FIGURE 6.8. Hidden unit activations for LW simulation 2, with 7 hidden units.

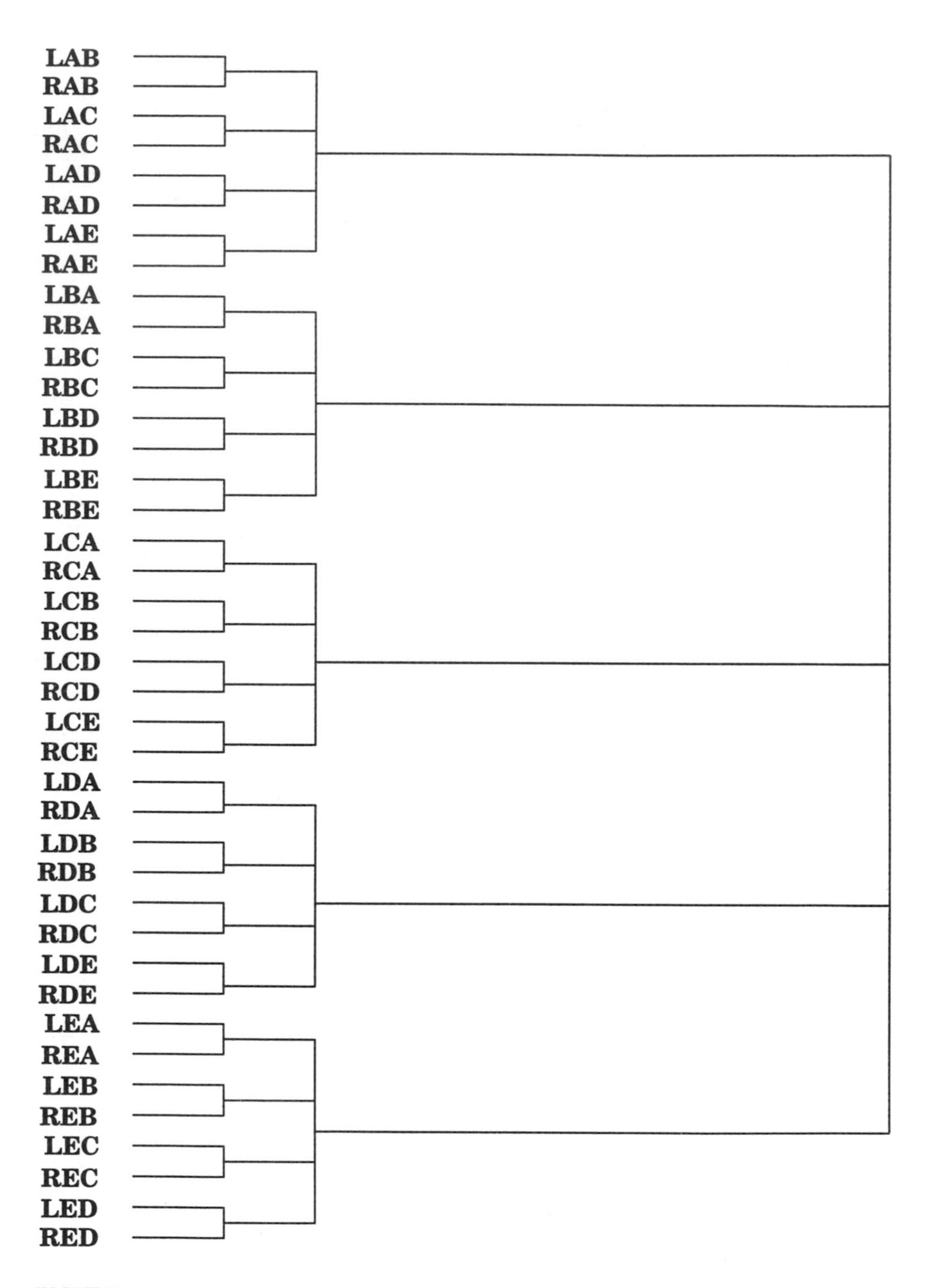

FIGURE 6.9. Raw input data for LW simulation 2.

the symbols of the assertion are symmetrical, whereas LDA and LAD do *not* cluster together because the symbols of the assertion are *not* symmetrical. A number of alternative ways of combating this failing of the coding present themselves. First, in line with the notion that the meanings of spatial terms can be carried by weight-encoded representations, one plausible, but not very subtle, method of increasing the efficacy of LW Simulation 2 would be a wholesale increase in the weight resources of the network. That is, instead of having an input vector of 10 bits, with only one bit coding for the spatial term, the input vector might be 20 bits long, with five bits simultaneously coding for the spatial term: This would have the effect of increasing the representational resources of the network over its lower weights. Instead, the length of the output vector might be increased, with two bits instead of one coding for a given symbol argument in a given output field: This would have the effect of increasing the representational resources of the network over its upper weights. Unfortunately, early exploratory simulations using such augmented networks failed to demonstrate any measurable increase in generalization performance over their unaugmented relations.

The pattern of results from LW Simulation 2 where, for example, the network consistently produced an output representation corresponding to LEA given REA as an input test assertion suggests that what is required, in order that the learning and applicability of the microprocedural meanings encoded in the weights be made more robust, is for the symbolic input assertions to go through some preliminary preprocessing. The processing should occur before they are mapped to their states of affairs so that, for example, the representations of the input spatial assertion LEA would cluster with a corresponding representation of RAE, and not with a representation of REA.

Accordingly, the next section details the execution of simulations designed to preprocess raw binary inputs with a view to warping the similarity space they occupy, such that patterns that need to be treated similarly by subsequent simulations will be forced closer together in this space. The problem to be solved is this: The form of the coding for input representations of the spatial assertions of the input set means that LW Simulation 2, for example, was not able to properly discriminate, when tested, between the four assertions: LAB (**A is on the left of B**), LBA (**B is on the left of A**), RAB (**A is on the right of B**) and RBA (**B is on the right of A**).

PREPROCESSING SIMULATIONS

One initially plausible method of preprocessing might seek to disambiguate assertions such as the four above by employing a variant on auto-association. That is, rather than mapping an input to an identical output, a network would be required to map a given input vector to its *complimentary* output vector. That is, given an input set comprising 80 spatial assertions, and using all four of the spatial terms *left*, *front*, *right* and *behind*, the network would be required to map each input assertion to its complimentary output assertion. For example, an input representation coding the assertion **B is behind E** would be correctly mapped to an output representation coding the assertion **E is in front of B**. The hidden unit representations extracted from such a simulation would then, of course, serve in lieu of the original binary representations. Unfortunately, exploratory simulations failed to demonstrate any real efficacy of this method of preprocessing, and an alternative method was tried and found to be more successful.

The method of preprocessing finally adopted sought to disambiguate the raw binary input data in a different fashion from that used in the early exploratory approach. Instead of an auto-association task, the simulation was designed to perform a classification task. Given an input set comprising the 80 possible assertions from the four spatial terms and the five symbols, a network was required to map a given input assertion and its complementary assertion to a unitary output. For example, an input representation coding the assertion in which **C is in front of D** would be mapped to a unitary output, comprising simply one unit on in the output. Additionally, the input representation coding the assertion in which **D is behind C** would also be mapped to the same unitary output. This procedure is illustrated in Figure 6.10.

The network, in effect, was designed to learn that two complementary inputs are different descriptions for the same states of af-

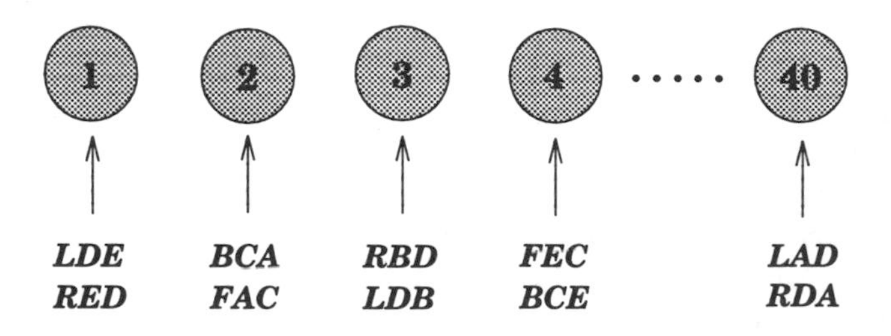

FIGURE 6.10. The classification task used in preprocessing.

fairs (without the states of affairs being explicitly specified). How to characterize the hidden unit representations extracted from this PP simulation is an intriguing question, but not one with which we should be greatly concerned. However, the fact that binary valued representations of LAB and RBA are forced to cluster together does not mean that somehow the network has devised a *new* spatial term, somewhere between **left** and **right**. Two consecutive network simulations were conducted for this classificatory task. The hidden unit vectors from the first run were extracted and used as the input vectors for the second run. In this way, binary-valued encodings over 14 units were compressed in two stages to continuously valued encodings over 12 units.

Results

A 14–13–40 feedforward architecture learned to an error tolerance of 0.075 after 11,925 cycles, with the momentum term set to 0.6 and the teaching constant term set to 0.05. The hidden unit representations from this simulation, when fed through a cluster analysis, did not show the kind of separation that I was looking for, in which instances of complementary pairs (such as LDE and RED) would be grouped together. Because of this lack of separation, the hidden units were extracted from this first simulation and used as the set of input vectors for a second. The 13–12–40 feedforward architecture learned to an error tolerance of 0.05 after 8525 cycles, which the momentum term set to 0.6 and the teaching constant term to 0.05. Extracting the hidden unit vectors and feeding them through a cluster analysis produced the dendogram shown in Figure 6.11.

As can be seen in Figure 6.11, the hidden vectors have clearly been clustered in terms of the sought-after regularity—that is, in terms of complimentary pairs of descriptions, such that RAB and LBA cluster together, BDC and FCD cluster together, LEA and RAE cluster together, and so on. Additionally, the dendogram shows that slightly larger clusters of four vectors bound the pairs of complimentary assertions. That is, the dendogram reveals quadruples of vectors clustered together, such as the pairs (LAB, RBA) and (RAB, LBA), at the top of the dendogram.

Consequently, the next set of simulations were executed using input data sets constructed using the hidden unit vectors extracted from the second preprocessing simulation. Running a simulation like this will yield interesting data on the relative

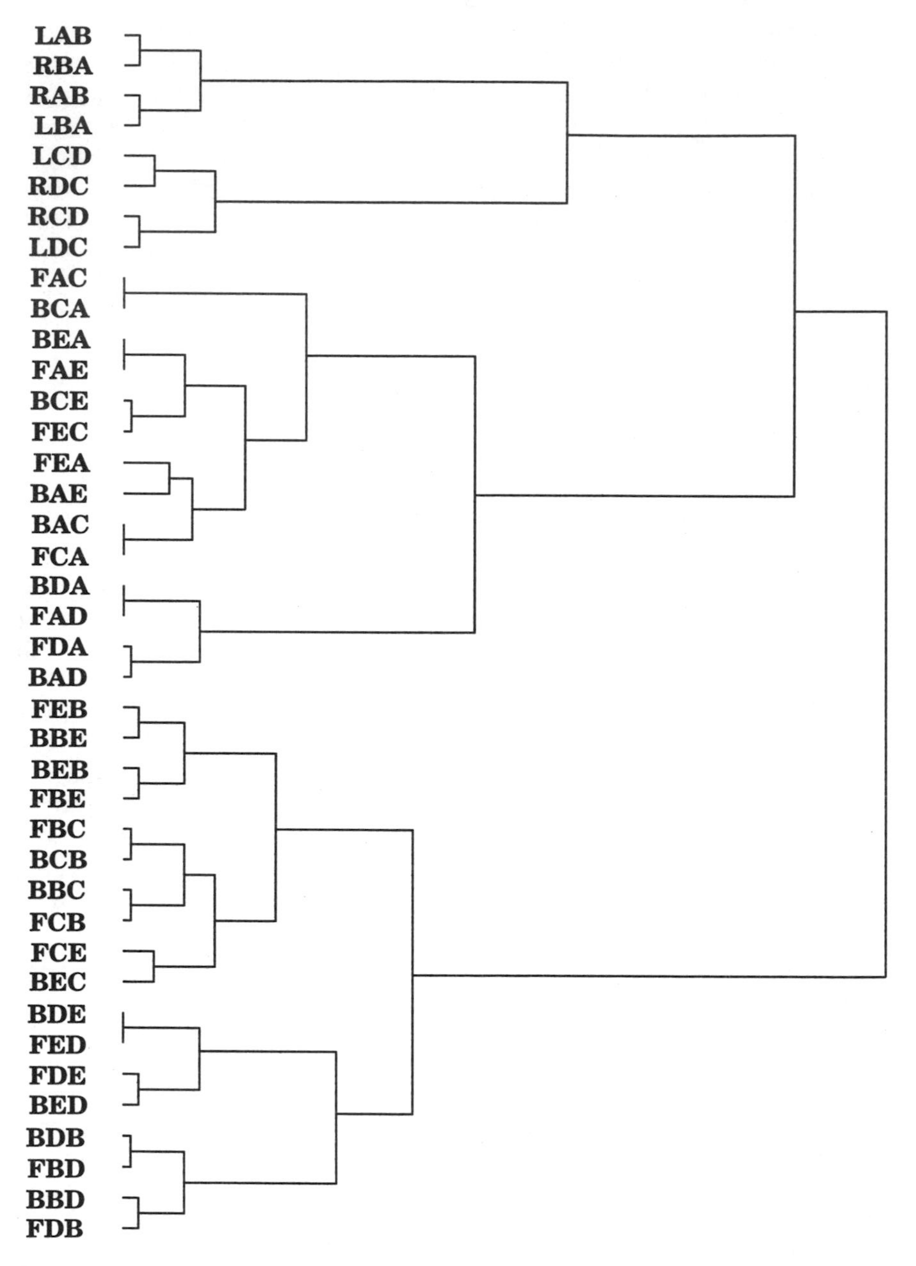

FIGURE 6.11. Representative sample of the hidden unit vectors from preprocessing.

generalization performance of a given network as a function of its input encoding (binary valued versus continuously valued). Also, of course, an improved measure of generalization, to reiterate the point, means that the putative meaning representations contained in the weights evolved after the network has learned to tolerance will be that much more robust.

LW SIMULATION 3

Replacing the binary encoded input set from LW Simulation 2 with the hidden unit vectors extracted from the second preprocessing simulation means that the efficacy of preprocessing can be ascertained immediately. Given an input set, coding over 12 units a continuously valued vector representation of spatial assertions, the simulated network was required to perform a mapping to an output set, coding over 10 units a state of affairs. The input set of vectors comprised 40 of the 80 hidden unit vectors extracted from preprocessing—those hidden unit vectors corresponding to spatial assertions constructed using only **left** and **right**. The output coding is exactly the same as that used in LW Simulation 2.

Initially, and consistent with the architecture of LW Simulation 2, a 12–7–10 configuration was trained on 30 of the 40 assertions extracted from preprocessing. A large number of variations (a) in number of hidden unit, (b) value of the momentum and teaching constant terms and (c) value of initial seed were tried in order that the mapping might be learned. In most cases, approximately 75% of the training set was learned, with the remaining 25% stubbornly refusing to learn.

Training was successfully completed, however, after the input training set of 30 continuously valued vectors was chopped in half, retaining all that were derived from **right** constructions. LW Simulation 3 was then trained solely on these 15 **right** assertions. After having learned to tolerance, the input set was expanded to include all 15 of the constructions derived from **left**, raising the total number of inputs to 30. This procedure was repeated for a number of simulations, with the numbers of hidden units varying from 6 to 10. The generalization performance of these networks is shown in Table 6.2.

The measure used to test the generalization performance of this series of simulations was deliberately strict: Each network had to produce an output activation of greater than 0.5 on the correct unit

Table 6.2.
Generalization performance of LW Simulation 3.

	6		7		8		9		10	
	L	R	L	R	L	R	L	R	L	R
LBC	√	√	×	×	×	ℵ	×	√	√	√
LAB	√	√	×	×	ℵ	ℵ	√	√	√	√
LCD	√	√	√	√	√	√	√	√	√	√
LDE	√	ℵ	√	ℵ	√	√	ℵ	√	√	√
LEA	×	×	√	×	√	ℵ	×	×	×	×
RCA	√	√	×	√	×	√	×	ℵ	√	√
REC	√	√	ℵ	√	√	√	√	√	√	√
RAD	×	×	ℵ	√	×	×	×	√	√	√
RDB	×	ℵ	√	√	√	√	√	√	√	√
RBE	√	√	√	√	√	√	√	√	√	√

in the correct field (either the *left* or the *right*) for that activation to be considered a correct response to the novel input. Moreover, the ℵ symbol in Table 6.2 refers to the case where the network correctly activated a unit in the appropriate field of the output vector, but also *in*correctly activated a unit in the same field. For example, looking at the performance of the network with eight hidden units when asked to generalize to LEA (i.e., generalize to the assertion that "**E** is on the left of **A**"), we see that the network correctly activated both **E** in the left field of the output vector, and **A** in the right field of the output vector. However, it also *in*correctly activated **E** in the right field of the output vector. Occasions where a failure to establish the correct referents of an assertion occurred are marked, to reiterate, with the ℵ symbol in the table. Every occurrence of the ℵ symbol serves to invalidate the network's response to the novel input, and generalization is deemed to have failed. That said, we can see that with a 12–10–10 configuration, the network achieved a 90% generalization to novel inputs, which is a very pleasing figure and a considerable improvement over the poor low 10%–20% generalization of LW Simulation 2.

Input to Hidden

A cluster analysis of the projective input to hidden weights is complicated for LW Simulation 3 because, of course, the input vector is continuously valued. This means that projective weight vectors cannot be assigned to discrete fields of the input and, hence, an interpretation of the dendogram shown in Figure 6.12 is difficult.

Although the dendogram does exhibit two major clusters, this cannot be taken to be a separation of **left** and **right** weight vectors. The reason for this skepticism can be seen in a cluster analysis of the hidden unit activations from LW Simulation 3, shown in Figure 6.13. This figure shows that the network has not achieved a partitioning of the hidden unit space corresponding to the difference between **left** and **right** assertions. The cluster analysis shows that the hidden units are sensitive to the prior preprocessing of the inputs, but does not appear to show any clear clustering of the kind found in LW Simulation 1.

Nonetheless, the efficacy of preprocessing the raw binary inputs has been demonstrated for network simulations carried out using **left** and **right** constructions. Although, as a caveat, we should say that the weights evolved during learning for LW Simulation 3 do not, after analysis, reveal any obvious regularity—which is to say, any clear separation of weight vectors respecting the distinction between **left** and **right**.

A natural extension of LW Simulation 3 is to employ the preprocessed inputs containing all four spatial terms. It may be the case that an analysis of a network simulation required to map **left**, **front**, **right** and **behind** over the same weights will reveal more about what that network has learned about the meanings of spatial terms, and also how it has learned such meanings. This endeavor is the subject of the next section.

Early exploratory simulations revealed a number of factors that affected the training of feed-forward networks on the task with which we are concerned. First, experience revealed that using con-

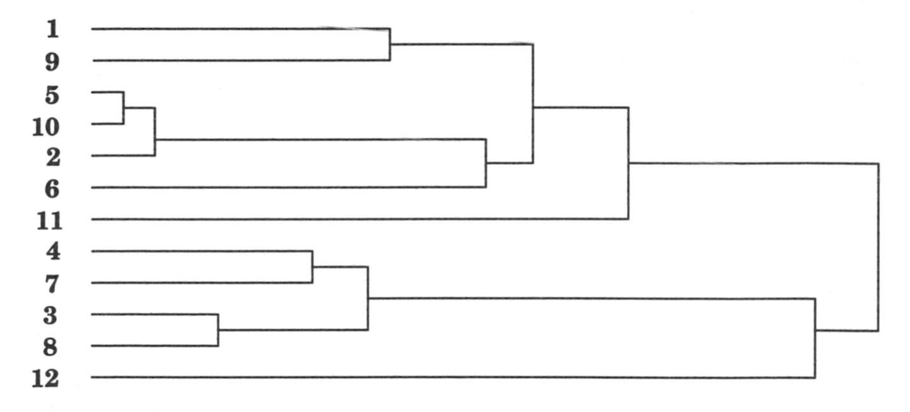

FIGURE 6.12. Dendogram showing the projective weight vectors of LW Simulation 3.

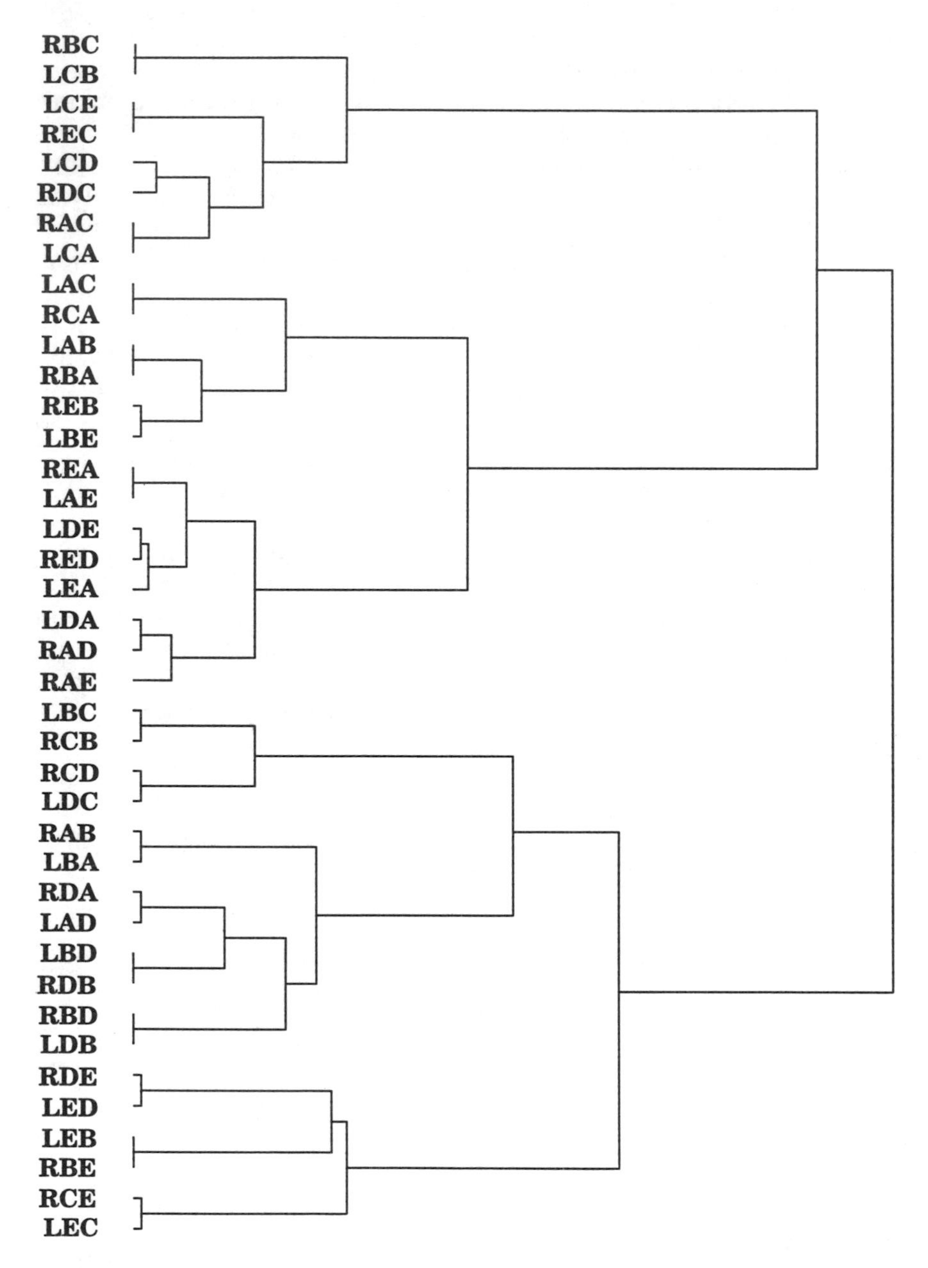

FIGURE 6.13. The clustering of the hidden unit activations of LW Simulation 3.

tinuously valued preprocessed input sets means that a given simulation is very sensitive to the relative ordering of that set of input assertions. Specifically, a network simulation required to map a *randomized* input set learned quicker than a network simulation required to map an unrandomized input set. Additionally, the generalization performance of a network trained on a randomized input set was found to be generally superior to that of a network trained on an unrandomized input set. These early findings are reflected in the next simulation, in which LW Simulation 4 was trained on an input set comprising randomized, continuously valued vectors.

LW SIMULATION 4

The input set for LW Simulation 4 comprised 60 randomized, continuously valued vectors extracted from preprocessing. The set of output training vectors was identical to that used in LW Simulation 1, where the four conceptual fields **left**, **front**, **right** and **behind** are coded over 20 units.

The pattern of learning of the three configurations of LW Simulation 4 remained faithful to expectations: The 12–10–20 configuration, for example, sank into a local minimum after 70,655 cycles, with the teaching constant term at 0.005 and the momentum term at 0.5, and with FAB, RAE and BAC remaining unlearned. In order that the network might learn the entire training set, a *decremental learning* strategy was employed. This strategy is very simple; a network is given an entire training set to learn, and learning is continued until the network is adjudged to have sunk into a local minimum. At this point, the inputs that the network has not learned are removed from the training set to form a partial training set, and the network is trained, with the same weights learned up to that point, on this partial training set. When the network has learned to tolerance on the partial training set, it is retrained on the entire training set once more.

Training the LW Simulation 4 using this decremental learning strategy resulted in the network learning the whole input set to tolerance after a further 1,400 cycles, with the network parameters up at 0.1 and 0.9, for the teaching constant and momentum terms respectively. The first measure of generalization yielded 80%, and the second, more stringent measure was at 65% (i.e., 13 assertions out of 20 completely correct). This is shown in detail in Table 6.3.

Table 6.3.
The generalization performance of LW simulation 4 with 10 hidden units. The ℵ symbol refers to the network activating the appropriate output units, but additionally activating *in*correct units on the output.

	LW Sim 4								
Assertion	**L**	**R**	**F**	**B**	**Assertion**	**L**	**R**	**F**	**B**
LAB	√	√	–	–	FAD	–	–	√	√
LBC	√	√	–	–	FBE	–	–	√	×
LCD	√	√	–	–	FCA	–	–	√	×
LDE	√	√	–	–	FDB	–	–	√	√
LEA	ℵ	ℵ	–	–	FEC	–	–	√	√
RAC	√	√	–	–	BAE	–	–	×	√
RBD	√	√	–	–	BBA	–	–	√	√
RCE	√	×	–	–	BCB	–	–	√	√
RDA	×	√	–	–	BDC	–	–	√	√
REB	√	√	–	–	BED	–	–	√	ℵ

INTERPRETATION OF MICROPROCEDURAL REPRESENTATIONS

In Jackson and Sharkey (1991), the stated position was that the input to hidden weights (of the class of LW simulations from which connectionist PS was developed as a explanatory framework) encoded the intensions of the training set of assertions, and the hidden to output weights encoded the extensional relations for the training set of assertions, a position that seemed to naturally account for both intensional and extensional aspects of the meaning of the set of assertions. Further down the line, this claim seems a trifle simplistic: In this section, I wish to spell out precisely why this is so, via a number of caveats.

Caveat 1

Consider how the claim would be affected by employing a network architecture employing *two* banks of hidden units, for example, instead of one. In such a case, the theorist would need to interpret (*a*) input to hidden1 weights, (*b*) hidden1 to hidden2 weights, (*c*) hidden2 to output weights, and (*d*) two sets of hidden unit activations. The intuitive appeal of distinguishing intensional and extensional meaning encoded either side of a bank of hidden units becomes obscured in such a case.

Additionally, the claim that the hidden to output weights encode extensional relations now seems to err in respect of what such relations comprise. To compute the entire function **F** corresponding to the intension of a given assertion, the network must, of course, deliver the extension for that assertion—which is to say, it must establish the correct referents on the output corresponding to that assertion. Both the input-to-hidden and the hidden-to-output weights participate in this mapping, and thus it seems more accurate to characterize both sets of weights as encoding aspects of intensional meaning rather than introduce an arbitrary distinction, based on architectural considerations between intensional and extensional meaning carried by separate microprocedural representations either side of a bank of hidden units.

In short, what seems more likely is that it is the network as a whole that is computing the function **F** corresponding to the intension of an assertion. This revised claim might seem to leave the notion of extensional meaning underspecified, and in a sense it does. This is not a criticism of the Microprocedural Proposal, however, as we will see in Chapter 7: That discussion can be pre-empted by stating the following axiom of a connectionist PS: Computational machinery, and specifically *microprocedural* representations, encode intensional meaning, whilst discourse representations, and specifically *structural analogue* representations encode extensional meaning."

Caveat 2

The analyses in this chapter of the simulations employing input training sets constructed using binary valued vectors, such as LW Simulation 1, reveal clear regularities related to the computational requirements of the mapping that the network was designed to perform. On the other hand, the analysis of the LW Simulation 4 did not reveal such clear regularities, yet this simulation also generalized well to its novel test set. In the light of these observations, consider the following: "Patterns of connection weights . . . are almost invariably undecipherable. These constitute not transient or episodic states that arise and pass away in the course of the net's operation ('thoughts') but rather its long term functionality or competence ('know-how')" (Haugeland, 1991, p. 85).

The results and analyses from the simulations in this chapter throw some interesting light on Haugeland's assertion that "patterns of connection weights [qua microprocedural representations] . . . are undecipherable." Referring back to an earlier section, consider Figure 6.4, showing the projective input to hidden weights of LW Simulation 1. The input set of vectors are, due to the discrete field encoding, semantically interpretable and, because of this, the clustering of the weights revealed by cluster analysis can be interpreted with respect to this coding. That is, the high dimensional space that the weights occupy reveal, after analysis, hyper-plane partitions that reflect the different meanings of the spatial terms coded in the input set of assertions. Now consider the corresponding dendogram of weights from LW Simulation 4, shown in Figure 6.14.

The input set of training vectors to this simulation are continuously valued vectors. Because the discrete field encoding is thus obscured, such vectors are prima facie, semantically uninterpretable. Thus, the microprocedural representations encoded in the weights of LW Simulation 4 cannot be interpreted with respect to this coding. The weight vectors show no sharp separation corresponding to the distinction between **left**, **right**, **front** and **behind**, as can be seen in the Figures 6.4 and 6.6. The use of a training set constructed from continuously valued vectors means that interpretation of the inputs to the network dips below the symbol surface, similarly, this results in the interpretation of the microprocedural representations also dipping below the symbol surface.

There is, however, one saving grace. The analysis of the representations developed over the hidden units, shown in Figure 6.15, is not hindered by issues of semantic interpretability. The dendo-

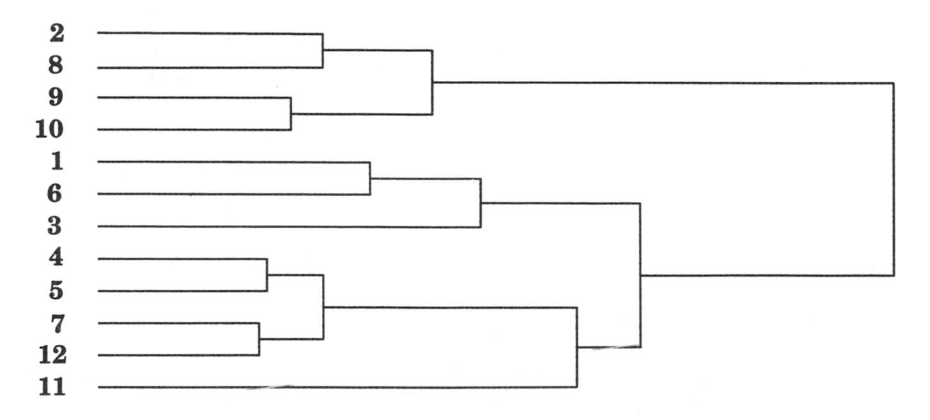

FIGURE 6.14. Projective weights of input for LW Simulation 4.

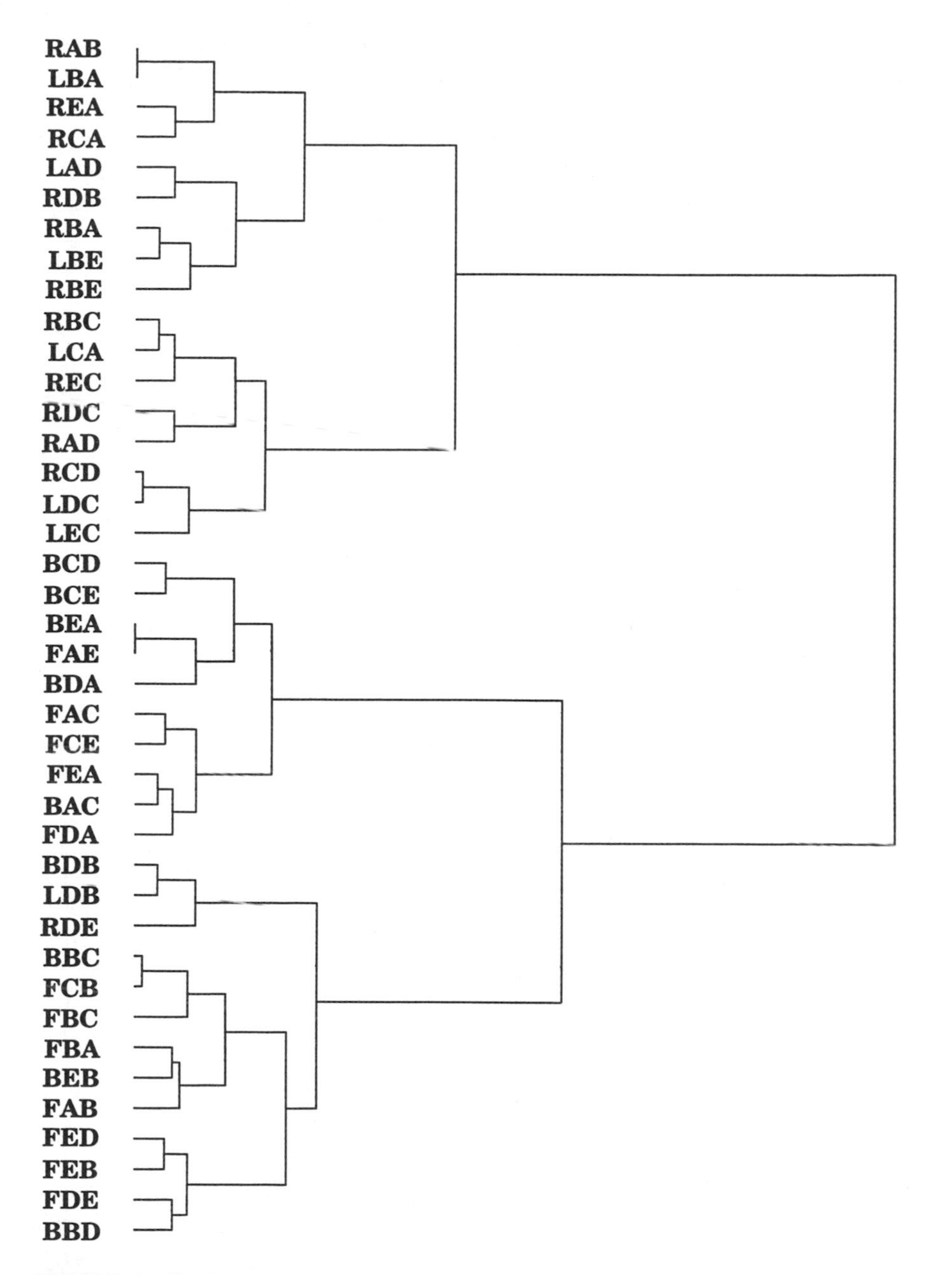

FIGURE 6.15. Representative sample of hidden unit activations for LW Simulation 4.

gram shows a clear separation of patterns into **left/right**, **front/behind**, which reflects the regularity of the microprocedural representations defined over the input to hidden weights, even if that regularity cannot be articulated on the symbol surface.

The clear separation of patterns into **left/right** and **front/behind** as shown in the dendogram, is clearly reflecting some aspect of the intensional meaning of the input assertions: If this were not the case, then one would expect clustering of hidden unit representations in terms of the **arg** fields, or some other similarity more to do with the symbols themselves.

Caveat 3

As I have already emphasized, one of the subsidiary concerns in formulating the Microprocedural Proposal was an analysis of the weight representations evolved during back propagation learning. Such an analysis, it was hoped, would reveal something of the nature of the representations of meaning that a given network simulation had encoded in order that it might perform the task required of it. However, the fact that the cluster analyses of the weights of both LW Simulation 1 and LW Simulation 4 show such marked differences, while still both exhibiting good generalization to their test sets, would seem to introduce an unexpected complexity.

Fortunately, however, I am in a position to offer some insight into why this should be so. The reason that this occurs is due to the fact that Euclidean space and what can be called *computational space*, although highly correlated are not, in actual fact, identical (see Sharkey & Jackson (1994b) for more details). Chapter 3 and Chapter 9 both have more to say about the notion of computational space, but for now, we can illustrate why this inequality (of Euclidean and computational space) is important by considering Figure 6.16.

Figure 6.16 depicts a simple computational space, constructed for the purposes of illustration (it is not derived from a particular learned network), with a decision boundary or *hyperplane*, implemented by weights, partitioning the space into decision regions (see Chapter 2 for how to calculate hyperplanes). The region to the right includes all of the **1**s and is the region in which, for example, a given network might output a 1, indicating a correct response. The region on the left includes all of the **o**'s, and is the region in which, correspondingly, a given network might output a 0, indi-

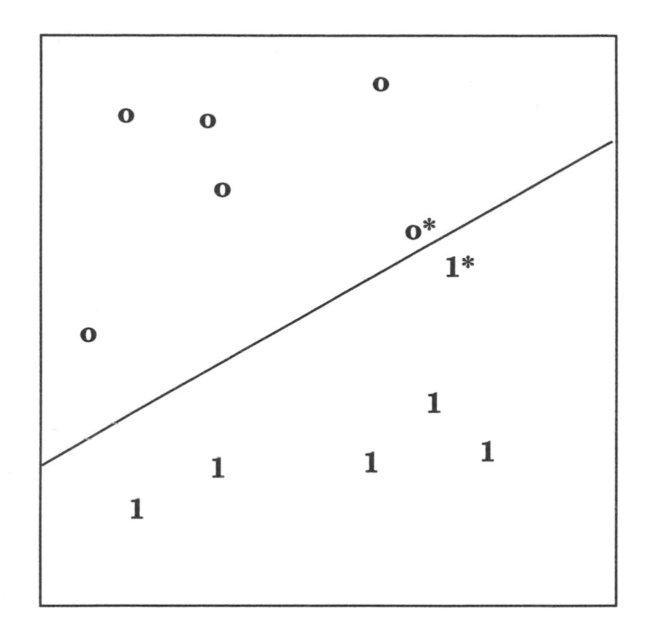

FIGURE 6.16. A simple computational space with hyperplane: Points in the space, either 1 or **o**, indicate representations.

cating an incorrect response. Thus, in terms of computational space, there is a clear separation of regions and consequently of representations. However, if a cluster analysis were to be performed on the data depicted in the figure then, because of the fact that hierarchical cluster analysis works by utilizing a Euclidean distance metric, the two points closest together in the space next to the hyperplane (indicated by*) would be clustered together, even though they lie in different decision regions. Thus, a cluster analysis would in actual fact, *obscure* the true nature of the computational space due to its use of a crude Euclidean distance metric for ascertaining similarity.

Remarks

Given these three caveats, I can draw on the novel vocabulary of a connectionist PS and offer the following description of the behavior of the learned LW Simulation 4. A given input vector serves to select which weights will participate in a given mapping: Specifically, a given input vector *engages* a microprocedural representation and the hidden unit representation then encodes some squashed result of engaging that microprocedural representation (where "squashed" refers, of course, to the operation of the sigmoid compressing the effective output of a given hidden unit to values between 0 and 1). The hidden unit representation, in similar fashion to the input vector, engages a further microprocedural representation, which results in the construction of the final output representation. Just as the hidden unit representations reflect the nature of the prior microprocedural representation that constructed them, so too the output unit representations similarly reflect the same.

The use of the phrase "engaging a microprocedural representation" was used, to recall, because neither of the two classically defined terms of compilation or execution seemed appropriate to describe what the LW simulations were doing. Does a given network simulation *execute* the microprocedural representation? In a sense it does, as it systematically produces a unique result over the hidden units when a given input vector is active. To my knowledge, no more precise notion of a *procedure executing* exists to supplement this account. Does the network *compile* the microprocedural representation? In a sense, it performs this task also, in as much as the representation can be described as participating in a translation of the input vector to a corresponding hidden unit vector. There are reasons, however, for supposing that the notion of engaging (a microprocedural representation) should not be interpreted too literally. *Execute* and *compile* are both terms for *classical* computational processes, but it is not entirely clear if engaging is (or should be thought of as) a term for a *connectionist* computational process. The implications of this point are explored in the next section.

COMPUTATION AND WEIGHT REPRESENTATION

The Microprocedural Proposal, first suggested in Jackson and Sharkey (1991) and outlined at the beginning of this chapter, makes one bold claim: Useful and legitimate representations in connectionist networks are not restricted to patterns of activation distributed over groups of units but, additionally, can be found in the weights evolved during learning. The notion of a weight representation is a novel one and has only slowly begun to find acceptance in the literature (cf. van Gelder, 1991a; Haugeland, 1991). However, where the notion has found favor, it has been taken up enthusiastically and, indeed, the possibility has been mooted that representations defined over weights, as opposed to representations defined over units, may well constitute paradigmatic *distributed* representations:

> Connectionist networks contain two quite different candidates for status as representations. One the one hand, there are patterns of activation across various sets of network units . . . On the other hand, there are patterns of connection weights among the units of the net. To the extent . . . that connection patterns are regarded as representations at all, they are the most compelling candidates in contemporary networks for illustrating a distinctive representational genus—to wit, distributed representations. (Haugeland, 1991, pp. 84–85).

In this section, I wish to draw some conclusions from the network simulations reported in this chapter based on the idea that weight representations illustrate a distinctive *genus* and point out some implications of the Microprocedural Proposal.

On the Notion of Computation

What is computation? The fact that both classical and connectionist theorists are in the business of devising explanations of mental phenomena couched in a computational vocabulary might seem to indicate that this question permits a definite answer. This is not, however, the case, a wealth of intellectual effort notwithstanding (cf. Clark, 1988b; Cummins, 1989; Cummins, 1991; Dietrich, 1990; Sloman, 1990). It is not my intention to enter into a debate about what is, and what is not, computation, but rather to take a common working conception of it and explore how the Microprocedural Proposal can contribute to it.

What is computation? Our inquiry will begin by taking an answer to this question to be *representation manipulation*: When a mechanism computes or performs computation, it executes some series of operations on a set of entities (the input) to systematically yield some other set of entities (the output). Such a conception of computation applies par excellence to the symbol manipulation of classical systems, computating by performing structure sensitive operations on complex structured representations. Is, however, such a characterization of computation adequate to encompass the nature of the information processing exhibited by connectionist networks and, moreover, as Boden (1988) asked, is the same theoretical vocabulary even admissible?

"Are non-sequential, cooperative and equilibrium seeking alterations of patterns of activity (as opposed to formally triggered syntactic manipulations of symbolic forms) really *computations*?" (Boden, 1988, p. 251).

If computation is defined in terms of representation manipulation then, in line with the theoretical force of the Microprocedural Proposal, the connectionist theorist must recognize that the manipulation of both unit *and* weight representations must be admitted into some (more suitably expansive) definition of computation, in contra-distinction to the thrust of Boden's assertion, that connectionist computation comprises only the manipulation of unit representations.

Of course, the representations under present discussion are not, as in the classical case, *symbols*. Rather, they are characterized as non-symbolic, non-classical representations, or subsymbols, or that old chestnut, *distributed* representations. However, let us be a little more precise: In fact, classical systems do not manipulate representations nor, indeed, do they manipulate symbols but, more accurately, they manipulate physically instantiated *tokens*. That is, classical computation comprises the manipulation of syntactically structured tokenings of physical objects. This fact is easily overlooked because, as Chalmers (1992) has recently noted, in classical systems, *tokens* and *representations* coincide (an important point—its implications for the grounding problem, and the role that it plays in the Subsymbolic response, are explored at length in Chapter 8). That is, the Classical symbol ATAHUALLPA serves simultaneously as an atomic computational token and as a referring representation, designating *Atahuallpa*.

What about the corresponding connectionist case? Is the computation performed by a network simulation defined over physically instantiated tokens also? Initially, one might be tempted to answer in the affirmative. However, whereas in a classical system computational tokens are entities such as ATAHUALLPA, in a connectionist network computational tokens are the individual units and weights that comprise the network itself. Now we come to a crucial point: Unlike classical systems in which tokens and representations coincide (ATAHUALLPA serves as both token and representation) in connectionist networks, tokens and representations are distinct entities. However (individual units and weights) lie at one level of organization, representations (which Chalmers solely referred to as patterns of activity over units, ignoring weights) lie at a higher level of organization, defined over those tokens.

It is with regard to this distinction between tokens and representations that the Microprocedural Proposal is able to make a contribution. Specifically, it allows a theorist to conceive of tokens qua microprocedures and representations qua microprocedural representations. Consequently, to conceive of connectionist computation as representation manipulation is to conceive of computation as token manipulation, a notion that now, in line with the Microprocedural Proposal, includes microprocedure manipulation. The LW simulations reported in this chapter can be characterized as performing computation in this sense: A given input vector selects which tokens qua microprocedures are to participate in a given mapping, which in turn determines the manipulation effected on the unit that the microprocedure directly precedes. The question naturally arises as to the nature of the relation that exists between tokens and representations. Or, to put it another way, the relation that exists between computation and representation. What seems to be the case is that a connectionist theorist must distinguish two levels of explanation: A level of computational explanation and a level of representational explanation, with the former falling below the latter.

The activation values of units in a connectionist network are individually *causally efficacious* elements, which is to say, if the value of even just one unit were to be changed, this would directly affect the nature of the computation performed by the network (cf. Sharkey, 1992). Each unit contributes to representations defined over a collection of such individual elements. Representation in a connectionist network is thus accurately described as *emergent* from computational activity at a lower level. Just as units function as individual elements, so too individual microprocedures each play a significant and causal role in the computation of the network of which they are elements and similarly contribute to microprocedural representations emergent from the collective behavior of many microprocedures. Connectionist computation thus comprises a low-level manipulation of tokens (including the manipulation of microprocedures) indirectly resulting in the manipulation of representations.

I expressed reservations earlier that the term *engaging* (a microprocedural representation) might not accurately be described as a computational process, and one can now see why. Connectionist computation is defined over tokens, not representations, and so to describe the engaging of a microprocedural representation as a computational process is erroneous. Quite *how* to describe the process by which a microprocedural representation effects the transformation of one activation vector to another is unclear (how much sense does it make to speak of representational processes?) although one can still apply the term to (individual) microprocedures. In brief summary, the conception of connectionist computation as straightforward representation manipulation is obscured by a number of points:

- The distinction between tokens and representations.
- The distinction between types of token, namely, between tokens qua units and tokens qua microprocedures.
- The emergent nature of representation from lower level computation.

Context Dependency and Systematicity

The notion of microprocedural representations emergent from a low-level computational activity defined over individual tokens qua microprocedures has some interesting implications for one of the classical theorist's favorite notions, that of *systematicity*. This notion is intimately tied up with the *context-independency* of mental representations. We will begin by spelling out just what these two terms mean.

Systematicity is a term, originated by Fodor and Pylyshyn (1988) that refers to the fact that systematic relations hold between cognitive capacities. For example, the individual able to entertain the thought that *Salieri respects Mozart* must by necessity (in order for the mind of that individual to be systematic) be able to entertain the thought that *Mozart respects Salieri*. Similarly for a large number of other cases, if an individual can infer **P** from **P&Q&R**, then she must necessarily be able to infer **P** from **P&Q**. The classical theorist is very explicit about systematicity and the need for the connectionist theorist to account for it:

> The problem that systematicity poses for Connectionists [is] . . . not to show that systematic cognitive capacities are *possible* given the assumptions of a Connectionist architecture, but to explain how systematicity could be *necessary*—how it could be a [psychological] *law* that cognitive capacities are systematic—given those assumptions. (Fodor & McLaughlin, 1990, p. 200)

The classical solution to the problem of systematicity turns crucially on the notion of *compositionality*. Specifically, the classical theorist proposes that the content of thoughts is determined, in a uniform way, by the content of the context-independent concepts that are the constituents of such thoughts—which is to say, the thought that *Salieri respects Mozart* has the same constituents as the thought that *Mozart respects Salieri.* More precisely, a solution to the problem of systematicity depends on the fact that a compound classical symbol, such as SALIERI RESPECTS MOZART, is constituted from the atomic symbols SALIERI, MOZART and RESPECTS. Moreover, the atomic symbol SALIERI expresses the same content in the context of RESPECTS MOZART as it does in the context of MOZART RESPECTS, picking out *Salieri,* which is an element of both the proposition *Salieri respects Mozart* and of the proposition *Mozart respects Salieri.*

The classical theorist holds that systematicity can *only* be explained by appealing to the compositionality of mental representations and, specifically, to the fact that mental representations must have context-independent (or context-insensitive) constituents. The thinking proceeds as follows: If a connectionist accepts that minds are systematic (which they are), then he or she must also reason either that (a) systematicity can be explained without concatenative compositionality and context-independent constituents of mental representations or, (b) that connectionist architectures can accommodate concatenative compositional structure, but in which case they would be classical architectures.

The error in the classical thinking has two parts. The first concerns the notion of compositionality and the assertion that connectionist architectures are unable to accommodate such compositional structure. This assertion is false. Granted, connectionist representations do not have a classical compositional structure, resulting from a concatenative combination of tokens, but they *can* have a *spatial* compositional structure resulting from a non-concatenative combination of tokens (cf. van Gelder, 1990, and see Chapter 7 for more discussion of spatial structure). This, of course, is not a novel point.

However, the second error in the classical thinking, and the main point that we wish to make, is novel, and it concerns the idea that the connectionist theorist is necessarily committed to viewing the representations of her network as context-dependent. Fodor and McLaughlin (1990), for example, take issue with the account of systematicity proffered by Smolensky (1987; 1988):

> COFFEE (and presumably, any other representation vector) is context dependent. In particular, the activity vector that is the COFFEE representation in CUP WITH COFFEE is distinct from the ac-

> tivity vector that is the COFFEE representation in, as it might be, GLASS WITH COFFEE . . . Presumably, this means that the vector in question, with no context specified, does not give necessary conditions for being *coffee*. (Fodor & McLaughlin, 1990, p. 192)

The phrase "any representation vector is context dependent" contained in this quote is, for our purposes, quite crucial. Of course, vector representations defined over units are quite properly described as context-dependent: It is precisely this property that it touted as one of the major advantages of connectionist networks, providing as it does a domain for generalization and extraction of prototypical information. However, the connectionist is also able to avail herself of context-*in*dependent representations in the weights of the network. Specifically, the tokens (microprocedures) over which connectionist computation is (at least in part) defined, can accurately be described as context-independent constituents of context-independent microprocedural representations, emergent from a low-level computational activity.

Tokens qua microprocedures thus express the same content in a number of different contexts. Of course, the context-independent microprocedures do not combine concatenatively to form complex syntactically structured representations as tokens do in classical systems; rather, they combine non-concatenatively to form an emergent representation with a spatial structure. The representations of meaning evolved in the weights of the LW simulations reported in this chapter are properly described as such context-independent representations.

I do not wish to claim that a connectionist theorist, solely by virtue of the fact that she is able to make recourse to context-independent representations, has thereby solved the problem of systematicity. Rather, I wish to more modestly point out that, by adopting the Microprocedural Proposal, the connectionist can begin to answer the classical challenge that systematicity must *necessarily* be exhibited by a connectionist architecture. She is able to do this because the distinctive genus of weight representations provides precisely the resources that the classical theorist charges are missing from such an architecture, namely, context-independent constituents of complex compounds.

Conclusions

The Microprocedural Proposal, and the notion of a connectionist PS that it engenders, constitutes an attempt to render the notions of weights, activation vector, procedure, intension and representation explicable within a unified non-symbolic, non-classical computational framework.

The simulations in this chapter have shown the efficacy of the notion of weight representation for encoding the intensional meaning of a corpus of simple spatial terms, and the analyses of the network simulations have illustrated that the regularities immanent in the weights evolved after learning can be systematically related to the differences in the meaning of such spatial terms, taking into account one provision: Namely, the fact that any regularity revealed by a cluster analysis is contingent on it being explicable purely in terms of Euclidean space (and a Euclidean distance metric), which is only a crude approximation of computational space. Further, the notion of microprocedural representations was shown to contribute to the debate concerning the nature of computation, and the context-independency of microprocedures was also noted with respect to the problem of systematicity.

Set in a larger context, the notion of meaning being encoded and characterized microprocedurally is merely one part of the process of constructing *discourse representations* on the basis of a symbolic input. The further challenge is to design and run connectionist network simulations that are able to learn microprocedural representations of meaning necessary and sufficient to construct more complex representations of extension. This is the subject of the next chapter.

7

Structure and Analogical Relations

Intensions on a Lewisian interpretation, to recall Chapter 4, are extension yielding functions. Which is to say, the intension corresponding to the meaning of a symbolic expression such as *on the right of* is a function that yields an extensional set construed as a mathematical model structure and consisting of objects, sets, functions, and so on. As we have seen, one of the major claims of a (formally grounded) symbolic procedural semantics is that, by using the notion of procedure, psychological illumination can be thrown on this darkly shrouded, mysterious concept of meaning-as-intension. In the procedural semantics of Johnson-Laird (cf. 1978), for example, the meaning of lexical items are viewed as procedures (implemented in code from the list processing language POP-10) that compute functions, the result of whose execution is the construction and manipulation of representations of extension termed *mental models*.

In contrast to such a symbolic account of procedure, the last chapter introduced a Microprocedural Proposal of meaning and used it to outline a connectionist PS. In this non-symbolic framework, representations of meaning are said to *evolve* as a consequence of the class of mapping that a connectionist network is computing where that mapping is akin to that found in formal species of semantics. After learning is completed, and if the generalization performance of the network is adjudged satisfactory, the weights of the network are thought of as microprocedurally encoding the intensions of a corpus of spatial assertions. Drawing by analogy on the terms of explanation in Johnson-Laird (1983), a connectionist PS can also be thought of as constructing representations of extension. That is, the result of engaging a microprocedural representation on the basis of an input spatial assertion is the construction of (a representation of) the extension of which that assertion is a true description.

A natural way of extending the microprocedural proposal would be to design a network simulation that was required to construct representations of extension from *sequences* of spatial assertions. That is, instead of mapping single assertions at a time, such as **B is on the right of E**, the network simulation would be required to map sequences of more than one assertion, such as **B is on the right of E**, **A is in front of B** and so on. In this way, the microprocedural representations of meaning evolved in the weights would be involved in constructing a more complex representation of extension. That is, the result of engaging a microprocedural representation on the basis of a sequence of spatial assertions would be the construction of a complex representation of the extension of which that sequence of assertions is a true description. Such an elaboration of the Microprocedural Proposal has two requirements. First, because the simulations reported in the last chapter were feed-forward networks over two layers of weights, their architecture was ill-suited to mapping sequences of assertions. Such a task calls instead for a more powerful architecture, and the species of *recurrent* network architecture devised by Elman (1990) seem appropriate for this purpose.

The second requirement for an elaborated connectionist PS concerns the nature of the representation of extension that is constructed on the output. As we will see, this problem is not as simple as it might first appear, and we will spend some time considering the various solutions to it in the literature. Indeed, whereas in previous chapter discussion focused on the notion of meaning (i.e., truth conditions) as intensions (i.e., procedures), in this chapter, discussion is narrowed on the accompanying notion of meaning as extension (i.e., reference). The reason for this shift is simple. Just as the term *intension* is usefully equated with procedure in computational psychology, so too should the term *extension* be equated with some appropriate computational device. Intensions are (or rather can be) realized computationally as procedures, and we have explored how and in what fashion this is possible. The question to be addressed now, however, concerns the corresponding computational realization of extension. To recall, this question was raised in the introduction to Chapter 5 where it was couched as the requirement on a satisfactory theory of psychological semantics that it (the theory) should postulate a form of representation that could take the place of the model structure in formal semantics. The latter part of this chapter, accordingly, will be concerned with what I have called the *collapse2 strategy*, which is to say, a method of constructing connectionist representations that, in a computational theory of meaning, are able to take the place of the model structures employed in formal theories of meaning.

THE MICROPROCEDURAL PROPOSAL ELABORATED

Two main simulations will be reported in this section, both using the SRN architecture. The two simulations can be characterized as performing a mapping from a sequence of symbolic strings to a representation of the states of affairs that the sequence of strings describes. That is, the recurrent network is being asked to map sets of multiple input assertions to an output representation where instances of symbols are in positions relative to each other that correspond to (i.e., that are true of) a given input sequence. Taking an instance of the coding from the first simulation, for example, a sequence of assertions such as **B is on the left of A**, **B is in front of C**, **E is on the left of B**, **D is in front of B** would be mapped to an output representation like that shown in Figure 7.1.

SRN Simulation 1

Given an input set of vectors, coding over 12 unit representations of spatial assertions, SRN simulation 1 was required to perform a mapping to an output, coding over 25 units a states of affairs. The set of input vectors used in the simulation code for assertions constructed from the corpus of lexical items used in the simulations of the previous chapter. Each input vector can be conceptually broken down into three fields, comprising the **op** field (containing two units, coding for the two spatial operators **left** and **front**) and

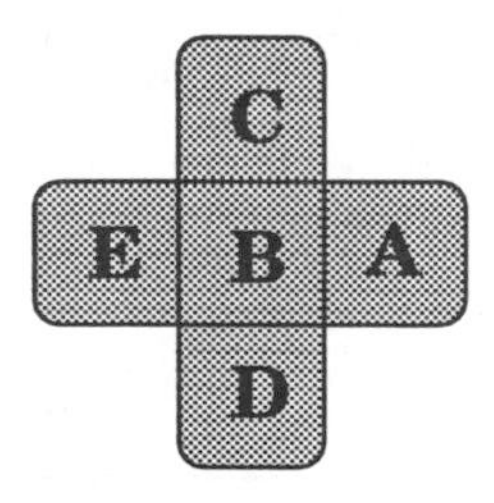

FIGURE 7.1. Diagrammatic output constructed from a typical RLW Simulation 1 sequence of assertions.

the **arg1** and **arg2** fields (each of which contains five units and each of which codes for the symbol arguments of the assertions).

The set of output vectors code for the states of affairs to which the sequences of input assertions refer. To reiterate, the output representation in, for example, LW Simulation 1 was limited in that it could only represent spatial relations between two symbol arguments at a time. In order that multiple relations between tokens might be represented, the richness of the output coding for SRN simulation 1 was increased by the addition of a fifth conceptual field, in addition to the **left** and **right**, and **front** and **behind** fields of the (conceptual) cross described earlier for all LW simulations in Chapter 6 (see Figure 6.3). This fifth conceptual field corresponds to a **center** position of the cross.

Training SRN simulation 1 was conducted using sequences of assertions. The network was provided with a target vector for each individual assertion of a given sequence, illustrated in Figure 7.2. In this way, the network was required to integrate a *subsequent* assertion with the output representation of a *consequent* assertion. There are six distinct types of sequences (of four elements each) that can be constructed using **left** and **front**: (a) **LFLF**, (b) **FLFL**, (c) **LLFF**, (d) **LFFL**, (e) **FFLL**, and (f) **FLLF**. The simulation was trained on an input set comprising four of these distinct types of sequence: specifically **LFLF**, **LFFL**, **FFLL** and **FLLF**. Correspondingly, the test set comprised (the novel) **LLFF** and **FLFL** sequences.

The output coding employed, with five conceptual fields as shown in Figure 7.1, means that the network is able to represent the following configurations of lexical items: (a) 20 1-place relations (i.e., between 2 symbol arguments), (b) 72 2-place relations

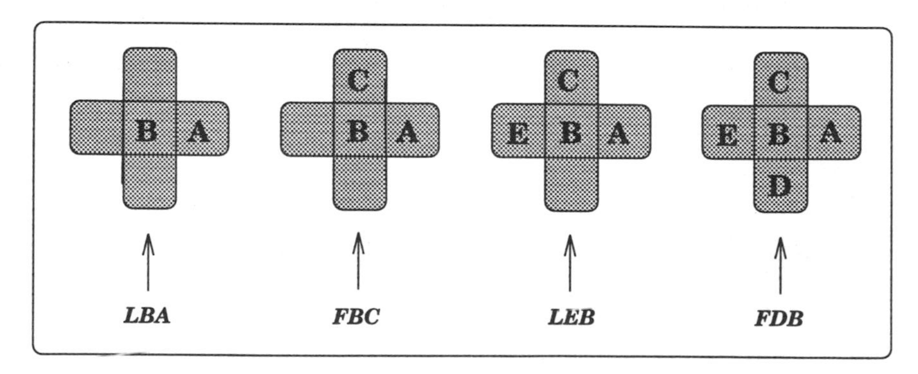

FIGURE 7.2. The incremental construction of the output representation from a sequence of assertions.

(i.e., between 3 symbol arguments) and (c) 120 3- and 4-place relations (i.e., between 3 and 4 symbol arguments, respectively).

Results Three different network configurations, with 20, 30 and 40 hidden units respectively, were run for SRN simulation 1. All were trained with a learning rate of 0.05 and 0.9. The 20 hidden unit configuration learned to a tolerance of 0.05 after 24,555 cycles; the 30 hidden unit configuration, after 19,467 cycles; and the 40 hidden unit configuration, after only 2,000 cycles. The results from testing are shown in detail in Table 7.1.

In the previous chapter, each LW simulation, when tested for generalization, was exposed to a certain number of novel inputs, where a "novel input" is taken to mean a binary (or continuously valued) vector similar to those on which the network was trained. In this first SRN simulation, however, the training inputs are *sequences* of vectors, which is to say, multiples of spatial assertions constructed from a finite set of *elements*. Consequently, when tested for generalization, the simulation should be exposed to novel sequences of assertions, where a "novel sequence" is taken to mean a sequence of binary vectors similar to those that the network was exposed to during training. Now we encounter an interesting question: Specifically, is it legitimate, or sensible, to expect the SRN to generalize, not just to novel *sequences* of inputs, but to novel *elements* of those novel sequences?[1]

In the literature, SRNs are mainly employed in tasks involving auto-association and prediction (cf. Elman, 1991; Cleeremans, 1993), and for such tasks, are predominantly tested for generalization on novel sequences of previously encountered elements. The task that this SRN simulation is performing (i.e., the incremental construction of an output on the basis of a sequence of inputs, what might be called *auto-accumulation*) does not conform to either of the auto-association or prediction task descriptions, however. Purely on the basis of this consideration, it would seem prudent to investigate the generalization performance of the simulation when tested on novel sequences constructed from (a training corpus of) previously encountered elements.

That said, network generalization for this first SRN was measured in two ways. For the first measure, if the network established the correct referents of a given assertion by constructing

[1] This kind of question is really part of the ongoing systematicity debate. I would direct the reader to recent papers by Hadley (1993) and Niklassen & van Gelder (1994) for more details.

Table 7.1.
Comparing the generalization performance of SRN simulation 1, with 20, 30 and 40 hidden units. The ∇ symbol refers to instances where the network "forget" a previously established referent.

	20					30					40				
	F	**R**	**C**	**L**	**B**	**F**	**R**	**C**	**L**	**B**	**F**	**R**	**C**	**L**	**B**
LBE	–	√	√	–	–	–	√	√	–	–	–	√	√	–	–
LAB	–	√	√	√	–	–	√	√	√	–	–	√	√	√	–
FCB	–	√	√	√	√	–	√	√	∇	√	–	√	√	∇	×
FBD	√	√	√	√	√	√	√	√	∇	√	√	√	√	∇	×
FCB	×	–	√	–	–	√	–	√	–	–	√	–	√	–	–
LAC	√	–	√	√	–	√	–	√	√	–	√	–	√	√	–
FDC	√	–	∇	∇	√	√	–	√	∇	√	√	–	√	√	√
LCE	√	√	√	∇	∇	√	√	√	∇	∇	√	√	√	∇	√
LDC	–	√	√	–	–	–	√	√	–	–	–	√	√	–	–
LBD	–	∇	√	√	–	–	√	√	√	–	–	√	√	√	–
FDE	√	∇	√	∇	–	×	∇	√	√	–	×	∇	√	√	–
FAD	√	∇	√	√	×	×	∇	√	√	√	√	∇	√	∇	√
FEA	√	–	√	–	–	√	–	√	–	–	√	–	√	–	–
LBE	√	–	√	√	–	√	–	√	√	–	√	–	√	√	–
FCE	√	–	√	∇	√	√	–	√	√	√	√	–	√	∇	×
LED	√	×	√	∇	√	√	√	√	∇	∇	√	√	√	∇	×
LAB	–	√	√	–	–	–	√	√	–	–	–	√	√	–	–
LCA	–	√	∇	×	–	–	√	√	√	–	–	√	√	×	–
FAD	√	√	√	×	–	×	√	√	∇	–	×	√	√	×	–
FEA	√	√	√	×	×	√	√	√	∇	√	×	√	√	×	√
FDE	√	–	√	–	–	√	–	√	–	–	√	–	√	–	–
LCD	√	–	√	√	–	√	–	√	√	–	√	–	√	√	–
FAD	∇	–	√	√	√	√	–	√	∇	√	√	–	√	√	×
LDB	∇	×	√	√	∇	√	×	√	∇	∇	√	√	√	∇	×
LAE	–	√	×	–	–	–	×	×	–	–	–	√	√	–	–
LDA	–	∇	×	×	–	–	×	×	√	–	–	√	√	×	–
FAB	×	√	×	×	–	×	×	√	√	–	√	√	∇	×	–
FCA	×	√	×	×	√	×	×	√	√	√	√	∇	∇	×	×
FEB	√	–	√	–	–	√	–	√	–	–	√	–	√	–	–
LEA	√	×	√	–	–	∇	×	√	–	–	√	√	√	–	–
FCE	√	×	√	–	√	∇	×	√	–	√	√	∇	√	–	√
LDE	√	×	√	√	√	∇	×	√	×	√	√	∇	√	√	√
LBD	–	√	√	–	–	–	√	√	–	–	–	√	√	–	–
LEB	–	√	√	√	–	–	√	√	√	–	–	√	√	×	–
FCB	–	√	√	√	×	–	√	√	√	√	–	√	√	√	√
FBA	×	√	√	∇	×	√	√	√	√	√	√	√	√	∇	∇
FDB	√	–	√	–	–	√	–	√	–	–	√	–	√	–	–
LED	√	–	√	√	–	√	–	√	√	–	√	–	√	√	–
FCD	√	–	√	√	×	√	–	√	√	×	√	–	√	√	×
LDA	∇	×	√	√	×	√	×	√	√	×	√	×	√	√	×

an output representation in which the units corresponding to those symbols exhibited an activation greater than 0.35, then generalization was deemed to have been successful. There is a caveat, however. For the first assertion of a sequence, the network had to construct a representation of a 1-place relation: Thus, generalization can be tested, *for that assertion* in terms of whether the network was able to establish the correct referents for the two symbol tokens of that 1-place relation. Therefore, we have two instances for which a measure of generalization can be derived. For the second assertion of a sequence, however, the network had to construct a representation of a 2-place relation: Thus, generalization can be tested *for the two assertions taken together* in terms of whether the network was (a) able to *maintain* the correct 1-place relation (two instances), and (b) able to integrate a successive symbol token so as to construct a 2-place relation (three instances). We now, therefore, have a cumulative total of five instances where a measure of generalization can be derived (and so on, for the remaining two assertions of a sequence). The result is that, for each complete sequence, there is a cumulative total of 14 instances where generalization can be measured. Adopting this measure yields a figure for SRN simulation 1 of 74% generalization.

This tortuously derived measure is fairly respectable. The second, a simpler and more stringent measure of generalization, tests each successively constructed *n*th-place relation. Taking the test set of five novel sequences together, the network had to construct four representations of each of the 1-, 2-, 3- and 4-place relations. Adopting this second measure of generalization yields a more disappointing figure of 45%. The reason for this poor measure of generalization can be partly explained by looking a little closer at Table 7.1, where we can see a pattern of results characteristic of the recurrent network architecture: namely—the selective "forgetting" of previously established referents. This forgetting is arguably largely responsible for the poor generalization performance of the network on the task.

The generalization performance for the first SRN can obviously be improved on. The network still, for example, exhibits the same "forgetting" behaviour associated with previously established referents. Couched in our favored vocabulary, the microprocedural representations of meaning evolved in the weights are not sufficiently robust to always construct and maintain a correct representation of extension.

SRN Simulation 2

In order to further explore the efficacy of the recurrent network architecture in performing the class of mapping being asked of it in SRN simulation 1, and in particular that the "forgetting" characteristic of the generalization performance of the SRN might be addressed, a second network simulation was designed. SRN simulation 2 was run in order (a) that a better measure of generalization might be obtained, but also (b) in order that the complexity of the task that a given network was being asked to learn might be increased. The increase in complexity had a number of aspects. First, all four spatial operators were employed in constructing the test set of assertion sequences, although, as a caveat, the number of symbol arguments from which the training corpus was constructed was reduced from five to three, using only **A**, **B** and **C**. Using only three symbols also had the effect of truncating the lengths of the sequences required to specify fully the relations between tokens to only two assertions. The plausible rationale was that this would have the effect of forestalling any "forgetting" that the network might otherwise exhibit.

Second, the nature of the output coding was changed. The reason for this was simple. The form of the output for the first SRN meant that it was only possible to train the network on certain types of sequence: Specifically, those sequences that described states of affairs *representable* by the output coding (i.e., those sequences composed of **LFLF**, **FFLL** and so on). SRN simulation 1 was not trained, for example, on a sequence such as the following: **E is on the left of A, D is on the right of A, A is on the left of C, B is on the left of D.**

The reason for this is obvious: Such a sequence would require a representation of extension in which a 4-place relation between symbol tokens could be specified, for want of a better description, along a *horizontal axis*. Using the output coding employed in RLW Simulation 1 would mean that, by only the *second* assertion in the sequence, the network would, so to speak, "fall off the edge of the world." In order to combat this failing, a more comprehensive form of output coding is evidently required, one that is able to represent all possible configurations of the three symbols that go to making up the corpus of input training assertions. Just such a coding scheme is illustrated in Figure 7.3, where an output is shown constructed on the basis of the sequence of assertions, **A is in front of C, C is on the right of B**. Using this coding scheme ensures

FIGURE 7.3. The conceptual array employed in SRN simulation 2, showing a sample output.

that instances of the six qualitatively different *conceptual configurations* derivable using the four spatial operators **left**, **right**, **front** and **behind**, and the three symbols **A**, **B** and **C** can be represented. These six configurations are shown in Figure 7.4.

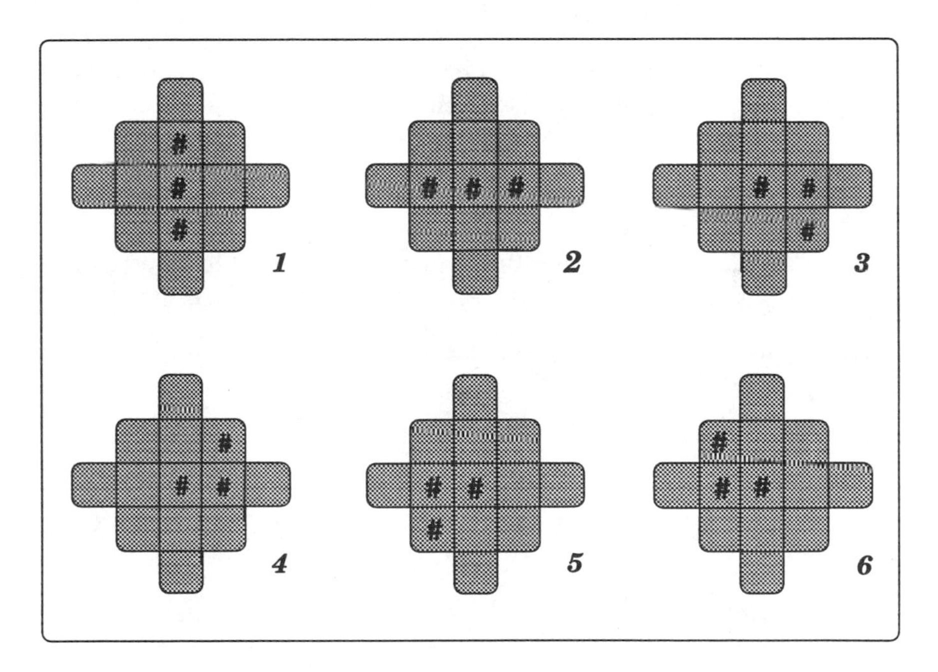

FIGURE 7.4. The six different possible output configurations of the coding scheme of SRN simulation 2.

Each conceptual configuration has 18 possible instantiations on the output. That is, there are 18 possible vectors all conforming to one conceptual configuration. Each of these instantiations is (obviously) dependent on the relative spatial positions of the tokens, but is also dependent on where and with what symbol instance construction of the representation began. This point can be illustrated by considering the following three sequences, which construct three different output instantiations of the same conceptual configuration, as illustrated in Figure 7.5.

- **A is on the left of B, C is in front of A**
- **B is on the right of A, A is behind C**
- **C is in front of A, B is on the right of A**

Using the four spatial operators **left**, **front**, **right** and **behind**, and the three symbol arguments **A**, **B** and **C**, means that it is possible to derive an exhaustive specification of 384, 2 element sequences of assertions. For the conceptual configurations shown in Figure 7.4, the first and second configurations are each exhaustively specified by 94 sequences, and each of the remaining four configurations is exhaustively specified by 48 sequences. SRN simulation 2 was trained on slightly more than three quarters of the possible input sequences, comprising 312 sequences: Conceptual (output) configurations 1 and 2 were each described by 72 sequences, and conceptual configurations 3 through 6 each were described by 42 sequences.

Results The network configuration of 10-40-39 learned to tolerance of 0.1 after 121,500 cycles, with values of learning rate and

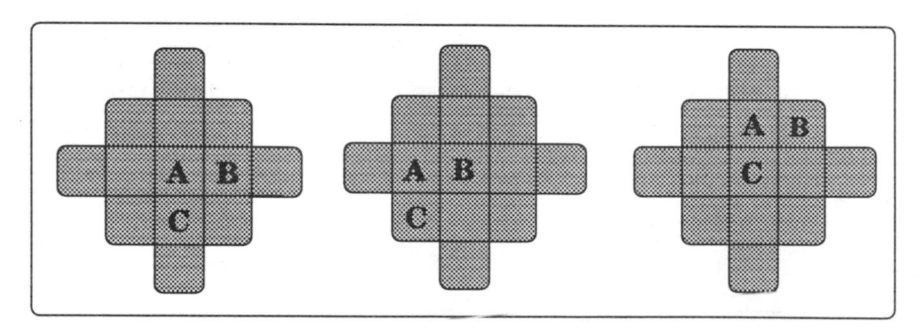

FIGURE 7.5. Three of the possible 18 output instantiations of one *conceptual configuration.*

momentum of 0.05 and 0.5 respectively. The number of cycles taken to learn is in fact an approximate figure because a good deal of decremental learning was necessary in order to pull the SRN up out of its local minimum.

The set of vectors used to test the generalization performance of SRN simulation 2 comprised 48 assertions describing conceptual configurations 1 and 2, and 24 assertions describing conceptual configurations 3 through 6. The generalization performance of the network to this test set of 72 sequences was 70%, using the first of the measures we have been employing. The second, more stringent test of generalization test each successively constructed output. Performance of the network when tested using this measure fell to 60%. Additionally, the network generalized well to the first element of the sequence, but much worse to the second element. Conceptual output configurations 1 and 2 exhibited 79% to the test set, using the second measure, and conceptual output configurations 3 through 6 exhibited only 50% to the test set.

The first of the requirements for an elaboration of the Microprocedural Proposal, mooted in the introduction to this chapter, has been explored using the SRN architecture. Both of the network simulations had varying degrees of success in performing the more complex task that they were designed to perform: mapping sequences of spatial assertions to a corresponding output. The most impressive generalization performance is seen from SRN simulation 2, which is able to generalize to 79% of a test set of novel sequences.

However, there are a number of reasons for supposing that the SRN architecture is not entirely suitable for elaborating the Microprocedural Proposal. First, the task that the simulated networks are required to learn is neither a prediction nor an auto-association, tasks on which the SRN architecture excels, and for which much more is known. The forgetting behavior of the network is our second concern. The SRN was extensively explored by Cleeremans (1993), who noted that one of their principal failings is the inability to retain information about events distant in the past (of the temporal input). The SRN would seem, in other words, to have a very *short* memory. This is backed up by the generally poor generalization of the SRN simulations and their forgetting behavior.

One possible response to these reservations, that adheres to a spirit of economy of processing, is to revert back to a feed-forward architecture, but to use input and output training sets extracted from the hidden units of prior simulations. Therefore, some of the work that a given recurrent network simulation would have to do in order to perform the mapping required of it is accomplished beforehand. In this way, the integration of the content of two successive assertions would not be accomplished temporally, as in the case of the RLW simulations, but rather *structurally*. This approach can be thought of as involving a species of structure-sensitive operation on non-concatentively compositional representations (cf. Baldwin, 1992; Blank, 1991; Chalmers, 1990; Meeden & Marshall, 1991; Niklasson & Sharkey, 1992).

The third reason for doubting the efficacy of the SRN simulations is the most important for our concerns. It is not an architectural reason, but rather it concerns the nature of the output representations that the networks had to construct. Specifically, I began to doubt whether the form of the coding employed, shown in Figure 7.3, was in fact capable of appropriately encoding the spatial relations for a given set of spatial assertions. The form of this problem, which is actually whether the output representations can adequately "take the place of the model structure," is the subject of the next section.

GROUNDING REVISITED

Recalling the beginning of the chapter, one of the requirements for an elaboration of the Microprocedural Proposal was identified as being a specification of the representations of extension constructed by the PS on the output. Or to couch the requirement in different terms, an elaborated connectionist PS requires a form of representation that is able to take the place in a computational, formal system of the model structure. In this latter part of the chapter, as promised, a method for constructing representations that are able to satisfy this requirement, the collapse2 strategy, will be detailed.

In order to appreciate fully the collapse2 strategy and, in particular, to realize the nature of the representations that it delivers, it is necessary to be aware of the complexities hiding behind the glib phrase "taking the place of the model structure." Revealing these complexities will lead us from discussion of grounding to real and projected worlds, to discourse representation theory and to psychological studies concerning integration of information from text, and from talk of structure and compositional schemes through to the systematicity of cognition and mental representations of natural language. A convoluted route to follow, and my apologies for that. Let's begin with a question:

How is representation of an environment external to an agent possible?

In many (if not most) psychological theories, the question of what makes a mental entity a representation of something has been largely ignored. A commensurate concentration on relations internal to a representational system blind the psychologist to how that representational system links up with the world inhabited by the individual. This problematic oversight I have, of course, already mentioned in chapters 5 and 6, along with many of the labels that the problem has attracted in a number of disciplines. It is, simply, the grounding problem. Pre-empting some of the material from Chapter 8, let us briefly consider two solutions to this problem.

The first is a *classical* response, and it goes like this. The mind of an individual is a symbol system (where the term *symbol system*, with all its theoretical accoutrements, is meant in the narrowly prescribed sense of Newell and Simon (1972) and Fodor and Pylyshyn (1988); see Chapter 9 for more details). All that is required in order to solve the grounding problem, is some means of connecting the symbol system to the world "in the right way," via some peripheral perceptual apparatus. This is the classical response, and its dreadful. The second, a *Hybrid* response (cf. Harnad, 1990b), takes the grounding problem much more seriously by contrast and makes use of the computational resources of connectionist systems in order to connect a symbol system to the world. Chapter 8 discusses these two responses in more detail, but in the meantime, I want to concentrate on a point that tends to be taken for granted: Namely, that it is the real world to which a symbol system must be connected. Quite what does this mean?

Real and Projected Worlds

Jackendoff (1984) stated that the two central questions of semantics mirror the by now familiar distinction between intension and extension:

- What is the information that language conveys?
- What is this information *about*?

The first question Jackendoff (1984) identified with the notion of intension, the second with the notion of extension. The information that language conveys consists of meanings: Discrete lexical items convey meaning, combine to form sentences that convey meaning, combine to form discourse structures that convey meaning, and so on. What the information is about, Jackendoff (1984) notes, is almost universally considered to be the real world—which is to say, the real world of people, events, states of affairs and so on. Is this, however, the case? Jackendoff (1984) had the following to say about the familiar example shown in Figure 7.6.

> The four dots are quite naturally seen as forming a square, although no linear connections are present on the page. Why? Furthermore, why these particular linear connections and not, say, an X, which is logically just as possible an organization? [This example shows] . . . that what is seen cannot be solely environmental in origin, since the figures are imbued with organization that is not there in any physical sense . . . This organization, which involves segmentation of the environmental input and unification of disparate parts, must be part of the *mind's encoding* of the environmental input in preparation for presentation to awareness. (Jackendoff, 1984, 57–58)

Jackendoff urged that if, as the example suggests, the world as experienced by an individual is dependent (to whatever extent) on mental (qua computational) processes of organization, then it is crucial to a psychological theory to distinguish between the source of environmental input and the world as experienced: The former he termed the *real* world, and the latter, the *projected* world.

> We have conscious access to only the projected world—that is, the world as organized by the mind, and we can talk about things only insofar as this organization has been imposed. Hence the information conveyed by language can only be about the projected world. (Jackendoff, 1984, 62)

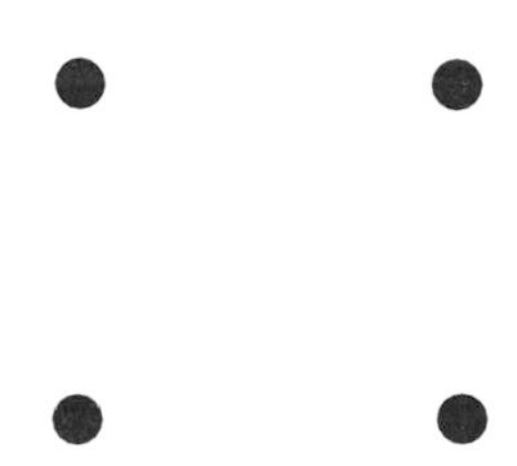

FIGURE 7.6. Do these dots represent a square or an X?

One might think, "So the mind internally represents aspects of the external environment? So what?" There would be no school of Representationalism in cognitive science (*pace* Fodor & Pylyshyn, 1988) without such representation. What is interesting about Jackendoff's muse is the way it is applied to the semantics of natural language. Specifically, it calls into question the notions of *truth* and *reference* in formal species of meaning theory. For example, the tradition in formal semantics is to regard reference as a relationship between expressions in a language and the things in the real world that these expressions refer to: Given Jackendoff's distinction between the real and the projected world, this notion begins to look a little suspect. In fact, what begins to seem the case is that the reference of lexical items and concatenations of lexical items (viz. sentences and larger discourse structures) is not the *real* world at all but the *projected* world.

In order to explicate this distinction, Jackendoff referred to real world entities in ordinary type, but enclosed references to projected world entities in # #. To illustrate, the discipline of physics, for example, holds that the real world contains (among other things), electromagnetic radiation of various wavelengths, frequencies and intensities. The properties of such radiation are, on Jackendoff's understanding, isomorphic with the #properties# of the theoretical construct #electromagnetic radiation#. Using such a notational device, it is perfectly satisfactory to posit #color# as the reference of color words, even though there is no such thing in the real world, because #color# is a consequence of the mind imposing its own organization on an unstructured environmental input.

Discourse Representation Theory

Jackendoff's conception of the reference (or extension) of linguistic expressions as projected world #entities# and #relations# has a parallel with proposals that have emerged from formal semantics, linguistics and psychology, all of which converge on the area of inquiry known as discourse representation (DR) theory. From the formal school can be cited the treatment of tensed discourse and nominal anaphora in Kamp (1979; 1981) and Heim (1983), and from linguistics the contributions of Karttunen (1976) and Sag and Hankamer (1984) are valuable, and from psychology, we should particularly note the contributions of Stenning (1978), van Dijk and Kintsch (1983), Kintsch and van Dijk (1978), Johnson-

Laird and Garnham (1980) and Johnson-Laird (1983). DR theory can be categorized by a number of general features that all of its disparate forms share, shown below. I will consider each of them in turn:

- Representation of the content of a discourse is essentially *non-linguistic*.
- Construction of the discourse representation is incremental.
- Processing discourse results in a model, containing entities (evoked by the discourse) and relations between entities.

Non-Linguistic Representation As a preliminary caveat I should say that the discourse representation, by being non-linguistic, is not also, therefore, non-*language-like*. In the present context of use, the term "language" is equivalent to the term "systematic representational system." Claiming that a representation is "non-language-like" is, consequently, meaningless: One would be claiming that a representation is non-representation-like. Rather, the claim that discourse representations are non-linguistic is meant to indicate that the representation is, in some sense, *non-symbolic*. We will have more to say about this in due course.

The best evidence for this first notable feature of discourse representations comes from psychological evaluation of the nature of the content extracted by subjects from texts. Barclay (1973), for example, presented subjects with sequences of sentences describing aspects of the two dimensional array shown below:

lion bear moose giraffe cow

He asked half of his subjects to memorize the sentences, and half to try to work out the array consistent with the sentences. After a subsequent, unexpected recognition test, the first group reliably identified the original sentences they had previously seen. The second group, however, were unable to reliably distinguish between the original sentences and new sentences true of the array. Similar results were reported by Potts (1973) on similar tasks. These results suggest that the second group of subjects had constructed a representation *analogous to* the array itself and that contained no information about the particular linguistic circumstance from which it was derived. Johnson-Laird and Garnham (1980) referred to representations with such characteristics as *mental models of discourse*.

The importance of such non-linguistic representations is illustrated in the failing of a psychological theory of text comprehension, proposed by Kintsch and van Dijk (1978), that does without them. In this theory, van Dijk and Kintsch urged that discourse is essentially represented propositionally (i.e., symbolically), although the theory contains no specific algorithm for a mechanistic extraction of propositions from texts. The theory requires that two aspects of a text (rather, of a set of propositions) be specified: (a) its microstructure, and (b) its macrostructure. The microstructure of the text comprises the referential links between propositions and is established on the basis of the co-occurrence of (proposition) argument, and the macrostructure corresponds to the general topic or theme of the text. For example, consider the following fictional passage of text: "A number of intense, seductive meetings between the Poet and the Dressmaker filled the early days of February, 1983."

In the (1978) version of their theory, Kintsch & van Dijk represented this piece of text propositionally, as shown below. The words (**number**, **meeting**) and (**intense**, **meeting**) would be adjudged co-referential—adjudged to have the same extension in (some putative model of) the discourse—because their two arguments, **meeting**, are identical.

- (number, meeting)
- (intense, meeting)
- (seductive, meeting)
- (between, meeting, poet, dressmaker)

The theory is sensitive to the fact that a chain of co-reference is not sufficient for a discourse to be *coherent* (cf. Garnham, Oakhill & Johnson-Laird, 1982). There must also be a unifying theme or topic, what Kintsch and van Dijk (1978) referred to as a macrostructure. Accordingly, the theory proposes a number of macro-rules that operate on the microstructure, under a schema corresponding to the reader's goals, to yield the macrostructure. The macrorules determine, for each proposition of the microstructure, whether that proposition is relevant or irrelevant to the macrostructure.

Its benefits notwithstanding, the theory suffers, however, from one terminal fallacy related to our point overleaf that the representation of the content of a text be essentially non-linguistic: That fallacy is the theories' reliance on a single form of propositional (qua symbolic) representation to represent both micro- and macrostructures. Illustration can be provided by considering following kind of passage: "Pete lost his copy of *Viz* the other day, and had to buy another. That's it over on the table there."

This contains two references to Pete's comic, "*Viz*" and "it." On the theory of Kintsch and van Dijk (1978), propositions containing these two referents have sufficient argument overlap for appropriate referential links in the microstructure to be established. Despite an overlap in argument, however, the two phrases are certainly not co-referential. The moral: A superficial propositional (i.e., symbolic) representation is necessary to capture the linguistic form of the sentences of a text, but that same propositional representation cannot do double duty and represent what those sentences refer to. In a subsequent formulation of van Dijk and Kintsch (1983), the force of this argument, and the paucity of a purely propositional form of representation, is recognized, and the new formulation of the theory actually incorporates a form of representation akin to a discourse representation, termed a *situation model*.

Incremental Construction The second general identifying feature of DR theory is that construction of the discourse representation is incremental, so that what has gone before (i.e., what has been constructed previously) serves as part of the context for disambiguation (if necessary) of subsequent discourse. The wish to accommodate such incremental construction was the motivation behind the SRN simulations, of course. Numerous psychological studies have shown that such incremental construction, either clause-by-clause or (more usually) sentence-by-sentence, is the only way to account for anaphora resolution (see Garnham, 1985).

A Formal Notation The third of the general identifying features of DR theory is that constructed discourse representations comprise entities and relations between (or conditions on) those entities. We can illustrate this feature of DR theory by considering a formal treatment of a small piece of discourse, adapted from Spencer-Smith (1987), that displays the semantic content of that discourse. Consider the following two sentences: (a) Persephone has a shell, and (b) She polishes it.

Boxes are often employed in DR theory to embody the semantic content of a discourse. The box is assumed to contain a model of a part of the world, which will expand (and contract) as the discourse unfolds. Both the name, *Persephone*, and indefinite, *a shell*, are alike in introducing *discourse referents* (cf. Karttunen, 1976) or *discourse markers* (cf. Webber, 1983). These stand as proxy for individuals and (or) entities, and to which subsequent anaphoric reference is made. Processing the name *Persephone* produces the reference marker x_1, and similarly, processing of the indefinite, *shell*, produces the reference marker x_2, resulting in the first complete discourse representation structure **drs1** shown in the Figure 7.7(a).

Below the reference markers x_1 and x_2 are the (partial set of) conditions that the world must satisfy if the first sentence is to be true. The name, *Persephone*, introduces (a marker that must be linked to) a particular individual. In contrast, the conditions on the indefinite specify only that its object should be something to which the concept *shell* applies. Moving onto the second sentence, this will initially expand **drs1** to produce **drs2** shown in Figure 7.7(b):

Because new singular terms *could* refer to something new in the discourse, the pronouns *She* and *it* introduce new discourse markers x_3 and x_4. DR theory assumes that non-systematic, pragmatic factors, such as Relevance (cf. Sperber & Wilson, 1986) operate in the resolution of anaphora, and that such resolution would determine that $x_1 = x_3$, and $x_2 = x_4$, resulting in the final discourse representation structure, **drs3**, shown in Figure 7.7(c):

Such a formal account of DR theory provides a perspicuous notation for displaying the semantic content of sentences of natural language. For those of a computational persuasion, however, mere notation is not enough: How are representations, comprising tokens of entities and relations between tokens, realized *computationally?*, is the gist of their inquiry. The theory of *mental models*,

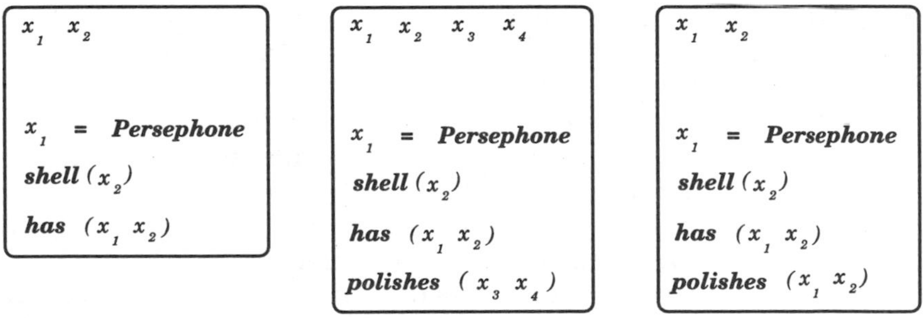

(a) drs 1	(b) drs 2	(c) drs 3
x_1 x_2	x_1 x_2 x_3 x_4	x_1 x_2
x_1 = ***Persephone***	x_1 = ***Persephone***	x_1 = ***Persephone***
shell (x_2)	***shell*** (x_2)	***shell*** (x_2)
has $(x_1\ x_2)$	***has*** $(x_1\ x_2)$	***has*** $(x_1\ x_2)$
	polishes $(x_3\ x_4)$	***polishes*** $(x_1\ x_2)$

FIGURE 7.7. Shows the three discourse representation structures resulting from processing of the two sentence "Persephone has a shell. She polishes it."

as proposed by Johnson-Laird (1983) represents the state-of-the-art in a psychological, computational realization of a discourse representation. Central to the notion of a mental model is what Johnson-Laird calls the *principle of structural identity*: A mental model is structurally analogous to the states of affairs it is a representation of. A number of questions present themselves. How is a representation structurally analogous to something? What precisely is analogue structure? What, in point of fact, does *structure* mean in this context? I will attempt to answer these questions in the next section.

THREE KINDS OF STRUCTURE

The debate between the classical and the connectionist schools in the cognitive science literature hinges, in large part, on what kind of *structure* the representations advocated by each of schools exhibits (see, for example, Fodor & McLaughlin, 1990; Fodor & Pylyshyn, 1988; van Gelder, 1990; Sharkey, 1992; Smolensky, 1987, 1988). For the classicist, the advocate is a *syntactic* structure, yet for the connectionist it is some kind of non-syntactic structure, what is often called a *spatial* structure. This label is not intended to imply that a connectionist representation is, internally, a spatially structured object. Rather, the label refers to the fact that the (non-formal, non-syntactic) structural similarities amongst connectionist representations are usefully understood as similarities of *location* in high-dimensional computational *spaces* (although, as we shall see, this conception may not be *entirely* accurate).

What has been hinted at in the prior section, in addition to these two types of structure, and what will emerge more fully from the following discussion, is the notion of a third type of structure: Analogical, or *analogue* structure. We will have considerably more to say about this type of structure, but for now we can say the following: Central to the concept of analogically structured representation is some notion of *significant similarity* between the structure of the representation itself and the thing represented. A representation exhibiting just such a significant degree of structural similarity I will term a *structural analogue* representation. Unfortunately, the term "analogue" is a polyseme, and recourse to a dictionary for disambiguation is likely to prove frustrating. The term "analogue" is sometimes touted as a spelling variation of "analog" (and hence is touted as having the same meaning) and sometimes it is not. For purposes of clarity, here and throughout only one of the senses of analogue is intended, that is, in the sense of *analogous* and *analogical*, and not in the analog-as-continuous sense. The two terms "analogue" and "analogical" will be used interchangeably.

The notion of analogue structure is a slippery notion to grasp, not least because, unlike syntactic and spatial structure, which are attributes of representations at a *functional* level, analogue structure is (almost certainly) an attribute of representations at a *virtual* level. Nonetheless, analogue structure is a valuable theoretical tool. For example, another way to think of the debate between classical orthodoxy and connectionist heresy—up a level, so to speak, from systematicity arguments and so forth—is in terms of whether the computational resources of connectionist networks are sufficient to construct, maintain and manipulate virtual representations. The classicists, headed by Fodor and others, would claim that, lacking a (purely formal) syntactic structure, the representations of a connectionist network cannot support such virtual representations, whereas the connectionists, headed by van Gelder and Clark and others, would claim that, in virtue of a distinctly connectionist notion of spatial structure, they can.

In line with this reasoning, the aim of the rest of this chapter is to contribute to the growing corpus of connectionist literature on systematicity (see for e.g., Baldwin, 1992; Chalmers, 1990; Chrisman, 1991; Niklasson & Sharkey, 1992), by detailing the theoretical background to, and the computational construction of, connectionist representations whose spatial structure at a functional level supports structural analogue representations at a virtual level. We will begin by clarifying the terms that have been introduced in this introductory exposition, including *structure*, *virtual* and *functional* levels of representation and, of course, *analogue structure*.

Syntactic and Spatial Structure

In Clark (1989), the vehemence of the exchanges between the classical and the connectionist schools is likened to that found in a Holy War. Moreover, the only proper response to such a rhetorical jihad, it is responsibly urged, should be a healthy ecumenicism. Being a philosopher of equally sound mind, van Gelder (1990) proposed in his seminal paper, "Compositionality: A Connectionist Variation on a Classical Theme," just such a ecumenical response to the classical claim that the functional architecture of the human mind must be a classically conceived architecture. He was able to do this by re-couching the debate in terms of the representational efficacy of different compositional schemes: Which is to say, in terms of the different ways of combining primitive objects to form more complex, compounds objects.

Compositionality As van Gelder (1990) detailed, a given compositional scheme requires, to be fully articulated, specification at two levels. First, at the level of (primitive and expression) *types*, and second, at the level of (primitive and compound) *tokens*. The level of type specification is more usually associated with linguistics: That is, with abstract grammatical structures and recursive rules that operate on (the bounded set of) primitive types to yield (the unbounded set of) expression types. The grammar of a natural language such as English, for example, differs from the grammar of another natural language, say German, at the level of type specification.

The level of token specification, on the other hand, is the level of concrete, physical instantiation. Using the example of English once more, expressions of this language can be instantiated in a variety of different ways: As alphanumeric characters, as sound patterns, as Braille, as systematic fluctuations of a magnetic field, and so on. Just as at the level of type specification, recursive rules specify how to take primitive tokens (qua instances of primitive types) and combine them to form compound tokens (qua instances of expression types) at the level of token specification also. It turns out, however, that the beguilingly simple process of combining primitive tokens to form compound tokens is more complicated than it might first appear. In particular, it seems it is possible to distinguish two very different compositional schemes: *concatenative* and *non-concatenative*.

Concatenative and Non-Concatenative Compositionality Concatenation is the most pervasive method for combining primitive tokens to form compound tokens there is, used in constructing natural and artificial languages, mathematical symbolisms and logic formalisms. Van Gelder (1990) distinguished (at least these) three species of concatenation: (a) spatial, (b) temporal, and (c) operational.

Spatial concatenation is employed in, for example, written text, where primitive tokens (viz. letters) are literally placed next to each other in order to generate more complex compound tokens (viz.

words). Temporal concatenation is found in, for example, spoken language, where primitive tokens (viz. phonemes) are temporally juxtaposed to form compound tokens, whereas the third species of concatenation, the operational genus, is illustrated, for example, by *pointers* in C code, which provide a way of linking (the locations of) tokens. About concatenation generally, van Gelder had the following to say:

> The essence of a concatenative mode of combination is [the] *linking* or *ordering* [of] successive constituents without altering them in any way . . . For a mode of combination to be concatenative, it must preserve tokens of an expression's constituents (and the sequential relations among tokens) in the expression itself. (van Gelder, 1990, p. 360).

When describing a representation as having a compositional concatenative structure, one is also saying that the representation will have an internal, formal structure of a particular kind, where abstract *constituency* relations find direct, concrete instantiation in the physical structure of the tokens. The name for this kind of structure is *syntactic.*

The classical school in cognitive science is ideologically wedded to concatenative compositionality and syntactic structure, and it is on the basis of such computational fundamentalism that the classical view of the cognitive architecture is constructed. That view goes as follows: Only a classical theory of mind acknowledges syntactically structured representations. *Precisely* in virtue of such syntactic structure, classically conceived mental processes are able to operate on mental representations by being sensitive to that structure. Which is to say, by having representations with similar syntactic structure *guarantees* that those representations will be treated similarly by mental processes sensitive to that structure. Only a theory acknowledging syntactic structural similarities among representations can explain how mental representations of similar things are treated similarly by mental processes; it can explain, in other words, the *systematicity of cognition.*

The classical rebuke of connectionism lies in such computational fundamentalism: Connectionist representations, according to the classicist (pace Fodor & Pylyshyn, 1988) lack a syntactic structure; they lack, in point of fact, any kind of significant structure at all. The lack of structure means that representations cannot be treated similarly by mental process sensitive to that structure, and hence cannot provide the basis for the systematicity of cognition and hence provide no real promise of explaining the cognitive architecture. The error in the thinking turns on the twin assumptions that, (*a*) there is no other method of combining primitive tokens to form compound tokens other than concatenation, and (*b*) there are no other kinds of compositional structure other than the syntactic structure resulting from concatenation. Both of these assumptions are wrong.

To recall, for a mode of combination to be concatenative, tokens of constituents must be preserved in the expression itself. There are no principled reasons, however, for why such preservation should occur, or for why it should be a defining feature of a compositional scheme: In illustration, van Gelder (1990, p.362) used the example of the Godel numbering scheme for formal languages. All that is required of a mode of combination is that there be available a systematic method of generating compound tokens, given (a bounded set of) primitive tokens, and a systematic method of decomposing those compound tokens back to their more primitive (constituent) tokens again. For a mode of combination to be non-concatenatively compositional (as opposed to concatenatively compositional), it must merely be the case that it be able to produce a compound expression given its constituents and decompose that compound expression back to its constituents again.

Representations in such a non-concatenative compositional scheme do not contain primitive tokens as (literal) parts of compound tokens and, accordingly, have no syntactic structure. This lack of constituency, however, does not also mean a corresponding lack of structure in the representation. Rather, when describing a representation as having a non-concatenative structure, one is saying that the representation will have an internal structure of a certain kind where abstract constituency relations find no direct, concrete instantiations in the physical structure of the tokens. The eminently plausible claim that van Gelder (1990) made is that whereas the classical school makes use of a concatenative compositionality, the connectionist school makes use of (or rather, should make use of) a non-concatenative (or merely functional, as it is also known) compositionality.

Characterizing the type of structural similarities between representations that results from a non-concatenative compositional scheme requires a departure from a formal conception. Connectionist unit representations are vectors in a certain high dimensional space where the bounds of that space are determined by the discrete activation values of a set of units. Moreover, the representations produced from such a non-concatenative compositional scheme are able to stand in structural similarity relations with each other. One initially appealing way to view such similarities, as I have mentioned, is as similarities of *location* in the computational space that the representations occupy. In this way, abstract constituency relations among the representations themselves are given in terms of Euclidean distance, where two structurally similar unit representations can be likened to neighboring points in the computational space defined by the vector values of a set of units.

It is this idea, that Euclidean distance relations between points in *n* dimensional space could capture systematic structural similarities, which serves to separate connectionist and classical approaches to theory building in cognitive science. For classical theorists, structural similarity relations are, as we have said, syntactic, whereas for connectionists structural similarity relations are spatial. Unfortunately, some confusion exists about precisely what spatial structure similarity relations are constituted by. Conventionally, as I have said, they are construed in terms of simple similarity of location in computational space, but is this the case? Consider the following, from van Gelder (1991b):

> The position of a representation in the space has a semantic significance; vary its position in that space, and you automatically vary the interpretation that is appropriate for it . . . Representations with similar sets of constituency relations end up as 'neighbouring' points in the space. The systematic nature of these spatial relations can be discerned using sophisticated techniques for the comparison of large groups of vectors such as cluster analysis. (van Gelder, 1991b)

Such an account of spatial structure, let us call it the relative point-distance (RPD) account, where only the Euclidean distances between *points* is considered, is incorrect. It is incorrect in that it does not accord with what *can* happen in a network. To reiterate, the conventional wisdom of the RPD is to consider that points in similar locations can be considered, for the purposes of computation, to be significantly similar in *structure (i.e., spatial structure). Thus, points in similar locations will have, to put it simplistically, similar* meanings.

Sharkey & Jackson (1994b), however, argued that the RPD explanation takes no account of weight representations in its explanation and, as a matter of empirical fact, it is only with respect to decision hyperplanes, implemented by weight representations partitioning the computational space of the network (see Chapter 2 for how to calculate hyperplanes), that points can be described as significantly and structurally similar, for purposes of computation. Here we have, in a sense, a working illustration of the benefits of the Microprocedural Proposal in which a theorist admits weight representations into his or her understanding; so allows a theorist to frame a more accurate explanation of an empirical phenomena, namely, the nature of the compositional structure of connectionist representations.

The compositional spatial structure of connectionist unit representations is, thus, not a simple notion best understood in terms of an RPD explanation, in terms solely restricted to the relative distance of one unit vector from any other. Rather, spatial structure results from a complex *interaction* of both units and weights, where both the relative, Euclidean distance between points *and* the absolute placement of points with respect to hyperplanes (i.e., with respect to weight representations) are equally important. We might call this the RPDAH explanation (for RPD, Absolute Hyperplane). It remains to be seen, and is in itself an important future research question, just how this RPDAH explanation of spatial structure, given in terms of the interaction of both unit representations and weight representations, will develop. Chapter 9 contains some interesting ideas on the subject.

The sophistication of explanations for something as abstract as spatial structure is, despite being an interesting topic in itself, slightly tangential to our present concerns. What is of more relevance is the manner in which representations acquire this spatial structure in the first place, from the mode of composition that constructed them. By distinguishing concatenative and non-concatenative species of compositionality, van Gelder (1990) showed that two systems of representations standing in systematic, structural similarity relations are possible: The first system has representations that are able to stand in syntactic structural similarity relations, whereas the second has representations that are able to stand in spatial structural similarity relations (howsoever it may be defined).

Of course, the challenge posed by the distinction that van Gelder has drawn between concatenative and non-concatenative compositionality is for the connectionist to actually design and build systems that perform systematic computation, with recourse to neither syntactically structured representations, nor manipulations of those representations sensitive to a syntactic structure. Couched in such terms, the challenge represents the opportunity to redefine the boundaries of *computation* itself. To reiterate the point, the collapse2 strategy detailed in the next section takes up the spirit of the first part of van Gelder's challenge in a different, though related, fashion to that of Hinton (1988), Pollack (1990), and Smolensky (1987)—namely, the construction of representations supporting (virtual) analogue structure relations. The second part of the challenge, involving the systematic manipulation of these representations on the basis of their spatial structure similarity relations, will be detailed in Chapter 8 and used in the Radical Connectionist response to the grounding problem. However, for now, we have a more pressing concern: Just what *is* this analogue structure I have been talking about?

Analogue Structure

> The game drifted on for a few moves, as they both lost concentration, then came alive again. He became aware, very slowly, very gradually, that he held some impossibly complex *model* of the contest in his head, unknowably dense, multifariously planed. (Banks, 1988, p. 49)

Kant (1787) argued that, like the fictional gameplayer invented by Banks above, individuals might employ modes of reasoning and inference using "intuition," "insight," "apprehension of relations between structures" and so forth, rather than rules of a hypothetical mental logic. Sloman's (1978) interpretation of this philosophy was that forms of reasoning and inference might be *analogue* or *analogical*, involving the manipulations of analogical representations, rather than *Fregean*, involving the manipulation of Fregean representations. A Fregean representation in this context can also be labeled as a *symbolic*, *linguistic*, *formal*, *propositional*, or *verbal* representation.

There are a number of misconceptions about analogue representations. The use of diagrams to illustrate analogical reasoning, for example, should not be taken to mean that analogue representations form part of the *imagery* debate (cf. Kosslyn, 1980). Images *are* undoubtedly analogical, in that structural relations between their parts correspond to the perceptible relations between the parts of the objects represented, but perceptible relations are not the only kinds of relation possible. As such images can profitably be regarded as a subset of analogical representations, analogical representations are associated with *intuitive*, *iconic*, *nonverbal* and *imagistic* modes of reasoning. They are often required to be continuous qua analog (as opposed to discrete) in nature, although there is no reason in principle why this should be the case. They are also often required to be *isomorphic* with what they represent, where "isomorphic" should be taken to mean "of the same form." Sloman (1978) showed that it is entirely possible to relax the full rigors of this requirement and still retain an analogical representation, for instance, when a two-dimensional image is used to represent a three-dimensional scene.

The biggest contrast that can be made between Sloman's analogue and Fregean representations concerns the nature of their structure. Specifically, Sloman asserted that Fregean representations bear no significant similarity to the structure of what they denote: This point can be illustrated with an example couched at the *type* level of specification, but the argument applies equally well at the level of *token* specification (as we will see):

> It will suffice to notice that although the complex Fregean symbol `the brother of the wife of Tom' has the word 'Tom' as part, the thing it denotes (Tom's brother-in-law) does not have Tom as a part. (Sloman, 1978, p. 164)

The moral is plain: A Fregean representation has a structure that is not analogous to the structure of the objects it represents. Correspondingly, in contra-distinction to a Fregean representation, Sloman (1978) offered the following definition of "analogical":

> Analogical representations have parts which denote parts of what they represent. Moreover, some parts of, and relations between, the parts of the representation represent properties of and relations between parts of the thing denoted . . . unlike a Fregean representation, an analogical representation has a structure which gives information about the structure of the thing denoted, depicted or represented. (Sloman, 1978, p. 165)

This definition could well have been coined to describe the nature of the representation that Barclay (1973) conjectured his experimental subjects had constructed in his psychological experiments. In addition, it bears a strong resemblance to the proposals we encountered earlier from the school of DR theory, namely the situation model in the text comprehension theory of van Dijk and Kintsch (1983), the discourse model in Johnson-Laird and Garnham (1980) and the mental model of Johnson-

Laird (1983). It is to the details of the latter theory that we now turn.

Mental Models The theory of *mental models* postulates a triumvirate of three kinds of mental representation:

- propositional representations;
- mental models;
- images.

Propositional representations correspond with Sloman's Fregean representations, which is to say, they are essentially linguistic (symbolic) strings of symbols. Mental models are structural analogues of states of affairs in the world, and (loosely) correspond to Sloman's analogical representations, whereas images (in the mental models framework) are described as *perceptual correlates* of mental models from a particular point of view.

Like the analogical representations we have discussed already and unlike Fregean representations, Johnson-Laird's mental models do not have an arbitrary syntactic structure, but rather one that plays a direct representational role because it is analogous to the corresponding state of affairs in the world. This feature of mental models is spelled out quite explicitly by Johnson-Laird in the *principle of structural identity*: "The Principle of Structural Identity: The structures of mental models are identical to the structures of the states of affairs, whether perceived or conceived, that the models represent." (Johnson-Laird, 1983, p. 419)

The mental models theory of structural analogue representation differs from the corresponding notion in Sloman (1978) in several ways. First, Sloman's definition allows for the possibility that some information in the analogical representation be not significant such as, for example, the color or the chemical composition of the inks used to draw the lines on ordinance survey maps. The principle of structural identity, however, is largely motivated by a sense of efficiency or *economy*: Every element of a mental model, including its structural relations, plays a representational role. There is nothing wasted in the structure of mental models.

Second, Johnson-Laird required a higher *degree* of analogical structure in his mental models than Sloman did in his analogical representations; for instance, Euler Circles are cited by Sloman as examples of analogue representations, where geometrical relations in a diagram can be taken to represent analogically the relations between sets of people. Johnson-Laird's principle of structural identity, however, did not allow such lax isomorphism. Euler Circles, though candidate analogical representations in a *weak* sense, do not have structures identical to the structures of the states of affairs they represent, and hence are not structural analogues in a *strong* mental models sense: Their structure is an artificial structure, invented by logicians, and hence arguably syntactic.

The representational efficacy of mental models hinges on their analogue structure. Johnson-Laird asserted that mental models differ from the various forms of propositional representation postulated by other psychological theories—which include the predicate calculus through to semantic networks—purely on the basis of their analogue structure and the principle of structural identity. Recycling some vocabulary from the previous section, we can describe propositional representations as classical symbol structures, constructed from a finite set of tokens by a concatenative mode of combination and hence exhibiting a syntactic structure. Distinguishing mental models from propositional representations is thus also to distinguish analogue structure from syntactic structure. Recent work in the connectionist literature (cf. Franklin & Garzon, 1990; Siegelman & Sontag, 1991) has shown, however, that connectionist networks are computationally equivalent to the abstract Turing machine underlying the classical notion of a symbol system. Hence, one has prima facie grounds for distinguishing mental models from connectionist representations, and hence for distinguishing analogue structure from spatial structure.

To couch this point slightly differently, syntactic structure and spatial structure, given the equivalence of computational power of classical and connectionist systems, are both sufficient to support virtual representations with an analogue structure. The fact that a psychological theory can be expressed in a form that makes use of only one kind of representational system, be it connectionist or classical, based on syntactic structure or spatial structure, does not mean that such a formulation is correct *at that level*. Rather, what it does mean is that, by describing the psychological account at a lower level, it is highly likely that the operational principles of the account will thereby be obscured. The argument can be extended to the representations themselves, which is to say, redescribing the virtual structure of, for example, a mental model at a lower level would almost certainly obscure the *functional role* of that virtual representation. The point can be illustrated by considering Figure 7.8, adapted from Port & van Gelder (1991)

The **F**, **C** and **R** dimensions define a space of possible representational systems. Location within the space varies as a function of

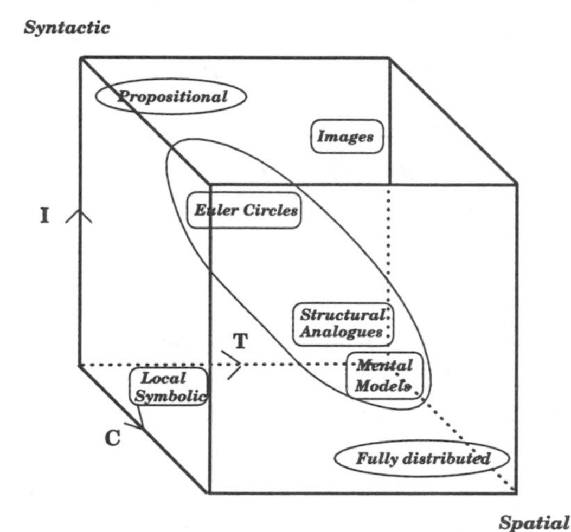

FIGURE 7.8. A possible space of representational systems, defined by the three dimensional of **R**epresentation, **C**ompositionality, and **F**unctional role.

properties of each dimension. The **R** dimension stands for representation, and representational systems can vary in this dimension as a function of properties of tokens: Whether they comprise a continuous or discrete set, whether the tokens are static or dynamic, and so forth. The **C** dimension is compositionality, and location in the representational space can vary as a function of this dimension, in terms of properties of compositional schemes: Whether the mode of combination is concatenative or non-concatenative and so on. The **F** dimension is the functional role of the representations in a given scheme. The functional role of a representation is concerned with how that representation is handled in the context of processing. As Figure 7.8 shows, connectionist and classical representational schemes are at far corners of the space. Can we similarly place analogically structured representational schemes in this space? An initial placement might look like Figure 7.8, where the location of analogically structured representations is shown as an ill-defined region in the middle of the space. To this preliminary placement, we should perhaps add a number of caveats.

First, the representational space bound by the three **FRC** dimensions was defined to distinguish representational systems at the level of the functional architecture. Analogue structure is exhibited by representations at a virtual level in, for example, the mental models theory, and hence the **FRC** space may be bound by dimensions (and properties of those dimensions) that are inappropriate to accommodate analogue structure. This point was conceded by Port & van Gelder themselves:

> "We do not claim these are a minimal set of dimensions . . . there are probably differing possible assumptions about semantics, and about the world itself [that the FRC conceptualization ignores]" (Port & van Gelder, 1991, p. 2).

Second, representations exhibiting spatial structure and analogue structure similarity relations seem to share a number of properties along the **F** dimension, which are not similarly shared by representations exhibiting spatial structure and syntactic structure similarity relations. In Blank, Meeden & Marshall (1991), the spatial structure of connectionist representations is described as having a *microsemantics*. Quite precisely what this term means is unclear (even when contrasted with the so called "macrosemantics" of syntactically structured representations) but, intuitively, one can see how the relations internal to a connectionist representation might be adjudged akin to the internal relations in a structural analogue representation. Another point worth considering is that, unlike syntactically structured representations that can only be manipulated by first decomposing the complex symbol, representations exhibiting analogue and spatial structural similarities can be operated on *holistically*, that is, without the need to first decompose the complex representation to more primitive constituent parts.

Third, in addition to the two properties of tokens of the **R** dimension detailed by Port and van Gelder (1991), there is a third property of this dimension relevant to the location of structural analogues in the representational space. It is a property previously mentioned by Sloman in his discussion of the structure of Fregean representations, and it is this: How are primitive and compound tokens paired with what they represent? Typically, tokens have no intrinsic relation to what they represent. Unfortunately, what this means as a consequence is that representational systems making use of both syntactic and spatial structure are, on the surface, vulnerable to the symbol grounding problem. That is, the tokens used at a functional level in both connectionist and classical computation are ungrounded and arbitrary.

The situation is slightly different, however, for a structural analogue representation, such as a mental model. Such a representation is constructed from tokens that *do* have an intrinsic relation to what they represent. However, the tokens that go to make up an analogical representation are a peculiar breed because they are themselves *virtual* entities. Quite how much sense it makes to speak of virtual tokens is unclear, given that the level of token specification is the level of concrete instantiation. However, the fact that theorists regularly make appeal to the notion of causal relations existing between states and events in virtual machines, where such relations depend on *physical* causation for their efficacy can, to an extent, legitimize their mention in any explanatory framework. Whatever the precise status of virtual tokens, there is no direct, physical correspondence between a token **A*** in a structural analogue and some world entity **A**, but there is a virtual structural correspondence.

Summary In this section, we have been concerned with the notion of the structure of representations, which is to say, with the internal configuration of a complex token resulting from a combination of primitive tokens. We also have seen how the connectionist and the classical theorist differ with respect to the type of structure that they ascribe to their representations. In addition to such functional level notions of structure, we also introduced the virtual level notion of analogue structure. A representation exhibiting a structure analogous to the structure of whatever it is a representation of, is characterized by Sloman (1978) as an analogical representation, as a mental model by Johnson-Laird (1983) and as a structural analogue by this book.

Structural analogue representations are ecumenical constructs. Being virtual entities, they can be supported by any functional architecture with appropriate computational resources, which is to say, they can be supported in principle by both connectionist and classical architectures. A tentative placement of analogue structure in the middle of a space of representational properties reflects its ecumenical nature, although it does seem to share something of an affinity with the connectionist's spatial structure over and above its similarities with a formal syntactic structure. The task of actually constructing representations that exhibit an analogue structure is taken up in the next section in the form of the collapse2 strategy.

THE COLLAPSE2 STRATEGY

Recalling chapter 6, and the first series of LW simulations, consider the nature of the output coding employed in those simulated networks, shown in Figure 7.9.

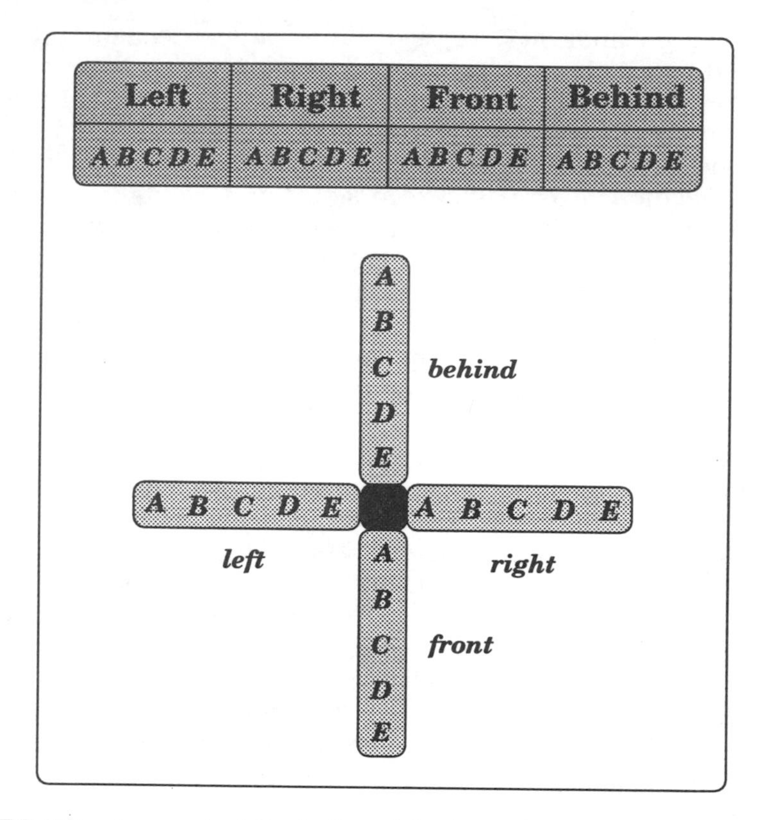

FIGURE 7.9. A conceptual illustration of the form of vector field coding used to construct output representations.

This vector field method of encoding produces representations that are structurally analogous to states of affairs in the world. That is, the similarity relations between functional representations can be seen to reflect structural similarities between the states of affairs being represented. In this chapter, a series of simulations executed over a recurrent network architecture, as opposed to a feed-forward network architecture, were detailed. Such recurrent network simulations required a commensurate increase in the complexity of the output coding, relative to that of the LW simulations of Chapter 6, in order that the network might exhibit a reasonably sophisticated behavior. In SRN simulation 2, for example, this increase in complexity resulted in a output coding where 13 locations of three symbol instances were encoded over 39 units. The form of this coding is reproduced in Figure 7.10(a), and an example of the outputs constructed from three different sequences of assertions is shown in Figure 7.10(b).

The particular form of this coding was chosen for a number of reasons. One was a resource requirement: The array had to be of a certain size to accommodate all the possible permutations of symbol instances and positions relative to each other that those symbol instances could assume. If this were not the case, as previously noted, then construction of an output on the basis of an input sequence would have "fallen off the edge of the world." The second reason for the choice of coding was simply that an array structure was apparently the most efficient and straightforward way of extrapolating from the simpler conceptual "cross" used in the LW simulations while still preserving an analogical structure. Let us consider the three output instantiations shown in Figure 7.10(b), and the three sequences of assertions that correctly describe them.

Each of the three sequences of assertions, on which construction was based, describe the same states of affairs, where some object **B** is both on the right of some object **C** and behind some object **A**. We can label each output instantiation using the following no-

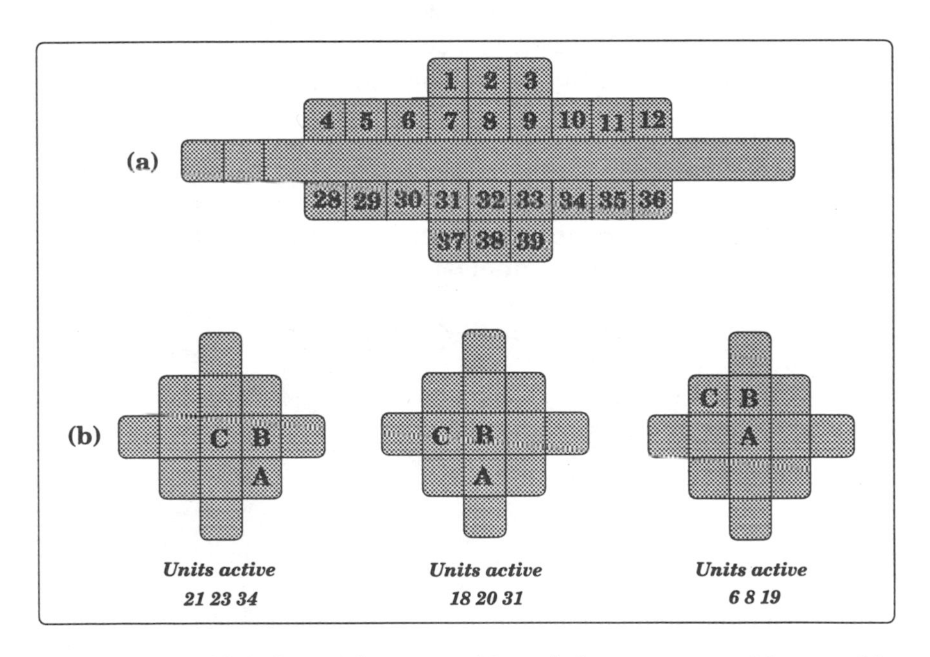

FIGURE 7.10. (a) Coding 13 locations of 3 symbol instances over 39 units. (b) Three output instantiations of resulting from the three sequences of assertions: (1) LCB, BBA, (2) BBA, RBC, and (3) FAB, LCB.

tational device: **.cba** refers to construction beginning with the **C** symbol, **c.ba** refers to construction beginning with the **B** symbol, and **cb.a**, to construction beginning with the **A** symbol. Using this device, and looking at the three different outputs that the connectionist PS constructs, one is suddenly struck by a very obvious flaw: What are being constructed are in fact *different* states of affairs. The three arrays in Figure 7.10(b), which we, an agency external to the representational system, can infer as equivalent, are described by three different 39 bit vectors: For the first array, units 21, 23 and 34 are active; for the second array, units 20, 28 and 31; whereas for the third, units 19, 6 and 8. Equally, consider the two output instantiations shown in Figure 7.11, and the two single assertions, which correctly describe them. The two assertions are: **A is on the left of B**, **B is on the right of A**.

Each of the two assertions, on which construction was based, describe the same states of affairs. However, like the outputs constructed on the basis of the three sequences of assertions described earlier, the two output instantiations are in fact different states of affairs: For the first array, units 19 and 23 are active, and for the second, units 16 and 20 are active.

The question we must ask, therefore, is to what extent are these 39-bit binary vectors supporting the appropriate analogical relations? This determination can be made by an appropriate cluster analysis. By feeding the exhaustive set of output vectors from SRN simulation 2 through a cluster analysis—that is, by feeding **.ba** and **b.a**, and the 11 other such conceptual "doubles," and separately, **.abc**, **a.bc** and **ab.c**, and the other 35 such conceptual "triples" through a cluster analysis—the lack of *significant simi-*

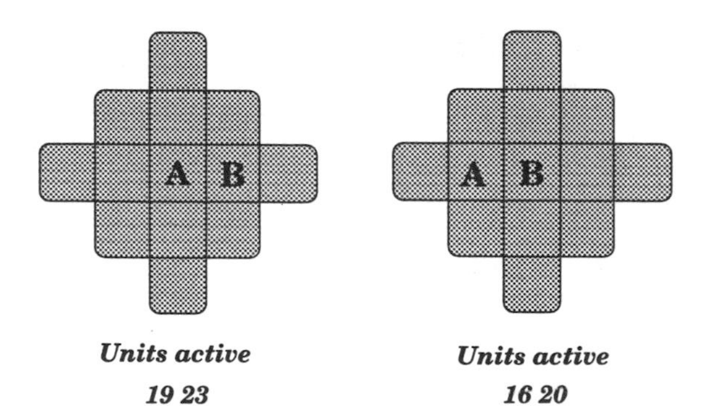

FIGURE 7.11. The two output instantiations resulting from the two assertions, (a) LAB, and (b) RBA.

larity between the (spatial) structure of these vectors and the structure of the states of affairs that they are representing, is plain to see. A portion of such an analysis is shown in Figure 7.12, and Figure 7.13.

The metric of similarity under discussion here is, of course, as I have already mentioned, a Euclidean one: Specifically, vector similarity is determined by the location of a given vector, in the 39-dimensional space occupied by the output, relative to all other vectors. It would appear from the two dendograms that the functional level representations generated by the form of the coding employed in the RLW Simulation 2 violate one of the features of

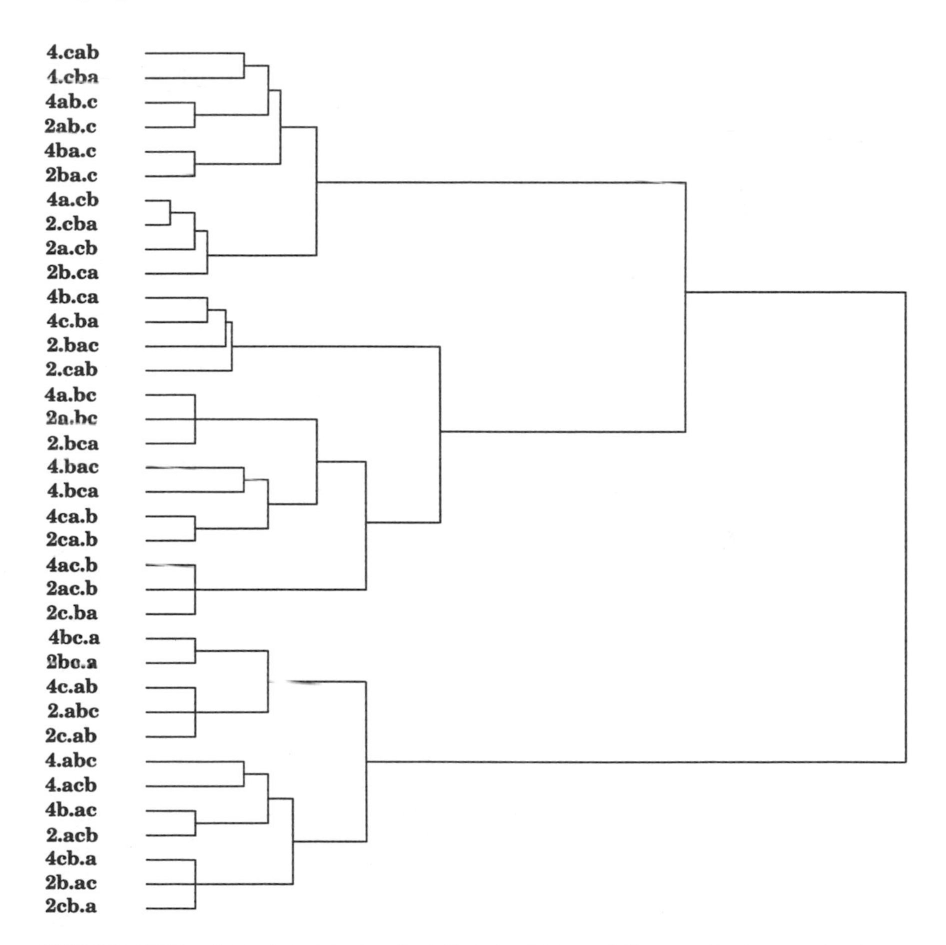

FIGURE 7.12. Dendogram showing the clustering of the raw 2-place outputs used in SRN simulation 2.

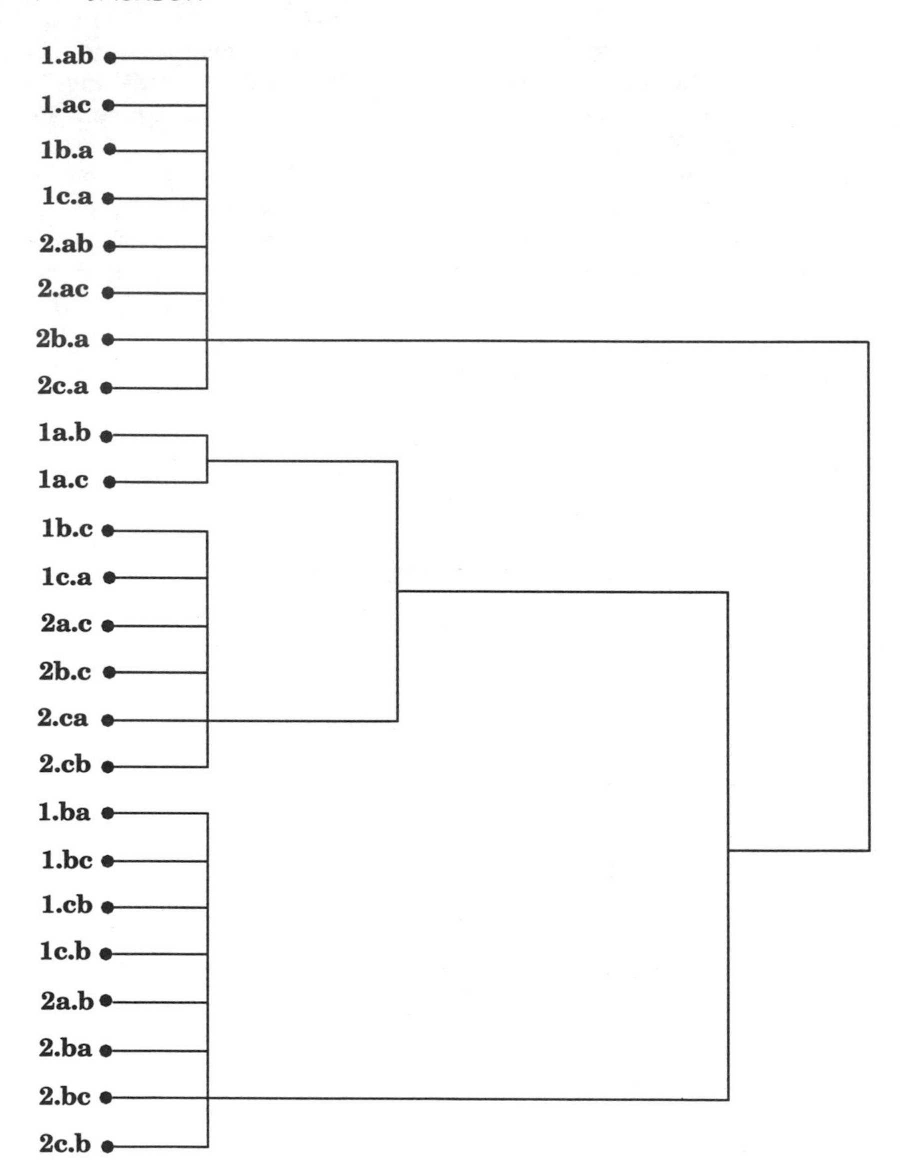

FIGURE 7.13. Dendogram showing the clustering of the raw 1-place outputs used in SRN simulation 2. The labels refer to where, and with what symbol, construction began.

an analogical representation, namely, that the representations should be non-linguistic, or non-symbolic. How so?

The SRN simulations, by necessity, had to begin construction of a given array at a certain point: The choice of where (and hence how) to begin construction of a given array, although important, was adjudged essentially arbitrary, and defaulted to the first symbol of the first assertion in a given sequence. With hindsight, this arbitrary choice can be seen more clearly, as a constraint on the representational efficacy of the coding. That is, by "anchoring" the construction of the array to units 19, 20 and 21 in the output vector, so as to meet the resource requirement, actually results in the intrusion of *symbolic* information, which obscures the analogical structure of the (virtual) representation. The raw, 39-bit output vectors employed in SRN simulation two consequently contain symbolic information about the spatial ordering of symbols within assertions, and the temporal ordering of assertions within sequences.

So, we have a problem: How to eliminate the unwanted intrusion of symbolic information? A solution presents itself almost immediately. What is required is for the spatial similarities between the raw, 39-bit output vectors to be *warped* so as to support the appropriate analogical similarities. Specifically, what this amounts to is the wholesale collapse of each conceptual output double, (such as **.ba**, **b.a**) and each conceptual output triple (such as **.bac**, **b.ac** and **ba.c**) to its most primitive structurally analogous parts. This is shown diagrammatically in Figure 7.14.

Using the form of the output coding used in RLW Simulation 2, with 39 units coding for the three symbols **A**, **B** and **C** over 13 lo-

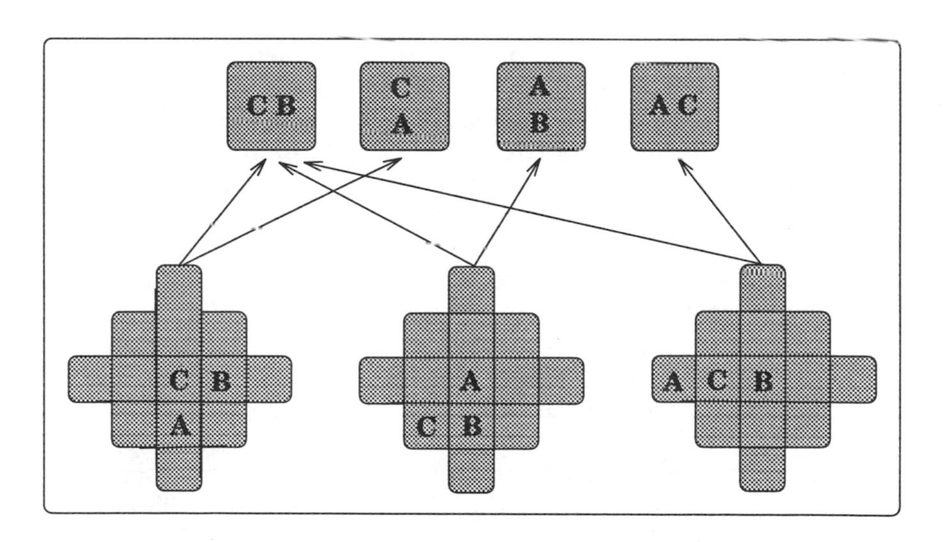

FIGURE 7.14. Illustration of the collapse2 strategy.

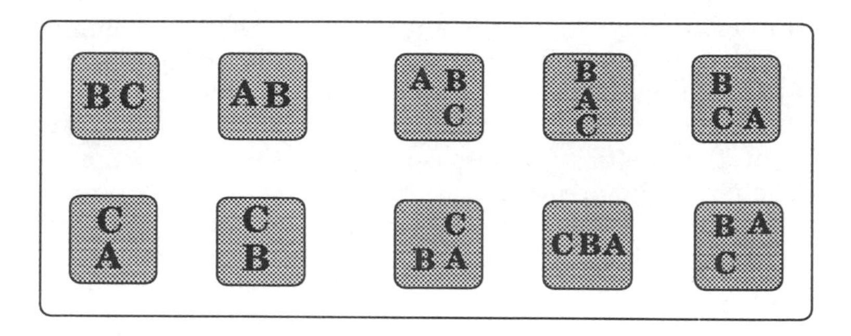

FIGURE 7.15. Examples of 1- and 2-place relations.

cations, means that 12 1-place, and 36 2-place relations can be represented, examples of which are shown in Figure 7.15.

For each 1-place relation, there are two possible output instantiations of the same states of affairs, corresponding to the two different symbols with which construction can begin: There are, consequently, 24 1-place output vectors. For each 2-place relation, there are three possible output instantiations, corresponding to the three different symbols with which construction can begin: Consequently, there are some 108 2-place output vectors. Preliminary simulations revealed that the complexity of the task might be lessened by splitting it up into two stages, so that the collapse2 strategy would involve an *incremental* collapse of each of the 36 conceptual triples.

Stage c1 In the first stage of such an incremental collapse, **c1**, the set of input vectors forming one conceptual double (i.e., **.bc** and **b.c**), plus the set of input vectors forming one conceptual triple (i.e., **.cba**, **c.ba** and **cb.a**), would be mapped to a unique, arbitrary output. This would have the effect of forcing each of the disparate vectors within a given double or triple closer together in the representational space over the hidden units.

Stage c2 The representations developed over the hidden units of the simulations in the **c1** stage would then be extracted and used to form the input set for the second stage simulation, **c2**, where each of the vectors from within a given double or triple would be mapped to their most primitive constituent parts. This incremental collapse, using the **c1** and **c2** stages, is illustrated below in Figure 7.16.

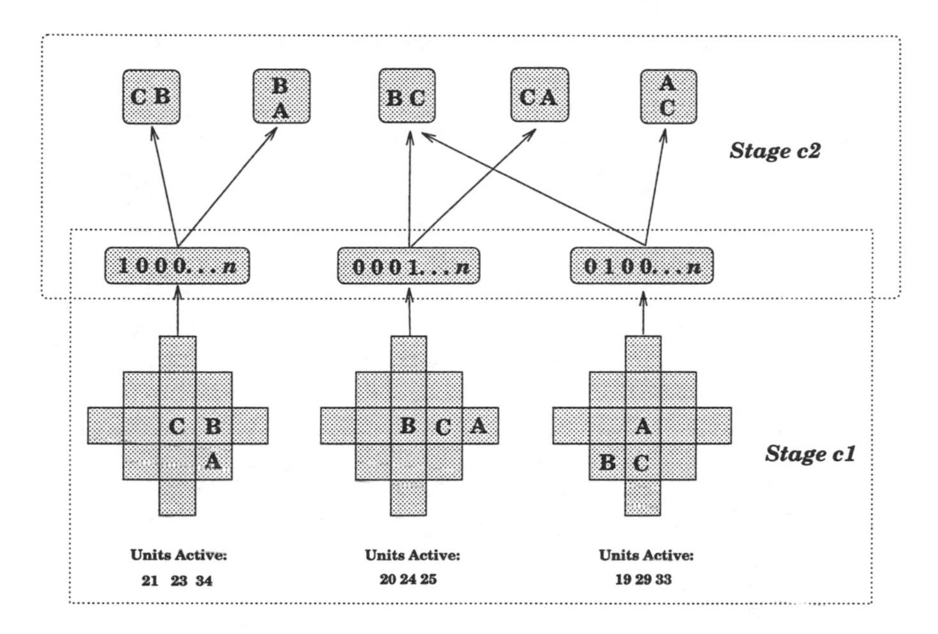

FIGURE 7.16. The rationale of the two **c1** and **c2** states in the collapse2 strategy.

Collapse2: Stage c1

Two sets of simulations were run for the **c1** stage of collapse2: The first set of simulations collapsed conceptual doubles such as **.ba**, **b.a**, and the second set collapsed conceptual triples, such as **.acb**, **a.cb**, **ac.b**. The following notational device will be used to identify these different sets of simulation: **D-c1** refers to the set of simulations collapsing conceptual doubles, whereas **T-c1** refers to the set of simulations collapsing conceptual triples. Using the form of the output coding employed in SRN simulation 2, with 39 units coding for the three symbols **A**, **B** and **C** over 13 locations, means that there are two output instantiations of each conceptual double, yielding 24 binary vectors, and three output instantiations of each conceptual triple, yielding 108 binary vectors.

For each conceptual double in the input set for the **D-c1** set of simulations, a given network was required to perform a mapping to a unique output: Consequently, the output training set comprised 24 binary vectors coded over 12 units. For each conceptual triple in the input set for the **T-c1** set of simulations, a given network was required to perform a mapping to a unique output comprising 108 binary vectors coded over 36 units.

In order to maximize the similarity of the representations developed over the hidden units for each of the two vectors in one conceptual double, and for each of the three vectors within one conceptual triple, the **c1** stage of collapse2 was executed over a number of progressively smaller network simulations, with the hidden units from a previous simulation being extracted and used to form the input set for a subsequent simulation. In order to keep track of all these simulations, the following notational device will be employed: Network simulation **D-c1.1** refers to the initial network configuration of 39-30-36 for the **D-c1 stage**, whereas network simulation **D-c1.5** refers to the final network configuration of 20-20-36 for the **D-c1** stage.

Simulation **D-c1.1** (39-30-12) learned to a tolerance of 0.05 after 800 cycles, with learning rate and momentum values of 0.05 and 0.5 respectively. Simulation **D-c1.2** (30–25–12) learned to a tolerance of 0.05 after 550 cycles, with comparable values of learning rate and momentum, whereas simulation **D-c1.3** (25–20–12) learned to tolerance after 625 cycles, with comparable learning rate and momentum values. Looking at the cluster analysis shown in Figure 7.17, we can see that the **D-c1** simulations have performed the required task, and that the hidden unit representations show a clear separation.

Simulation **T-c1.1** learned to a tolerance of 0.025 after 9,450 cycles with the learning rate and momentum set at 0.05 and 0.7 respectively. Simulation **T-c1.2** (30–27–36) learned to the same tolerance, after tinkering to avoid a local minimum after 9,569 cycles, with learning rate and momentum initially set at 0.035 and 0.3 respectively, subsequently raised to 0.1 and 0.9. Simulation **T-c1.3** (27–24–36) learned to tolerance after 11,450 cycles, with values of learning rate and momentum of 0.05 and 0.7 respectively. Simulation **T-c1.4** (24–20–36) learned to tolerance, after tinkering to avoid the local minimum once again, after 3,802 cycles, and simulation **T-c1.5** learned to tolerance after 3,225 cycles, with learning rate and momentum values of 0.075 and 0.7 respectively.

As can be seen in the cluster analysis shown in Figure 7.18, the 36 conceptual triples have been cleanly separated by the **T-c1** stage simulations. Using the notational device introduced earlier, the labels of the dendogram, such as **4.cba**, **4c.ba**, **4cb.a**, refer to the three vectors of a given triple.

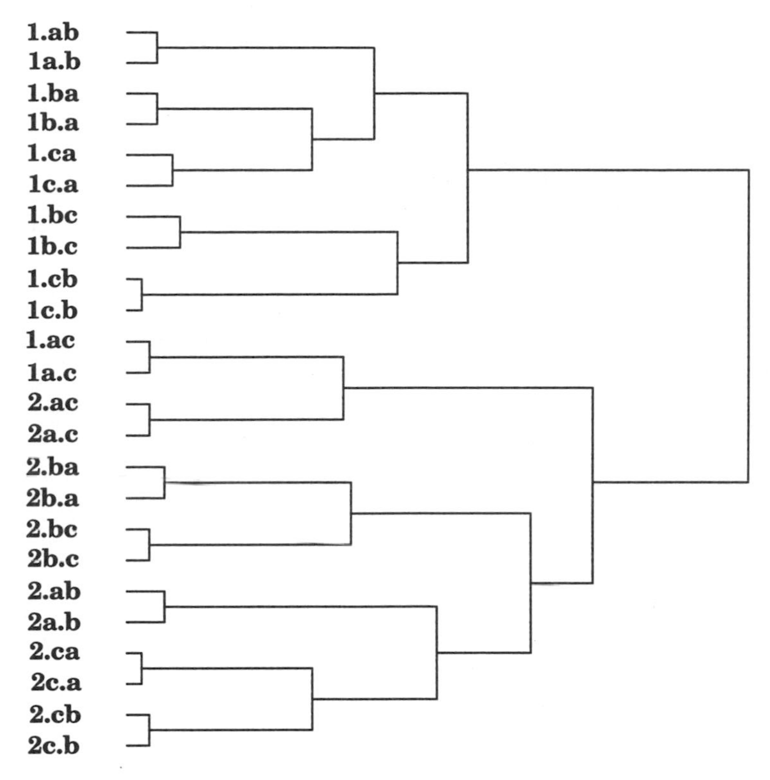

FIGURE 7.17. Cluster analysis of the hidden units from the final **D-c1.3** simulation, showing a clear clustering of conceptual doubles.

Collapse2: Stage c2

The input training set for the **c2** stage of collapse2 comprised the 24 hidden unit vectors extracted from the **D-c1.3** simulation, plus the 108 hidden unit vectors extracted from the **T-c1.5** simulation, coded over 20 units. Following the rationale of the **c1** stage simulations, the **c2** stage also comprised a number of simulations, in order to maximize the similarity of the representations developed over the hidden units. For each of the 132 input vectors, simulated **c2** networks were required to perform a mapping to an output vector, coded over 12 units, representing the appropriate most primitive states of affairs composed from the the three symbols **A**, **B** and **C**. These 12 states of affairs are shown in Figure 7.19.

Employing the same rationale as that used for the **c1** stage simulations, a similar notational device will be used to keep track of

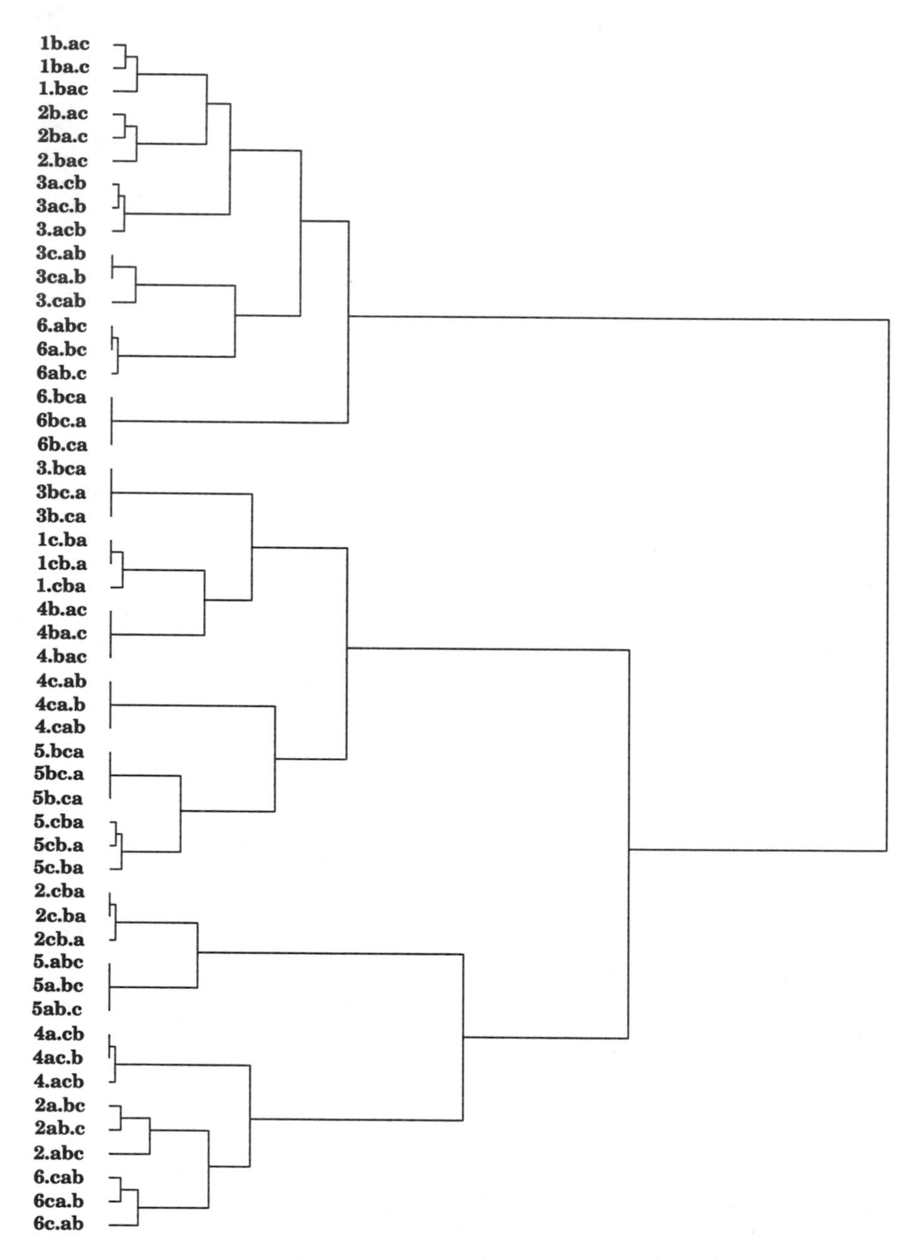

FIGURE 7.18. Cluster analysis of a portion of the hidden units extracted from the final **T-c.15** simulation, showing a clear separation of conceptual triples.

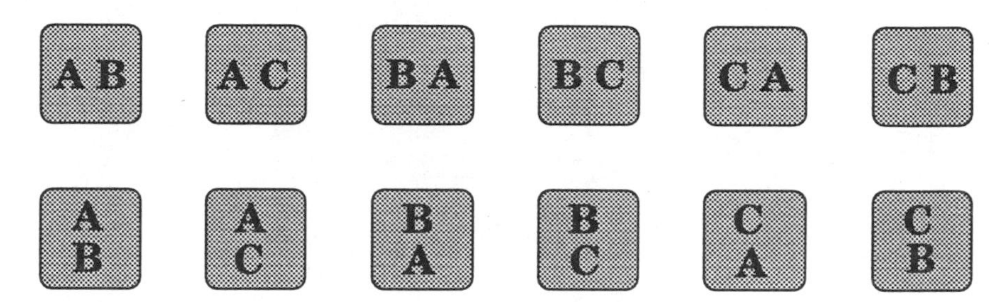

FIGURE 7.19. The 12 most primitive states of affairs of the three symbols **A**, **B**, and **C**.

the **c2** stage simulations: Thus, the **c2** stage comprised five network simulations, **c2.1** through **c2.5**.

Training the initial **c2.1** (20–18–12) simulation proved very difficult to achieve, and early attempts by the network to learn the entire input set were unsuccessful. An incremental learning strategy, in which the **c2.1** simulation was initially trained to tolerance on only a small number of inputs, the number of which was progressively increased, proved more productive. Using such a strategy, the **c2.1** simulation learned to a tolerance of 0.3 after some 160,085 cumulative cycles, with values of learning rate and momentum varying between 0.01 and 0.03, and 0.1 and 0.9 respectively. The hidden units from simulation **c2.1** were extracted after learning was completed, and were used to form the input set for the second **c2** simulation. This network simulation, **c2.2** (18–18–12), learned to a tolerance of 0.03 after 5,325 cycles, with learning rate and momentum values of 0.05 and 0.5 respectively. The third simulation **c2.3** (18–18–12) learned to a tolerance of 0.03 after 4,183 cycles, with values of learning rate and momentum of 0.05 and 0.5, subsequently raised to 0.15 and 0.9 respectively. Simulation **c2.4** and **c2.5** learned to a tolerance of 0.03, with values of learning rate and momentum of 0.075 and 0.7 respectively, after 1,625 cycles, and 2,025 cycles.

Extracting the hidden units from the final, **c2.5** simulation, I chose one of each of the vector instantiations of one conceptual configuration and fed the resulting 48 vectors, coded over 18 units, through a cluster analysis. This analysis produced the dendogram shown in Figure 7.20.

As can be seen, the dendogram shows that the hidden units have partitioned their representational space into 12 locations, corresponding to the 12 primitive states of affairs. It is worthwhile,

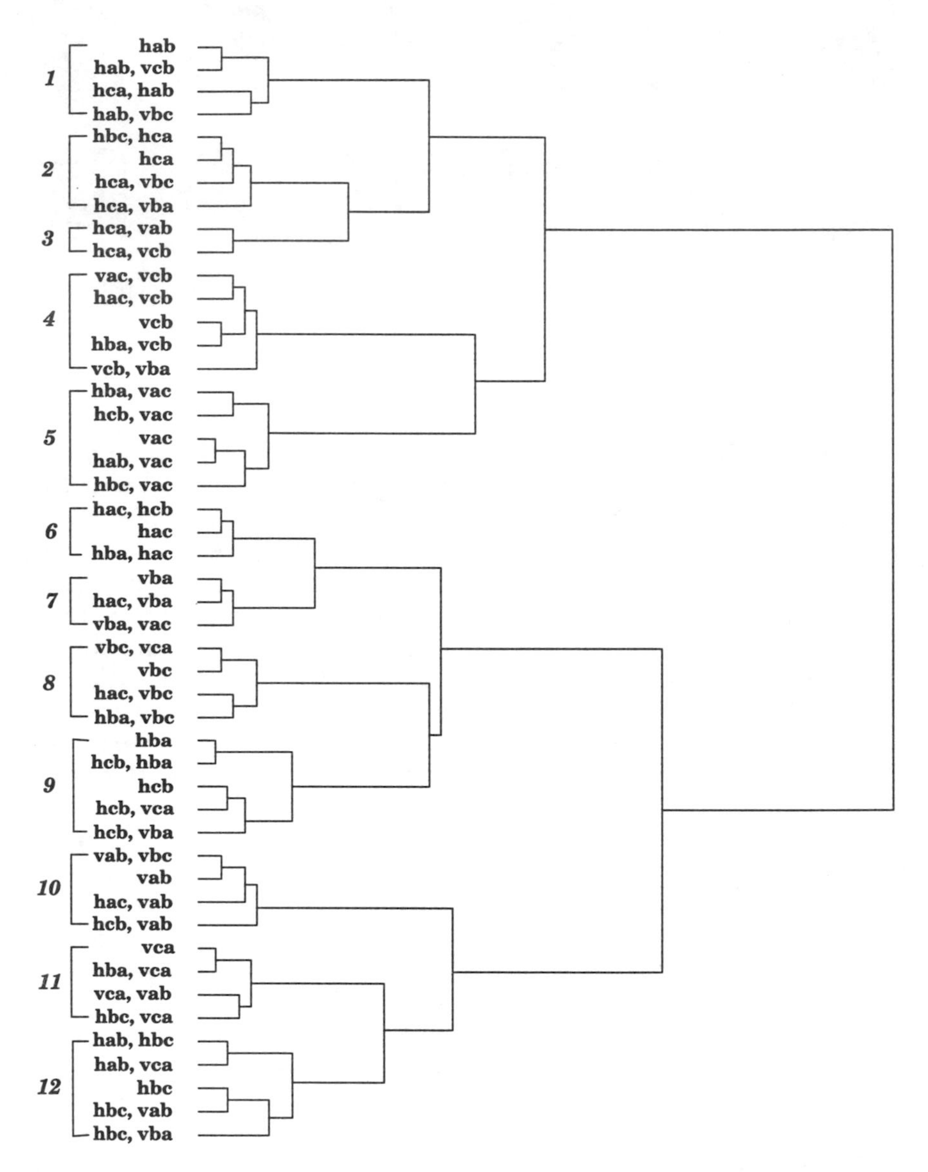

FIGURE 7.20. Cluster analysis of the hidden unit representations extracted from the collapse2 series 1 simulation **c2.3**. The labels refer to what primitive relations the hidden unit representations are encoding. For example, the label **HBA,VCB** refers to a representation where **B** is on the left of **A** in the horizontal plane, and **C** is behind **B** in the vertical plane.

at this point, to consider *why* this is a pleasing result. Surely, the objection might go, if the **c2** simulation is mapping the continuously valued **c1** vectors to 12 output units, then the partitioning we see is not surprising. Let us turn this objection on its head and consider how else one might expect these hidden unit representations to cluster, if not in the manner seen? Consider Figure 7.21.

Is Figure 7.21(a) more similar to Figure 7.21(b) than to Figure 7.21(c)? If it is, its clustering should reflect this, but on what basis are (a) and (b) more similar than (a) and (c)? Should Figure 7.21(a) cluster with 7.21(d) (with which it is similar in virtue of shared tokens in similar relative positions), rather than 7.21(e) (with which it is not similar, in virtue of shared tokens in dissimilar relative positions), even though the latter is a three-place relation, like 7.21(a), rather than a one-place relation? Should representations encoding two-place relations cluster together to the exclusion of one-place relations? The latter kind of clustering would reflect the most obvious *perceptual* similarity between representations, yet this is not what we find.

On the contrary, what we find is that the cluster analysis reveals that the pair-wise Euclidean distances between output representations is not sensitive to whether that representations encodes a one-place or a two-place relation. The dendogram shows that two-place relations cluster around loci formed by one-place relations. However, any two-place relation could equally well cluster around one of two loci, corresponding to the two one-place relations that it encodes; hence, one has grounds for supposing that the clustering of two-place relations around *particular* one-place relation loci is arbitrary. One can see this particularly clearly in cluster 3, which does not have a one-place relation as a loci. The main contributory factor for this discrepancy involves the previously mentioned point that Euclidean space and computational space, though highly correlated, are not identical. To reiterate the point, a cluster analysis

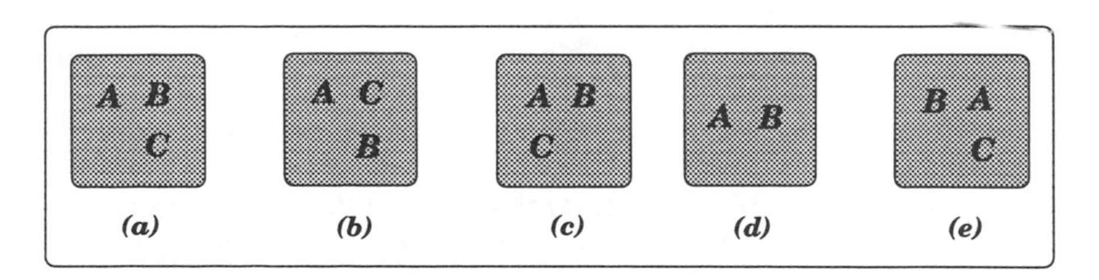

FIGURE 7.21. Is (a) more or less similar to (b) than to (c)? Or to (d)?[[A a21]]

using a Euclidean distance metric is only ever a crude approximation of similarity relations in computational space.

SUPERPOSITIONAL STRUCTURAL ANALOGUES

The collapse2 strategy was designed to banish the unwanted intrusion of symbolic (or linguistic) information into the analogical array structures used in the SRN simulation 2, where such symbolic information takes the form of information about the spatial ordering of symbol-instances within assertions, and the temporal ordering of assertions within sequences. Comparing the Euclidean distances between these "collapsed" vectors, and displaying such information in a dendogram, such as shown in Figure 7.20, we can see that simple *perceptual* similarity (between original binary vectors) is no longer in evidence, and that a more subtle *functional* similarity metric has been created, in response to the task requirements of the collapse2 strategy.

The collapse2 strategy delivers non-concatenatively compositional unit representations over the hidden units of the last, **c2.5**, simulation. Such unit representations result when a distributing transformation superposes constituent inputs into a non-concatenative, molecular compound. This means that the constituents of the molecular compound are, themselves, *superposed*, and hence unavailable to inspection on a symbol surface. What does it mean to say that constituents are superposed?

The notion of superposition is a relatively new one (cf. van Gelder, 1991a), and hence not without its critics. An adequate proposal for what counts as a superpositional representation would seem to have three components. Thus, **R** might be said to be a superposed representation of individual content items c_1, $c_2 \ldots c_n$ if, (a) **R** represents each c_i (b) **R** is an *amalgam* of representations $r_1, r_2 \ldots r_n$ of content items $c_1, c_2 \ldots c_n$, and (c), **R**, r_1, $r_2 \ldots r_n$ each uses the same resources in representing a given content c_i (cf. Aizawa, 1992). This kind of proposal will, undoubtedly, require substantial revision as an understanding of connectionist representation grows. For now, it is sufficient for our purposes.

Despite the fact that constituents of molecular unit representations (resulting from a distributing transformation) are superpositional, the unit representations extracted from the **c2.5** simulation of collapse2 are still structural analogues—which is to say, they still encode analogical relations. The novel feature of these unit representations, thus, is the fact that analogical relations between parts of the compound whole—between the primitive constituents—are analogical relations between *superpositional* parts. Hence, the label that I have chosen to give them, superpositional structural analogues, or SSAs. These novel form of analogical representations are investigated further in Chapter 8.

CONCLUSIONS

This chapter set out to investigate the nature of the representations of extension constructed by a formally grounded connectionist PS, with the ultimate aim of producing connectionist representations that, in a computational account of meaning, would be able to "take the place of the model structures" employed in formal theories of meaning. To this end, the literature on formal discourse representation theory, psychological theory on the representation of the content of texts, and the classical and connectionist notions of compositional structure were examined. From this discussion, two important notions emerged: the first, *analogue structure*, and the second, structural analogue representations. Analogue structure was detailed and compared to the the two other prominent notions of structure in cognitive science, and similarities between the connectionist notion of spatial structure and analogue structure were noted and placed in the context of a bounded space of representational properties. The rationale and the implementation of a method of constructing analogical representations, the collapse2 strategy, was specified. This strategy produces representations, called SSAs, extracted from the hidden units of the final in a series of collapsing networks, in which the spatial similarity relations support virtual analogue structure relations.

The *real* test of the efficacy of the SSAs constructed by the collapse2 strategy, however, is whether a subsequent, simulated network is able to learn the microprocedural representations (i.e., weights) that enable a mapping from a symbolic string to the structural analogue representation to be accomplished, and whether the microprocedural representations thus evolved in such a mapping are able to generalize to novel strings and construct the appropriate representation on the output. This endeavor represents a nontrivial extension of the microprocedural proposal and, as we will see in the next chapter, a re-orientation to the symbol grounding problem. Detailing this extension will be addressed in the next chapter.

8

Meaning and Grounding

Riding on the back of the theory of computability and the Church-Turing thesis, the doctrine of *functionalism*, first articulated by the philosopher Hilary Putnam (1960), is an empirical conjecture that the workings of the mind are intrinsically computational. It also supposes that detailed knowledge of a neurophysiological substrate underlying that mind is unlikely to contribute substantially to an understanding of the computation of which the mind comprises: In slogan form, the doctrine asserts that the mind can be studied independently of the brain. In the classical school of cognitive science, functionalist explanations of mental process are essentially elaborations of the operations and principles of the theoretical Turing machine and its digital successor, the von Neumann machine.

Imagine if you will a Turing machine. The machine takes as input a tape, on which are written a string of 1s and 0s. The machine goes into operation, writing and deleting symbols, and shifting the tape back and forth. Eventually, it will produce a second string of 1s and 0s, constituting the output for the function it has computed. For example, suppose the first string was 11010, and the second string 1110. In what sense does this mechanical derivation of 1110 from 11010 constitute an explanation? How might such a mechanical process be relevant to mental activity? In line with the doctrine of functionalism, the explanatory power of such a process lies in the specification of the function that was computed in mapping 11010 to 1110 (which in fact, in this case, is the function of addition: 11010 represents 2, punctuation marker, 1, punctuation marker, and 1110 represents 3, punctuation marker). This simple function is what is termed *implementation independent*, which is to say, the function can, in principle, be computed by any number of appropriate machines (given that they have the power of the abstract Turing machine).

At this point an interesting, and important, question arises: How does the classical theorist get at this function, without prior knowledge of it, from simply the input/output profile?[1] The answer is that he or she does *not* get at it: Until the input and output strings are given an interpretation by an agency external to the Turing machine (usually, in such a case, by the theorist who devised the machine in the first place), the symbol strings over which the computation is defined, and the function the machine has computed over those symbol strings are both, although perfectly precise, literally meaningless. (However, there are some who would disagree with the strength of this interpretation. Boden (1988), for example, argued that the inherent procedural consequences of a symbolic program of instructions imbue it (the program) with a kind of *causal* meaning.) This simple example serves to illustrate the essence of the *grounding problem*.

Throughout this book, I have alluded to this grounding problem without fully specifying just what it involves. To reiterate those earlier allusions, I mentioned that Wittgenstein encountered it at the terminus of his philosophical analysis in the *Tractatus* (1922), the Vienna Circle of philosophers thought they had solved it by appealing to ostensive definition, Carnap thought meaning postulates could address the problem, Johnson-Laird, Herrmann & Chaffin (1984) referred to the problem as a "symbolic fallacy," Hofstadter (1985) has argued that recognition of the problem requires that theorists wake up from a *Boolean dream*, and Searle (1980a; 1980b; 1984; 1987) has used the problem, couched as a lack of *intrinsic meaning* to demolish GOFAI (Good Old Fashioned Artificial Intelligence, a phrase due to Haugeland, 1985) in his *Chinese Room* thought experiment

The grounding problem was also referred to, in slightly more detail, in a number of different contexts in Chapter 5. In the first, the context was the last of the four inherent difficulties in assimilating a formal semantics with a corresponding psychological semantics—namely, the specification of the intensions of basic lexical items. By analogy, just as the functions computed by an abstract Turing machine are meaningless (viz. ungrounded) without interpretation by an agency external to the machine, so too the functions corresponding to the intensions of basic lexical items postulated by the formal semanticist, although they remain primitive and unanalyzed, are similarly meaningless.

---

[1] In fact, the Classical theorist must get at two things from the I/O profile: first, what function was computed, and secondly, what procedure computed that function.

The second of the contexts in which the grounding problem was raised was the last of the three problems faced by a theorist devising a theory of lexical semantics: Namely, the specification of how words are connected to the world. How is the word *left* connected to what it refers to in the world? What is the nature of this relation, and how might a theorist go about reconstructing it in a computational model? The third of the contexts in which the grounding problem was mentioned was the theory of meaning promulgated by Fodor, Garrett, Walker and Parkes (1980), based on Carnap's notion of a meaning postulate. The failure of that account of meaning to recognize the theoretical force of the grounding problem is, as we will see, endemic to the thinking of the classical theorist.

In this chapter, attention will be focused exclusively on this pervasive and mysterious grounding problem. To begin, I will present a detailed exposition of its most current formulation, drawing on both the intuitions contained in Searle's seminal (1980a) paper, "Minds, Brains and Programs," and the subsequent refinements of the ideas contained therein (Searle, 1980b; 1984; 1987), and also Harnad's characterization of the problem in Harnad (1990b; 1992). A number of responses to the grounding problem will be identified and discussed: The Classical response, the Computational Neuroethological response, the Hybrid response, and the Subsymbolic response. Two important points arise from a consideration of these various responses. The first is the recognition that the label the grounding problem is too narrow to encompass the observed phenomena, and might more informatively be redubbed, the *representation* grounding problem (cf. Chalmers, 1992). The reason for this new term will become more obvious as we proceed but, principally, it is intended to reinforce the point that the term *representation* should not be restricted to a purely classical interpretation.

The second consideration concerns the nature of the now redubbed representation grounding problem itself. Arguments will be presented that urge and qualify the reasons for a distinction to be made between two types of grounding: Extensional grounding and intensional grounding. On the basis of this distinction, another response to the grounding problem will be presented, the *Radical Connectionist* response: Radical because the response invokes the idea of the two types of grounding mentioned above, and connectionist because the response is couched in a non-classical vocabulary. In the last section of the chapter, the Radical Connectionist response, drawing on and combining the notions introduced in the last two chapters, of microprocedural representations of intension, and SSAs, will be illustrated in a simulation of intensional grounding.

GROUNDING ARTICULATED

We can distinguish two formulations of the grounding problem: The first, referred to in the guise of the *Chinese Room* argument was proposed by Searle (1980a), where the problem is referred to as the lack of intrinsic meaning in symbol systems, and the second is due to Harnad (1990b), the originator of the more specific *symbol* grounding problem, in which the problem is couched in terms of the need to ground symbolic function in non-symbolic function. We will tackle each of these formulations in a straightforward manner.

In the Chinese Room, Searle took issue with one of the core assumptions of GOFAI, that an appropriately programmed symbolic machine literally has (or is) a mind. More specifically, he challenged the view that if a computer is able to pass the Turing Test in Chinese—is able, that is, to take Chinese symbols as input and output Chinese symbols indistinguishable from those that a human speaker of Chinese would produce, given the same inputs—then it can be said to *understand* the meanings of the Chinese symbols in the same way that a human speaker of, for example, Punjabi understands the meaning of Punjabi symbols.

Articulating his reservations about this assumption led Searle to imagine himself doing everything that the computer might do in producing the Chinese symbols as output given Chinese symbols as input: He imagined himself, in other words, simulating a run of a program for "understanding" Chinese. Locked in the now famous *Chinese Room*, Searle received pieces of paper through a hole in the wall on which are written Chinese symbols. In with him in the Chinese Room is a book of rules, written in English for manipulating the characters purely on the basis of their shape (these rules, of course, represent the program), and Searle returned the manipulated Chinese characters back through the hole in the wall.

From without, the Chinese Room appears to be passing the Turing Test, whereas from within, all that is going on is the mindless manipulation of the characters. The essence of the Chinese Room argument is that if Searle-instantiating-a-program understands no Chinese in such a situation, which he didn't, of course, then neither would a computer-instantiating-a-program understood any Chinese in such a situation either. The Chinese symbols, and the manipulations of them, are systematically *interpretable* as having a meaning (most notably by those outside the Chinese Room) but the meaning is not intrinsic to the system of Chinese writings. Rather, it is parasitic on the interpretation provided by an external agency, in just the same way that the symbols on this page have no intrinsic meaning, but are parasitic on the interpretation provided by the reader.

Harnad's (1990b) formulation of Searle's intuitions about intrinsic meaning involved him postulating how a machine might pass a more stringent test of understanding. This Total Turing Test would test not only the *linguistic capacities* of a given machine, but also its *robotic capacities*, which is to say, its sensorimotor abilities to interact with an environment external to it. This Total Turing Test, Harnad argued, is a more accurate reflection of the capacities of human speakers of language, and hence a more accurate test of the capacities that a machine would need to acquire in order that an observer might ascribe a mind to it.

A machine that possessed a linguistic capacity but no robotic capacity would be unable to pass the Total Turing Test, yet GOFAI is concerned with the project to build precisely such a machine. Harnad urged the theorist to stop and consider just what this project involves: Specifically, the project is akin to asking an English speaking individual to learn the meanings of Chinese symbols on the basis of a Chinese-to-Chinese dictionary, without any prior knowledge of that language. This project is daunting, perhaps even impossible, because all that would result from such an endeavor, Harnad argued, is a meandering from meaningless *definiens* to meaningless *definiendum*. It is a systematic, and systematically interpretable meandering to be sure, but a finally pointless endeavor because the search would never halt on the intrinsic meaning of a symbol. Harnad (1990a) characterized such a meandering as like being in a hermeneutic hall of mirrors. Any computational system that makes use of Classically conceived symbols, and that defines the computation performed by that system as the manipulation of such symbols on the basis of their shape (i.e., on the basis of their syntactic structure), Harnad characterized as "hanging from a sky hook": it is, in other words, fundamentally *ungrounded*. This is the essence of Harnad's symbol grounding problem.

For both Searle and Harnad, the main thrust of their arguments concerned the nature of the computation performed by a classical symbol system. Classical theorists, of course, view the structure-sensitive manipulation of symbols, which comprises their understanding of computation, as a major advantage, providing as it does, the basis for the systematicity of such systems. In the context of Harnad's symbol grounding problem, however, the fact that computation is defined in terms of a syntactic structural manipulation is not an advantage at all, but rather a debilitating drawback. And, in fact, the following axiom is cited in support of the avowed "polemical aims" in Searle (1987).

Syntax is not sufficient for semantics

On the basis of the discussion of structure in Chapter 7, this axiom can be paraphrased slightly, so as to read:

Syntactic structure is not sufficient for semantics

As a connectionist, this elaborated axiom is intriguing. The possibility that it raises is that computational systems, whose definition of computation makes no recourse to syntactic structure sensitive manipulation, may not be as vulnerable to Harnad's symbol grounding problem (or the grounding problem more generally) as those computational systems that do make such recourse. Connectionist systems, as we have seen, can be thought of as performing a spatial structure-sensitive manipulation, and hence may exhibit just such a degree of invulnerability. The implications of this observation will be explored more fully in later sections, but for now, I wish to concentrate on three of the responses to the grounding problem that I mentioned in the introduction. These responses, I should add, are not all formally worked out solutions to the grounding problem (particularly the Classical and Computational Neuroethological responses) but are rather candidate solutions that I have extracted from the relevant literature.

The Classical Response

The Classical response to the symbol grounding problem is brief and cursory (see, for e.g., Fodor et al., 1980). To ground a symbol, the response goes, involves merely connecting up that symbol to the world *in the right way*. The thinking of the classical theorist that engenders this response is that cognition, comprising the manipulation of symbols in virtue of their syntactic structure, is an autonomous functional module that need only be hooked up to peripheral (and non-cognitive) perceptual mechanisms in order to "see" the world of objects to which its symbols refer. This response, in a sense, is already implicit in the notion of a *physical* symbol system (cf. Newell & Simon, 1972), as the syntactic structure of classical symbols is assumed to correspond to real physical structures in the brain. Given that such physical (qua neurophysiological) structures in the brain are part of a protein mechanism causally interacting with an external environment, then classically defined symbols are *already* connected to the world. Of course, this is in the wrong way because classical symbol systems derive a great deal of their explanatory force from the fact that they are implementation independent, which is to say, they can be simulated on any number of different machines, be they carbon or silicon. If a theorist is tempted to make this recourse to the symbol grounding problem, which is similar to the *Brain Simulator* reply to the Chinese Room argument then, of course, she gives up the implementation-independence of symbolic function.

The form of response presented here is, I believe, an accurate distillation of classical thinking. Most importantly, it reveals the failure of the classical theorist in cognitive science to appreciate the full force of the symbol grounding problem. Such disregard for the problem is revealed in the fact that even where a classical theorist grudgingly accepts the significance of the symbol grounding problem, they typically provide only the most vague of responses to it, involving some manner of peripheral hookup to sensory modules, as revealed in the number of replies to Searle's (1980a) paper, such as the *Systems reply*, the *Robot reply*, the *Brain Simulator reply* and so forth. With regard to the particular form of the Classical response outlined here, the crucial phrase is in the right way. This glib statement grossly underestimates the problem of formalizing just what it is about the external environment that the brain reflects in its physical structure. This problem of formalization is, arguably, coextensive with the problem of cognition itself, and it does no good at all to try and trivialize it.

The Computational Neuroethological Response

Computational neuroethology is a term originated by Cliff (1990) and refers to the emerging discipline where ethology (the study of behavior) meets neurophysiology. The computational neuroethological theorist seeks to construct computational models of the neural mechanisms of animal behavior. Cliff (1990) cited the modeling of the compound eye of the fly *Limulus* by Prakash, Solessio and Barlow, (1989), and Arbib's (1987) work on the computational frog *Rana computatrix* as good examples of this kind of approach.[2]

In computational neuroethology, the computational model the theorist is seeking to build is embedded in a simulated environment that serves to close the external feedback loop from motor output to sensory input. In this way, output of the simulated nervous system is then expressed as observable behavior. The relevance of the computational neuroethological response to our present grounding concerns, is its insistence that only by linking a computational system to an environment external to that system via a sensori-motor system, can symbols in the model be grounded. The argument by Lakoff (1988) with reference to connectionist computational models, and in response to Smolensky (1988), went like this:

> The neurons in a topographic map of the retina are not firing in isolation just for the hell of it. They are firing in response to retinal input, which in turn is dependent upon what is in front of one's eyes. An pattern of activation in the topographic map is not merely then a meaningless statistical object in a dynamical system: it is *meaningful*. (Lakoff, 1988, p. 39)

Similarly, if the units in a connectionist network are located relative to a sensori-motor system, then patterns of activation over that network of units are, to borrow Searle's terminology, intrinsically meaningful. Or put another way, no interpretation by an agency external to the representational system is necessary to give the activation patterns meaning. Equally, however, if the units that comprise a connectionist system are *not* located relative to a sensori-motor system, then patterns of activation over that network of units are not *intrinsically* meaningful, and one must admit the necessity of recourse to an external agency for interpretation. Lakoff's views were directed, equally, against (*a*) Smolensky's new computationalism and the distributed representations advocated by that school, and also (*b*) the classical school and the syntactically struc-

---

[2] Computational neuroethology has much in common with autonomous systems research and the esoterically named "Artificial Life" approach to robotics. See Varela and Bourgine (1992) for more details.

tured symbols that it advocates. Both kinds of representations, on a computational neuroethological view, require grounding in a sensori-motor system. Cliff (1990) provided the following summary of the computational neuroethological perspective:

> The representational contents of the constituent states of a computational process not connected to a sensori-motor system are dependent upon the intentions of the person who designs and observes the process . . . On the other hand, the contents of the states of a computational process which *is* connected to a sensori-motor system are, in a sense, objective. They do not depend upon the intentions of a designer or observer. They depend, rather, upon the role that they have in the sensori-motor system. (Cliff, 1990, p. 13)

The Hybrid Response

The Hybrid Response, formulated by Harnad (1989; 1990a; 1990b; 1992) is touted as a specific solution to the *symbol grounding* problem, the nature of which is exemplified in the Chinese Room argument. More specifically, the Hybrid response recasts the Chinese Room argument as an attack against *symbolic functionalism*: The doctrine that the computational workings of the mind are intrinsically, and solely, symbolic. The Hybrid response performs this feat of escapology by positing, in addition to symbolic functionalism, a *robotic functionalism*: The doctrine that the computational workings of the mind include some components that are not intrinsically symbolic. Now let's be a bit more precise.

The Turing Test outlined in the last section was formulated as a gauge of a very specific kind of machine intelligence, namely, *linguistic capacity*. If a human speaker was able to communicate, via a screen, with a machine, and was unable to distinguish whether the responses she was seeing originated with that machine or with another human speaker, then there would be no grounds for supposing that that machine did not possess "intelligence," or that it did not "understand," or that it did not "have a mind." Searle's Chinese Room argument is directed precisely at this notion of a Turing Test for machine intelligence. Specifically, he argued that it is in principle possible for a machine to pass such a test, and yet be doing no more "understanding" than a small pebble. The moral is the mind cannot comprise only symbolic computation; cannot, in other words, comprise solely a linguistic capacity.

Now, the Hybrid Response goes one step further: In place of the Turing test, consider the Total Turing Test for machine intelligence. This would require a machine to be indistinguishable from a human in both linguistic and robotic capacities. Robotic capacity, in this context, is a label for a range of sensori-motor interactions with the external environment, of which Harnad cited the *discrimination* and *identification* of objects, events, and states of affairs in the world as principal. Yet now the force of the symbol grounding problem appears to diminish. The Chinese Room argument worked against the Turing Test because Searle himself could do everything symbolic that the machine could do and yet still be obviously failing to understand. However, is Searle able to do everything symbolic *and* robotic that a machine would have to do to pass the Total Turing Test, just as easily as he was able to do everything in the purely symbolic case? Harnad conjectured that he is not, and for the following reason:

> Machines that behave as if they see must have sensors—devices that transduce patterns of light . . . So Searle has two choices. Either he gets only the output of those sensors . . . in which case he is not doing everything that the candidate machine is doing internally (and so no wonder he is not seeing) . . . or he looks directly at the objects that project onto the devices sensors . . . but then he would in fact be seeing! (Harnad, 1991, 50)

It would appear that sensory transduction is sufficient to foil the thrust of the Chinese Room argument. Harnad is quite explicit about the nature of the sensory transduction qua robotic capacity that serves to ground a symbolic capacity: Specifically he names the ability to discriminate, identify and manipulate what are called *non-symbolic* representations. Discrimination involves the machine making similarity judgments about pairs of inputs, and constructing *iconic* representations. These icons are nonsymbolic transforms of distal stimuli, which is to say, analogs of the projections occurring on sensory surfaces (where "analog" should be contrasted with "discrete," and not confused with "analogical"). Merely being able to discriminate inputs does not make for a robotic capacity, however. Additionally, the machine must also be able to identify the iconic representations. That is, it must be able to extract those invariant features of the sensory icons that will serve as a basis for reliably distinguishing one icon (one member of a category) from another (non member of that category). Harnad referred to such perspicuously filtered versions of icons as *categorical* representations.

Harnad was not urging that iconic or categorical representations "mean" anything—that they have, in Searle's terminology, any internal content. The categorical representations particularly are just an inert taxonomy of the possible space of shapes that the iconic representations encode. Somehow, that taxonomy must be linked up to representations on the symbol surface such that the properties of, principally systematicity, can be exhibited on that surface.

Accordingly, here is how the Hybrid Response views grounding to work (i.e., how "meanings" might be attached to symbols). The same mechanism that allows objects to be sorted into categories on the basis of their sensory projections also allows the assignation of a unique, arbitrary name to each category. This is the essence of the Hybrid Response to the symbol grounding problem: Assign an elementary symbolic representation to each element of the category taxonomy and string such elementary representations together into more complex propositional structures about further category membership relations.

The example Harnad used to illustrate is the word *zebra*. If some appropriately configured machine (what Harnad would call a TTT robot) is able to discriminate and identify horses and stripes on the basis of iconic and categorical representations, then it is also able to assign an unique name to each of the categorical representations, so as to yield the two elementary symbols **horses** and **stripes**. The machine, in principle, is now in the position to use the complex symbol string: **zebra = horses + stripes**. The symbol **zebra** is what Harnad calls a *grounded* symbolic representation, implying that **zebra** is grounded in the iconic and categorical representations underlying **horses** and **stripes**.

Objections I have two objections to the Hybrid Response: The first is a fairly specific criticism of what Harnad considered robotic capacity to comprise, whereas the second is a more global concern. With regard to the first objection, consider what kind of machine it would be that possessed *only* those robotic capacities that Harnad envisioned accompanying a linguistic capacity. Consider, that is, a machine that was only able to discriminate and identify energies impinging on its sensory surfaces. Are we to accept that the complex social, feeding, territorial and mating behaviors of all those animals that possess no linguistic capacity is the expression of only discrimination and identification? If a chimpanzee is capable of constructing and using about 30 different tools, is one to suppose that such complex behavior is mediated solely by perspicuously filtered icons of sensory projections?

What seems to be the case is that Harnad's conception of robotic capacity is an impoverished one, unable to perform all that is evidently required of it. Using terms that will be explained in due course, a *Cartesian* automata does not become a *Craikian* automata merely by the addition of a mechanism capable of extracting and labeling invariants of sensory projections. Rather, a Cartesian automata is upgraded to a Craikian automata by the addition of computational machinery sufficient to construct and manipulate representational models of the world. There is more to robotic capacity than simply iconic and categorical representations: In addition, there is the computational machinery necessary and sufficient to construct analogical models of the world.

The second of my objections to the Hybrid response is short and to the point: why ground *symbols*? Pause for reflection: What is significant about symbols in this context?

Implicit in the Hybrid response to the symbol grounding problem is the belief that *only* symbolic representations are in principle capable of exhibiting the systematicity characteristic of mental representations, and hence of providing sufficient representational resources to build an artificial intelligence. As we saw in Chapter 3 and 7, however, connectionist representations, suitably constructed, are in principle equally capable of exhibiting such systematicity. So the question can be asked again, in a slightly different form: Why particularly concentrate on grounding classical symbolic representations, given that the problem applies equally to *all* classes of complex structured representations, which includes suitably constructed connectionist representations?

The answer, in the context of Harnad's muse, is that there is no principled reason to so concentrate, other than a blind adherence to the notion that cognition must involve the class of complex structured representations subsumed by the classical label. This is why, of course, the grounding problem is referred to as the *symbol* grounding problem in the Hybrid response (and not simply the grounding problem). However, if one accepts that classes of complex structured representation, other than those classes resulting from a concatenative compositionality, are possible, and if one accepts that such representational systems have, in principle, all of the features necessary for that system to exhibit, for example, systematicity, then one must also accept a primary outcome of such observation, namely, that the label the *symbol grounding* problem is too narrow to encompass the observed phenomena, and might more informatively be re-dubbed the *representation* grounding problem.

The reason for this should be clear: At the heart of the grounding problem (as opposed to the symbol grounding problem) is the lack of intrinsic meaning in any representational scheme that is not connected up, in some fashion, to the world. On the basis of the understanding outlined above, the term *representational system* now subsumes representations constructed from both concatenative and non-concatenative compositional schemes, which is to say, both classical and connectionist schemes. Accordingly, one cannot talk about the symbol grounding problem in the context of connectionist representations, rather, one must talk about the *representation* grounding problem, and Chapter 9 does just that.

Classical theorists respond to the Hybrid symbol grounding problem in poor fashion: Either they try to deny the significance of the problem, or in grudgingly accepting it, offer only the vaguest of solutions, postulating some peripheral, non-cognitive "add-on" sensory modules. The Hybrid theorist, by contrast, argues that a non-classical component is in actual fact *necessary* to a solution of the symbol grounding problem, and that a linguistic (or symbolic) capacity must be grounded in a robotic (or nonsymbolic) capacity to interact with the external environment. The computational neuroethological theorist, by contrast, is off at something of a tangent, arguing that sensorimotor interaction with the environment (viz. a robotic capacity) is necessary to ground the representations of any computational system, be it classical or connectionist. The computational neuroethological responses thus shares with this book a belief in the need to ground both classical and connectionist representations. The implications of this requirement are explored in the next section.

REPRESENTATION GROUNDING

To reiterate a somewhat obvious point, the symbol grounding problem as formulated by Harnad is expressly concerned with the lack of intrinsic meaning of *symbols*: that is, with classically conceived, syntactically structured entities. However, as we noted in discussion of the Hybrid response, a concentration on classical representations blinds a theorist to the fact that the grounding problem applies equally to all classes of complex, structured representations. The Computational Neuroethological response recognizes this fact, of course, and thus arguably is the only one of the three responses discussed so far that address the broader *representation* grounding problem.

Accordingly, the question we will consider next is whether connectionist representations are as vulnerable to the *representation* grounding problem as classical symbols are to the *symbol* grounding problem. We can begin this inquiry by reiterating and paraphrasing Searle's first axiom.

Structure is not sufficient for semantics

On the basis of this axiom, we can distinguish two avenues by which the relative vulnerabilities of the classical and the connectionist schools to the representation grounding problem can be investigated: (*ai*) a structural avenue, and (*b*) a semantic avenue. Let us consider the first of these.

The gist of Searle's axiom is that manipulation of symbols on the basis of a formal syntactic structure is not sufficient for ascribing meaning to that symbol. As we saw in Chapter 7, the formal structure of a representation is an attribute at the level of *token* specification. The manner in which primitive tokens combine to form more complex tokens determines the type of structure the resulting complex token will exhibit. In a classical system, tokens combine concatenatively and thus subserve a syntactic structure manipulation, whereas in a connectionist system, tokens combine non-concatenatively and subserve a spatial structure manipulation.

Referring back to Chapter 7 once again, however, computation defined in terms of both syntactic and spatial manipulation is equivalent in terms of computational efficacy: Anything that a computational system employing representations exhibiting syntactic structural similarities can do, a computational system employing representations exhibiting spatial structural similarities can (in principle) do also. The conclusion we must draw in this case is that we cannot ascertain any difference in the relative vulnerabilities of classical and connectionist systems to the representation grounding problem in purely structural terms. Our first avenue is, in fact, a cul-de-sac. What about the second, semantic avenue? Is this able to help in distinguishing the relative vulnerabilities of the classical and the connectionist schools to the representation grounding problem? In Chalmers (1992), the conclusion is that it can.

The argument that Chalmers put forward turns on an ambiguity in the way that the term "symbol" has previously been used. Specifically, Chalmers made the distinction between (structural) *tokens* and (semantic) *representations*, urging that whereas in classical systems, no important difference is recognized between these two entities, in connectionist systems, the difference between tokens and representations has crucial theoretical significance. About this distinction, Chalmers had the following to say:

> In symbolic systems, representations and tokens coincide. Every basic representation is an atomic computational token, and other representations are built by compounding basic representations. A LISP atom, for example, serves simultaneously as a computational token and as a representational vehicle. In connectionist systems, by contrast, representations and tokens are quite separate. Individual nodes and connections, the computational tokens, lie at a completely different level of organization from representations . . . the level of computation falls below the level of representation. (Chalmers, 1992, p. 10)

The tokens that Chalmers referred to in a connectionist system are the individual units. One might initially balk at the notion of computation defined as the manipulation of units but, in fact, the notion is entirely coherent. Recalling Chapter 6, we stated that no completely satisfactory definition of computation actually exists. We have, of course, the *formal*, classical notion of computation, as the structure sensitive manipulation of syntactic tokens, but there is no equally robust characterization for connectionist systems. However, we can reiterate some of the relevant points mentioned in Chapter 6. Just as Chalmers distinguished tokens and representations, defined over single units and groups of units respectively, so too one can distinguish microprocedural tokens and microprocedural representations, defined over single weights and groups of weights respectively.

The representations that Chalmers referred to in his (1992) work are always unit representations, which is to say, distributed patterns of activation over groups of units. In addition to these unit representations, of course, the Microprocedural Proposal detailed in Chapter 6 points to the advantages of also considering weight representations, which can serve to enrich Chalmers' arguments, rather than to detract from it. In point of fact, the notion of microprocedural representations plays an integral part in the reorientation to the representation grounding problem presented in the next section. For now, however, using Chalmers' distinction between tokens and representations means that we can distinguish a fourth response to the grounding problem.

The Subsymbolic Response

The Subsymbolic response (cf. Chalmers, 1992) acquiesces to Searle's axiom that structural manipulation is not sufficient to give symbols meaning. It begins its departure from the thrust of the classical and the Hybrid responses, however, by examining how the axiom applies to symbols-as-tokens and symbols-as-representations. With regard to symbols-as-tokens, the full force of the grounding problem is undiminished: No amount of structural manipulation (even where that manipulation is based on spatial, as opposed to syntactic, structural similarities) can imbue a symbol-as-token with meaning. Because of the concept of computation that the classical theorist adheres to, where tokens are also representations, the fact that structural manipulation is not sufficient to imbue a symbol-as-token with meaning also means that such manipulation is not sufficient to imbue a symbol-as-representation with meaning. In the Subsymbolic response, the fact that "structural manipulation is not sufficient for semantics" is not a problem. The symbol-as-tokens over which connectionist computation is defined are not *intended* to be representations, that is, to possess meanings. They are, instead, purely structural objects (i.e., units).

With regard to symbols-as-representations, the situation is quite different, however. A classical symbol-representation is an atomic, primitive entity, with no structure internal to it. There is nothing to the classical symbol-representation PERSEPHONE that makes it intrinsically refer to *Persephone* any more so than it might refer to *Antigone* or *Eteocles*. Internal to the computation defined over it, the symbol-representation PERSEPHONE is a featureless chunk, with only an arbitrary label to distinguish it from other tokens. To recall Chapter 7, a similar point was also made by Sloman (1978), in a slightly different fashion, in the context of his distinction between Fregean and analogical representations, when he asserted that Fregean representations have a structure that bears no relation to the structure of what it is they are representing. To put the global point in another way, the symbol-representation PERSEPHONE is not *causally efficacious* in the computation defined over that representation.

In contrast, the symbol-representations in a connectionist network are not atomic, primitive entities. The specific patterns of activation comprising a connectionist symbol-representation is responsible for making it represent what it does. Consequently, there is something to the connectionist symbol-representation PERSEPHONE that makes it intrinsically refer to *Persephone* and not *Antigone* or *Eteocles*. That something is *precisely* the internal structure of the symbol-representation that allows it to play a causally efficacious role in the computation defined over the constituent tokens of the representation. In line with the Computational Neuroethological response, the internal structure of the connectionist symbol-representation PERSEPHONE is essential, in other words, to the behavior of the computation defined over it, and not just any arbitrary pattern of activation will do.

It would seem that by dissociating tokens and representations, the Subsymbolic response is able to divert the full force of Harnad's symbol grounding problem. As formulated by Searle and Harnad, the problem refers to the fact that tokens have no intrinsic meaning, hence are ungrounded and consequently require some means by which they can be made intrinsically meaningful. The Subsymbolic response, by advocating representational vehicles that are not also primitive tokens, and that possess complex internal structure, offers a different perspective on the symbol grounding problem. The prospect, as Chalmers (1992) urged, that is raised is of intrinsic meaning being carried, not by primitive symbols qua tokens, nor indeed by the manipulation of such tokens, but rather by the internal structure of computationally, causally efficacious representations.

There is, however, an issue that I have not yet considered in detail. When Searle asserted that classical symbols lack intrinsic meaning, when Harnad similarly urged that symbols are ungrounded qua meaningless, and when Chalmers asserted that the internal structure of distributed patterns of activation can carry meaning, it is crucially important to be aware of precisely what it is that they are asserting: What, in other words, is the nature of the meaninglessness that they all ascribe to classical symbols? We will consider this question in the next section.

A REORIENTATION TO THE GROUNDING PROBLEM

In addition to Searle's first symbol grounding axiom that we have already mentioned, there is a second that we must now consider. It asserts that:

> Minds have contents: Specifically, they have semantic contents.

As well as the term "intrinsic meaning" that Searle used to refer to the lack of meaning in classical symbols, he also used this second term, "content" or "semantic content" more or less interchangeably. About intrinsic meaning qua semantic content, he had the following to say:

> What I have in mind when I say that the operation of the brain is causally sufficient for intentionality, [is] . . . that it is the operation of the brain and not the impact of the outside world that matters for the content of our intentional states, at least in one important sense of content. (Searle, 1980b, p. 452)

Other than Searle's insistence that only the brain, and more specifically, that only the *causal properties* of that protein mechanism, are sufficient to carry intrinsic meaning (an insistence that has resulted in him being labeled a *carbon chauvinist*), what this quote reveals is a significant difference between the notion of meaning comprising Searle's understanding of the term, and the notion of meaning comprising, for example, the computational neuroethologist's understanding of the term. Thus, when Lakoff stated that: "The nature of the hookup to the body will make . . . an activation pattern meaningful, and play a role in fixing its meaning" (1988, p. 39), he was referring to a notion of meaning that involves an explicit recourse to an environment external to a computational system. By contrast, Searle's understanding of meaning eschewed any appeal to an external environment: His notion of meaning is one that exists entirely internal to a given system. Not that Searle ignored the kind of meaning to which Lakoff alluded, only that where he does make reference to the role of the external environment in grounding, it is dismissively, as in the context of the Robot Reply.

Consequently, we can begin to see that what Searle might take grounding to mean appears to be something rather different from the notion of grounding as used by the Computational Neuroethological response, for example, involving as they do, two different notions of meaning in each case. In the first case, that of Searle and Harnad, a grounded symbol would possess meaning in virtue of properties internal to a computational system, whereas in the second case, that of Lakoff, a grounded symbol would possess meaning in virtue of properties external to a computational system. For the sake of clarity and completeness, I should add that Lakoff would probably argue that a grounded symbol would possess meaning in virtue of an *interaction* with properties external to a computational system. In the context of our discussion, such an interaction, being unspecified, does not substantially alter my argument. In philosophical circles, Searle's notion of intrinsic meaning qua semantic content is referred to as *narrow content*, whereas Lakoff's notion of meaning is referred to as *wide content*.

There is another characterization of the two types of content of which we can avail ourselves, however, that involves the two terms I have spent a good deal of this book discussing, *extension* and *intension*. As we have seen, these two terms have evolved in a lumpy fashion from purely formally conceived entities into important parts of a psychological theory of meaning. The extension of a symbol qua representation is the meaning that the representation possesses in virtue of its relation to some entity in the world (although whether that world is, following Jackendoff (1984), real or projected, is a different matter), whereas the intension of a symbol qua representation is the meaning that the representation possesses in virtue of its relation to its extension (in a simplified sense, no pun intended on "sense"). Chapter 6 was concerned with how intensional meanings could be realized as weight-encoded microprocedural representations, whereas Chapter 7 concentrated on how to construct representations of extensional meaning.

Given these two complementary notions of meaning, we can characterize Searle and Harnad as referring to intensional meaning, whereas we can characterize Lakoff as referring to extensional meaning. The reorientation to the grounding problem that I am pursuing takes this distinction as its starting point. The distinction between wide and narrow content, or between intrinsic and extrinsic content is not, of course, original. What is original, however, is the use of the intension and extension distinction and the greater sophistication and range of application of these terms in a treatment of the grounding problem.

A Representational Vehicle for Intrinsic Meaning

The implications for the grounding problem of the difference between extensional and intensional meaning is recognized in Chalmers (1992), although in an imprecise way, prompting him to postulate two types of representation grounding, what he termed *causal* and *internal* grounding.

> If one is concerned with how our symbols can possess . . . extensional content, then one will engage in the project of *causal grounding*. This is the enterprise of hooking up computational systems to

> the external environment, and thus connecting representations directly to their referents . . . When one asks the question, 'How can representations be grounded?' about internal content, a different project suggests itself. This is the project of *internal grounding*: Ensuring that our representations have sufficient internal structure for them to carry intrinsic content. (Chalmers, 1992, p. 20)

Although this characterization of causal and internal grounding seems superficially plausible, there are, I believe, a number of reasons for supposing that it is not. Taking causal grounding first, this notion does not differ substantially from the type of grounding advocated by the computational neuroethologist. However, we should make a note of the laxness of Chalmers' phraseology. Specifically, he was being lax when he asserted that causal grounding is in the business of "connecting representations directly to their referents." Assuming an Antirealist position (pace Jackendoff) in preference to a Realist position (pace Gibson), one can say rather more accurately, that it (i.e., causal grounding) *should* be in the business of constructing representations of extension (i.e., #extensions#) via the mediation of a sensori-motor system, on the basis of energies impinging on sensory surfaces. Given that the only knowledge a computational system can possibly have of an environment external to it is the representations it has constructed of that environment, it makes little sense to speak of "connecting representations to their referents." Strictly speaking, a given computational system has no knowledge that there are discrete, real-world referents (or extensions) until it has constructed #referents# (or #extensions#): All it knows is an undifferentiated bombardment of energies on sensory surfaces.

Perhaps we are being picky. When, however, we consider the second type of grounding that Chalmers identified, that of internal grounding, we must reserve a particular skepticism. Chalmers asserted that symbols qua representations can be grounded in distributed patterns of activation over units, and specifically, can be grounded in the rich, internal structure of such distributed patterns of activation. This is an altogether more suspect notion, and I will illustrate why via a number of grounding caveats.

First, the project of internal grounding does not seem to be much of a project at all, merely an observation, and an incomplete one at that. The Subsymbolic response seeks to ensure that representations have internal content merely by ensuring that representations have a certain degree of internal structure, which is to say, merely by ensuring that those representations are connectionist. Internal grounding, for the Subsymbolic response begins and ends with the statement that connectionist representations have a complex internal structure sufficient to carry internal content. As well as being an impoverished project, internal grounding is an ill-specified one also. Precisely how does a complex internal structure carry something as ephemeral as internal content, and just what is this internal content anyway? is the gist of this caveat. In the Radical Connectionist response, both of these concerns are addressed: Internal content is more accurately expressed as intensional meaning, and a network simulation is detailed in which the complex internal structure of a weight representation is evolved to carry this meaning.

Second, following on from the first caveat, the proposal of the Subsymbolic response to ground symbol-representations in distributed patterns of activation, rather than being beneficial, is in fact an unnecessary constraint on a connectionist response to the grounding problem because it precludes a role for weight representations. By concentrating his attentions on representations defined over unit activations, Chalmers has ignored the potential benefits of also considering representations defined over weights. Just as the structure of representations defined over unit activations can systematically represent features that it is intended to represent, so too, weight representations are similarly able to systematically represent those features they are intended to represent. They are able to do this, of course, because representations defined over units and weights both exhibit a non-concatenative, spatial structure. In addition, of course, a recourse to the notion of weight representations means that the response to the grounding problem can incorporate the benefits of context-independent constituents of representations.

Third, grounding representations in a spatial structure (whether exhibited by a representation defined over units or weights) is also an *undifferentiated* one. Specifically, it precludes the possibility that the grounding of intrinsic qua intensional meaning might be accomplished by employing something in addition to spatial structure. Although the syntactic structure of classical representations is not a viable candidate, of course, nonetheless, the analogue structure that we discussed in Chapter 7 might very well be. In fact, whereas Chalmers urged that an appropriate vehicle for intensional meaning is a spatial structure, the Radical Connectionist response admits the possibility of a role for analogue structure. It is to the specifics of this Radical Connectionist response to the representation grounding problem that we now turn.

THE RADICAL CONNECTIONIST RESPONSE

The Radical Connectionist response to the representation grounding problem postulates, in line with the subsymbolic response, two kinds of grounding: *Extensional* grounding, and *intensional* grounding. Although there are similarities in some respects to the projects of causal and internal grounding, there are significant differences due to the novel conception of the representation grounding problem itself to which the Radical Connectionist response is addressed. In addition, there are also elements of the Hybrid response common to the Radical Connectionist response, particularly, the notion that symbolic activity is grounded in robotic activity.

Extensional Grounding

Answering the question of how representations can be grounded, when asked about the notion of extensional meaning, requires a theorist to explicate the process of extensional grounding: that is, how a representation has meaning in relation to an external environment. In evolutionary terms, extensional grounding is a very old problem with an ancient solution: Arguably, there is no evolution of a protein mechanism beyond a certain point without the ability to represent the world, and hence the capacity to ground representation extensionally.

Consider a lowly unicellular organism, such as paramecium. This simple organism swims through its aquatic environment by the coordinated beating of its cilia. If it bumps into an obstruction, it reverses the direction of its ciliary beat, backs off, and swims away in a new direction. This kind of behavior might seem to indicate that paramecium has solved the problem of extensional grounding, that is, its avoidance behavior might seem to depend on an ability to make a decision based on a perceptually derived model of the world. This is not, however, the case. Rather, mechanical stimulation, such as bumping into an obstruction, causes the membrane of the organism to depolarize. The ensuing sequence of electro-chemical events leads the cilia of paramecium to reverse direction. When the membrane repolarizes a few moments later, the cilia resume their normal beating, and paramecium moves off (Quinn & Gould, 1979).

Paramecium is an example of what Johnson-Laird (1983) terms a *Cartesian automata*. Such automata have not solved the problem of extensional grounding: Rather, there is a direct, physically mediated causal link from stimulus to response. The responses of a Cartesian automata are not based upon an internal representation of the world, but upon a direct, causal interaction with it. In contrast to Cartesian automata, the natural world also contains a vast multitude of protein, self-replicating machines, that Johnson-Laird (1983) called *Craikian automata.* These kinds of automata have solved the problem of extensional grounding that the simpler Cartesian automata could not. With the evolution of nervous systems, organisms developed sensory transducers sensitive to stimuli from distant objects. These transducers convert impinging energy into electrical nerves impulses, and these impulses in turn are the input to a computational process that leads to construction of representations of the world. A Craikian automata uses the constructed model of the world to guide its behavior through its environment: It avoids some obstruction **A** in virtue of its internal representation **A***. The richer and more veridical the internal model, the better the chances of the organism finding its way safely through its environment.

The process of constructing internal models by a Craikian automata is the process of extensional grounding. It is an altogether more complex capacity than a Hybrid robotic capacity. The end result of the kind of sensorimotor function comprising a Hybrid robotic capacity is merely an inert taxonomy of categorical representations. For the reasons I have mentioned, a Hybrid robotic capacity is unable to generate the rich variety of behaviors exhibited by all those self-replicating machines with which our world abounds, and is consequently impoverished. Specifically, it incorporates no mechanism by which internal models of the world, both necessary and sufficient to guide the behavior of that machine safely through its environment, might be either constructed or manipulated.

In contrast to the Hybrid notion of robotic capacity, the Radical Connectionist response proposes the mechanism of extensional grounding: The end result of this construction process is an internal model of the world, what I have called a structural analogue representation, and what Johnson-Laird (1983) termed a mental model. As we saw in Chapter 7, the structure of the internal model, unlike a classical representation that is an arbitrary syntactic structure, is an analogue structure, playing a direct representational role. This discussion glosses over an important point that we have not yet discussed, and that is the relation of the collapse2 strategy to extensional grounding.

Collapse2 and Extensional Grounding One way to view the collapse2 strategy is as an abstraction of a latter stages of extensional grounding. Consider the kinds of raw input to collapse2, which take the form of triples of conceptual configurations such as **.bac**, **b.ac**, **ba.c**. What might such triples correspond to in extensional grounding? Perhaps it might be clearer if, instead of symbols, we used something a little more concrete. In such a case, the three conceptual triples above might look like the following:

.bac *banana*^{1} *apple*^{1} *cherry*^{1}
b.ac *banana*^{2} *apple*^{2} *cherry*^{2}
ba.c *banana*^{3} *apple*^{3} *cherry*^{3}

The first example, *banana*^{1} *apple*^{1} *cherry*^{1} might correspond to a visual scene where an unripe banana is on the left of a Granny Smith, which is on the left of a syrupy Morello cherry, whereas the the second example, *banana*^{2} *apple*^{2} *cherry*^{2} might correspond to the visual scene where a large plantain is on the left of a red Coxe's Pipin which is on the left of another syrupy Morello cherry. Such representations accord with the Hybrid notion of an iconic representation, inasmuch as that the representations serve to discriminate one input from another. As such, we might view the inputs to collapse2 as iconic representations. The first stage of collapse2, performing a mapping over these iconical representations delivers, over the hidden units, something very like the Hybrid categorical representations. That is, the network is being asked to extract those invariant features of the given triple **.bac**, **b.ac**, **ba.c** that distinguish it from any other given triple by creating representations over the hidden units that encode appropriate category relations.

Just like categorical representations, the hidden unit representations extracted from the learned first stage of collapse2 are in inert taxonomy, and without any additional machinery, do not constitute the basis for (anything remotely like) a plausible robotic capacity. However, the second stage of collapse2 comprises, arguably, the minimum of additional machinery serving to separate a Cartesian automata from a Craikian automata. That is, the machinery required to construct representational models of the world, whose structure is analogous to structures of states of affairs in the world. In essence, the second stage of collapse2 takes categorical representations as input to a mapping in which representations over the hidden units are developed, the spatial structure of which support structural relations analogous to those that hold between objects in the real world. The ability to construct such internal structural analogue representations is precisely what is missing from the Hybrid conception of robotic capacity.

The Role of Transduction The Hybrid response arrives at a single notion of meaning at the interface of two, symbolic and non-symbolic, types of representation: At this interface, meaning is seen to arise and is grafted onto the underside of a symbol system. The Radical Connectionist response, by contrast, arrives at two notions of meaning. The first, extensional meaning, is found at the far end of extensional grounding, able to deliver structural analogue representations of the environment external to that mechanism, and also able to manipulate those constructed representations on the basis of their analogue structure.

After having been quite scathing about the Hybrid response, I would now like to praise it, in a round-about fashion. There is one crucial factor that mitigates viewing the collapse2 strategy as a real contribution to an understanding of extensional grounding (or a extended robotic capacity, if you prefer). That factor is transduction, which is to say, the actual physical conversion of one form of energy to another. If there is no transduction then, strictly speaking, extensional grounding does not operate, because it has transduction at its core. It is certainly possible to simulate aspects of this process of extensional grounding but, to coin a Harnadian phrase, simulated transduction is not transduction in the same way that a simulated wave, for example, is not wet. For a Cartesian automata to be upgraded to (even the simplest kind of a) Craikian automata, it *must* be the case that it actually has physical components that are able to transduct impinging energies, whether that energy is in the form of EM radiation, pressure waves or whatever. This is the real contribution of Harnad's (1990b) Hybrid approach: namely, the realization that simulated transduction *is not enough*. In the next chapter of the book, I consider the process of extensional grounding in more detail, and from a slightly different perspective. For now, however, I have other fish to fry, namely, the upgrading of a Craikian automata to what I have called (in deference to the genius of the man) a *Fodorian* automata.

Intensional Grounding

In addition to the Cartesian and Craikian automata we have already discussed, we can identify a third type, which I have termed a Fodorian automata. Fodorian automata are able, like Craikian automata, to construct internal models of the world via sensory transduction of energies impinging on the sensory surface and some mechanism of extensional grounding. The representations it constructs, consequently, can properly be thought of as carrying extensional meaning—which is to say, meaning that the representations have in relation to properties external to the automata. Fodorian automata differ crucially from Craikian automata, however, in being able to construct models of the external environment in the *absence* of information impinging on sensory surfaces. That is, they do not *require* the mediation of sensory transducers, a sensori-motor system, indeed the process of extensional grounding itself, in order to construct internal models of the world. Moreover, they are able to interrogate and manipulate the models that, independent of sensory input, they have constructed. The capacity that serves to distinguish a Fodorian automata from a merely Craikian automata, and the one that allows them to construct internal models of the world without recourse to a sensory input, is, of course, a *linguistic* capacity.

Specifically, as well as solving the problem of extensional grounding, Fodorian automata have also solved the problem of *intensional* grounding. Thus, "How can representations be intensionally grounded?" requires the theorist to explicate the process whereby a representation has meaning in virtue of relations to internal models of the external environment, constructed by a prior extensional grounding. The process of intensional grounding in a Fodorian automata ensures that representations have intensional meaning: Which is to say, ensures that they carry a meaning defined internal to the automata. Chalmers urged that representations be grounded internally in distributed patterns of activation. The Radical Connectionist response, however, views things slightly differently, due to the fact that it has at its disposal a greater number of *representational genera* (cf. Haugeland, 1991), specifically, both *unit* representations and *weight* representations. Intensional meaning, defined internally to the machine with respect to the extensional meaning encoded by a structural analogue representation constructed by a prior extensional grounding, is carried by the complex internal structure of microprocedural (i.e., weight) representations.

If one were to seek a unifying idea behind these three responses to the representation grounding problem—the Hybrid response, the Subsymbolic response, and the Radical connectionist response—he or she might very well assume that crucial to *any* account of grounding is a *non-classical* structure. Chalmers advocated a spatial structure, and so does Harnad (after a fashion). The Radical connectionist response, however, is motivated by the spirit of a healthy ecumenism, and, supported by a tradition of ideas extending back to Craik (1943), and drawing on proposals from philosophical, formal and psychological semantics, advocates an analogue structure. That is, analogue structure plays a part in the Radical Connectionist response because the intensional meaning carried by weight representations is defined with respect to the analogue structure of the representations constructed by a prior process of extensional grounding.

There are a number of factors that mitigate viewing the Radical Connectionist response in absolute terms. The first is the fact that analogue structure is often distinguished by degrees. For example, we saw in Chapter 7 that Sloman's notion of analogical representation required a far weaker concept of analogue structure than Johnson-Laird's notion of a mental model. The second point is related, and concerns the fact that the differences between spatial and analogue structure are not at all clear, even now. Representations exhibiting both spatial and analogue structural similarities combine non-concatenatively, and both play a direct representational role in that they both are able to systematically represent semantic features of whatever it is they are a representation of. As was conjectured in Chapter 7, analogue and spatial structure may both be subsets of a more general class of non-syntactic structure.

However, the Radical connectionist response, and the two notions of grounding it postulates, can be stated in slogan form as follows:

Extensional Grounding—Concerned with how the world is related to the computational representation of that world via transducers.

Intensional Grounding—Concerned with how the computational representation of the world is related to the computational representation of language.

A MODEL OF INTENSIONAL GROUNDING

Intensional grounding is concerned with how the computational representation of an external environment is related to the computational representation of language: Or put another way, it is concerned with how intensional meaning is related to extensional meaning. We have spent some time considering the role that collapse2 plays in extensional grounding, and we will take the SSAs that it constructs as models of the external environment. Thus, we have one of the two requirements for a model of intensional grounding. The second requirement, the computational representation of language and the notion of a linguistic capacity, both referring to the philosophical notion of a *propositional* representation, we will take up here.

In the computational literature, there is often a confusion made between descriptions of things, and things described: Language itself is a prime example of such a thing. As van Gelder (1990) has noted, the use of concatenative compositionality, and the syntactic structure resulting from such composition, is pervasive not to say endemic, in many fields of inquiry: Certainly, all modern linguistic descriptions of language are couched in classical terms. A classical description of language does not mean, however, that the thing described—language—is a classically conceived system of representation, making use of concatenative compositionality and syntactic structure. It is equally as plausible that language is a representational system that comprises tokens combining non-concatenatively and exhibiting a spatial structure.

Such considerations underlie the thinking of the connectionist theorist. Whereas descriptions of language may well be labeled classical (with all the theoretical accoutrements that such a label implies) nonetheless, language described—which is to say, language as it exists in the head of an individual—may not be classical. In line with this kind of reasoning, the second requirement for a model of intensional grounding is a non-classical encoding of selected elements of a representational system that is widely described in classical terms: What we require, in short, are non-symbolic symbol strings.

Recursive Connectionist Data Structures

Consider, if you will, two examples of our familiar spatial assertions:

- **B is on the left of C**
- **A is in front of B**

What is required for the model of intensional grounding with which we are concerned are non-concatenative representations of such strings that are nevertheless recursively constructed so as to preserve the symbolic information in the description. The Recursive Auto Associative Memory (RAAM) of Pollack (1990) lends itself very naturally to such an endeavor. Specifically, we will make use of a sequential RAAM network, the architecture of which is shown in Figure 8.1.

The single assertion **B is on the left of C** would be presented to the RAAM network as a bracketed tree structure such as **C(L(B nil))**: Figure 8.2 illustrates the formation of the final complex representation from its constituent elements.

Let's take a single input. On the first processing step, the symbol **B** is input to the network as a terminal symbol over the **branch 1** units, along with the **nil** symbol over the **branch 2** units (in this case, **nil** was represented by a uniform value of 0.5 over all units). The auto-associative nature of the RAAM network requires that **B** and **nil** be reproduced on the **branch 1*** and **branch 2*** output units respectively. In performing this mapping, the network develops a compressed hidden unit representation of the combination of **B** and **nil**, which can be labeled $R_{(B,nil)}$ (R1 in Figure 8.2).

On the second processing step, $R_{(B,nil)}$ has been passed down to the **branch 2**. The **L** symbol is now presented as input, along with $R_{(B,nil)}$, and the RAAM network is required to produce **L** and $R_{(B,nil)}$ on the **branch 1*** and **branch 2*** output units respectively. Once more, the hidden units develop a compressed representation, which can be labeled $R_{L(B,nil)}$ in order to perform this mapping (labeled R2 in Figure 8.2). On the third processing step, the input

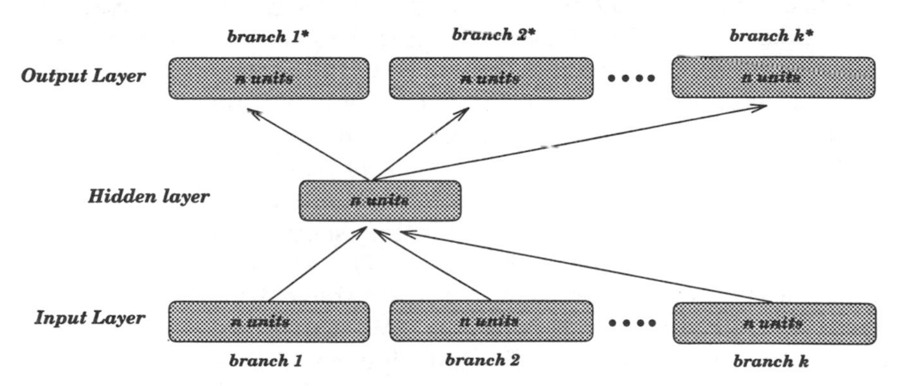

FIGURE 8.1. The architecture of a sequential RAAM network used to construct inputs for the model of intensional grounding.

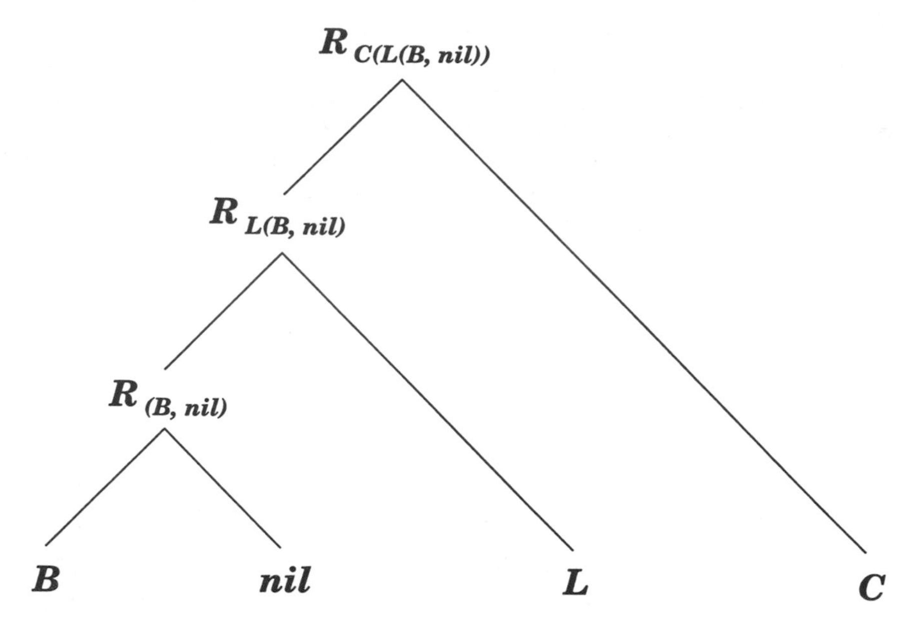

FIGURE 8.2. The sequential construction of a recursive connectionist data structure representing the assertion "B is on the left of C."

comprises **C** as a terminal symbol over the **branch 1** and $R_{L(B,nil)}$ over the **branch 2** units. Performing auto-association causes the hidden units of the network to develop the final compressed representation, $R_{C(L(B,nil))}$ (labeled R3 in Figure 8.2), which represent the entire assertion.

For sequences of assertions, the process is exactly the same. The sequence **B is on the left of C**, **A is in front of B** would be presented to the RAAM network as a bracketed tree structure, such as **B(F(A(C(L(B,nil)))))**. Figure 8.3 shows the construction of the final complex representation from its constituents.

Using the three symbols **A**, **B**, **C**, and the four spatial terms **left**, **right**, **front** and **behind** means that it is possible to construct 24 bracketed tree structures representing single assertions, and 384 bracketed tree structures representing sequences of assertions, giving a total of 408 possible inputs. The architecture of the sequential RAAM network used to encode these 408 inputs was the same as that shown in Figure 8.1. This sequential RAAM (with 7 **branch 1**, and 25 **branch 2**, units (i.e., a (7–25)–25–(7–25) network) learned to tolerance after some 11,375 cycles) with values of learning rate and momentum of 0.075 and 0.7 respectively. For the 24 single assertions, 24 hidden unit representations corre-

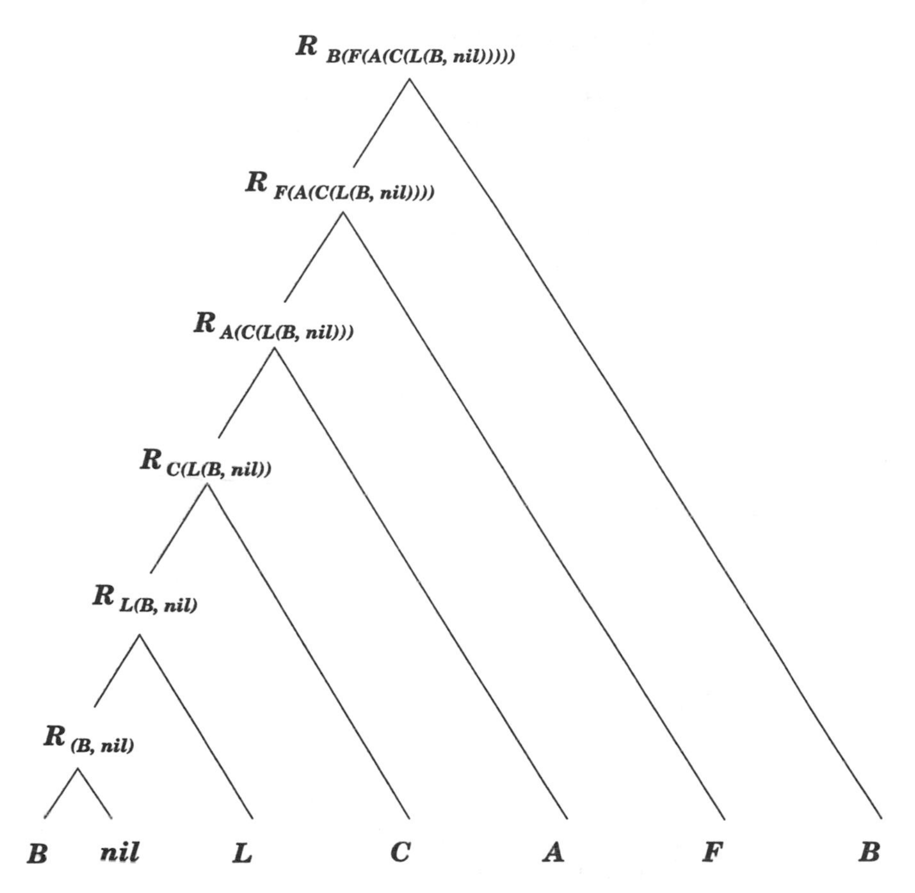

FIGURE 8.3. The sequential construction of a recursive connectionist data structure representing the sequence of assertions, **B is on the left of C, A is on the left of B.**

sponding to R3 were extracted, and for the 384 sequences of assertions, 384 hidden unit representations corresponding to R6 were also extracted.

Using the sequential RAAM network has allowed symbolic (linguistic) strings to be encoded in a manner that preserves the symbolic nature of the information that they carry, but that are themselves not symbolic representations (where symbolic means concatenatively combined, syntactically structured representations). Quite how to characterize such representations is unclear; one possibility might be to view them as *virtual* symbolic representations, where virtual syntactic structural relations are supported by functional level spatial similarity relations.

Intensional Grounding Simulations

This series of simulations was trained with a set of inputs comprising RAAM encoded spatial assertions, and a set of outputs comprising the SSAs extracted from the hidden units of the second stage of collapse2. The total corpus of possible inputs to a given grounding simulation, coded over 25 units, comprise some 24 RAAM encodings of single assertions, and some 384 RAAM encodings of sequences of assertions, making a total of 408 possible inputs to the grounding simulations.

The total corpus of possible outputs of the grounding simulations, coded over 18 units, comprise the 12 SSAs of one-place relations, and the 36 SSAs of two-place relations, extracted from the hidden units of the second stage of collapse2. Each SSA encoding a 1-place relation is described by two different RAAM encodings, and thus can be constructed by a given grounding simulation on the basis of those two inputs.

The SSA encoding two-place relations are also described by a number of different RAAM inputs, a different number for each different class of structural analogue, and hence can also be constructed on the basis of those different inputs. Those classes can be characterized using the vocabulary introduced in Chapter 7. Consider the output array employed in the second SRN simulation in Chapter 7, shown in Figure 7.3. This was described using the notion of six *conceptual configurations* derivable using the four spatial operators **left**, **front**, **right** and **behind**, and the three symbols **A**, **B** and **C**. Referring back to that description, for conceptual configuration 1, for example, the grounding simulations are required to construct the same SSA from some 16 different RAAM encodings. This is shown diagrammatically in Figure 8.4(a). For conceptual configurations three through six, the grounding simulations are required to construct the same SSA, as shown in Figure 8.4(b), on the basis of some eight RAAM encoded sequences of assertions.

Early exploratory grounding simulations discovered that mapping RAAM encoded inputs to their corresponding structural analogues was not a task that a network could accomplish quickly: Using an entire training set, comprising some 356 of the possible corpus of 408 inputs, exploratory simulation training times indicated that training to tolerance would require in excess of several tens of millions of cycles. In order to combat this problem of high computational expense, a number of smaller simulations were conducted, each of which had a training set constructed using

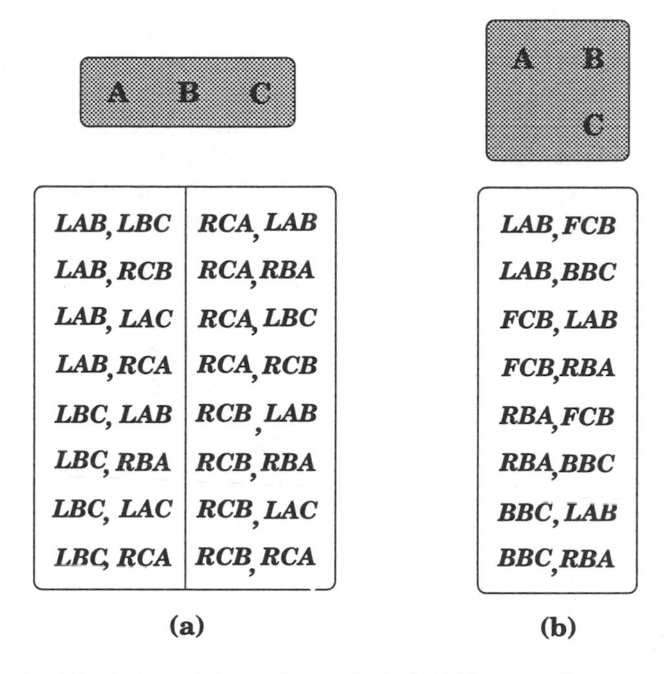

FIGURE 8.4. Illustration of two SSAs and the RAAM encoded sequences associated with them.

only those SSAs belonging to one conceptual configuration (of two-place relations) and, separately, those SSAs of one-place relations.

Testing Generalization The procedure for testing the generalization performance of the various grounding simulations was not a straightforward one. Because the set of both input and output vectors were continuously valued, the outputs produced by a given network were also similarly continuously valued. This meant that merely comparing the outputs of the network produced via testing with a target would first be very laborious, but second, and more importantly, quite possibly misleading. Indeed, this was found to be the case: Particular units in a given output vector were found to differ from their target quite substantially when the activation value of each constituent unit in a given vector was considered individually.

In order to avoid this difficulty, the continuously valued output vectors produced by a given test set of RAAM inputs was first decoded before a measure of generalization was established. The decoding was accomplished by employing the hidden-to-output weights of the the final collapse2 simulation, **c2.5**. In this way, the set of output vectors extracted from a given grounding simulation, coded over 18 units, would be decoded via the **c2.5** hidden to output weights, to emerge on a symbol surface, coded over 12 units. I could then use a standard measure of generalization, in this case, whether the activation value of a given unit was greater than 0.45. This procedure was performed to test the generalization of all of the grounding simulations.

Intensional Grounding Simulation 1 IG Simulation 1.1 comprised 84 RAAM encoded inputs, mapping to six SSAs, all describing two-place relations, and all belonging to conceptual configuration 1. The simulation was required to construct each SSA on the basis of 14 different RAAM encoded input sequences: Twelve inputs in all were reserved to test for network generalization. A 25–24–18 configuration learned to a tolerance of 0.05 after 200,875 cycles, with values of momentum and learning rate at 0.5 and 0.05 respectively. Using the decoding scheme explained above, this network exhibited some 64% generalization to a test set of 12 RAAM encoded assertions.

IG Simulation **1.2** and **1.3** each had a training set comprising 84 RAAM encoded inputs, mapping to 12 SSAs belonging to, for IG Simulation **1.2**, conceptual configurations 3 and 4, and for IG Simulation **1.3**, conceptual configurations 5 and 6. These two network simulations were required to construct each SSA on the basis of 7 RAAM encoded input sequences, with 12 inputs reserved to test for network generalization. For IG Simulation 1.2, a 25–24–18 configuration learned to tolerance of 0.05 on some 79 items of the training set of 84 inputs after 415,527 cycles, with values of momentum and learning rate of 0.5 and 0.05 respectively. Generalization of IG Simulation 1.2 to its test set, after decoding, was 58%. RGP Simulation 1.3, using the same configuration and same network parameters, learned to a tolerance of 0.075 on the entire training set after 967,175 cycles. This simulation exhibited only 32% generalization to a test set of 12 novel RAAM encoded sequences.

These measures of generalization, particularly the measure for IG Simulation **1.3**, are quite disappointing. However, early exploratory simulations revealed that the generalization performance of a given grounding simulation was greatly improved if, along with the RAAM encodings of sequences of assertions, the training set also contained a representative sample of the RAAM encodings of single assertions. Accordingly, the next series of simulations were trained on a data set composed of both RAAM encoded sequences and also RAAM encoded single assertions. The task that the network must perform is, one might think, only made harder by requiring it to construct SSAs of both one- and two-place relations, but this is not the case.

Intensional Grounding Simulation 2 IG Simulation **2.1** comprised some 94 RAAM encoded sequences, mapping to 12 SSAs. Six of the SSAs correspond to the two-place relations of conceptual configuration 1, whereas the remainder correspond to the structural analogues representing the six one-place relations, which are the most primitive states of affairs making up conceptual configuration 1. Fourteen inputs were reserved to test for network generalization, 12 testing two-place relations, and 2 testing one-place relations.

The same 25–24–18 configuration learned to a tolerance of 0.075 after 456,786 cycles, with values of learning rate and momentum of 0.05 and 0.5 respectively. Using the decoding scheme described previously, and testing the generalization performance of IG Simulation **2.1** with a test set comprising RAAM sequences of both one- and two-place relations, yielded a measure of generalization of some 81%, a pleasing improvement on IG Simulation **1.1**.

IG Simulation **2.2** comprised some 104 RAAM encoded input sequences, mapping to 24 SSAs: 12 belonging to conceptual configurations 3 and 4, describing two-place relations, and 12 describing one-place relations. Sixteen inputs were reserved to test for network generalization, 12 describing two-place relations and 4 describing one-place relations. The same 25–24–18 configuration learned some 101 inputs to a tolerance of 0.075 after 1,281,974 cycles. After decoding, IG Simulation **2.2** exhibited 68% generalization to its test set.

The generalization performance of these series 2 grounding simulations to a novel test set, particularly IG Simulation **2.1** at 81%, is an improvement over the series 1 simulations, but obviously the measures could still be improved on. In light of this, it is worth considering that, in terms of generalization efficacy in connectionist research, unaugmented back propagation algorithm over a standard, feed-forward architecture is known to produce notoriously variable figures of generalization on a given task. Indeed, a number of authors have devised augmented algorithms that result in considerably better performance. For example, the RuleNet architecture (McMillan, Mozer & Smolensky, 1991) has been reported to increase generalization efficacy in the range of some 300% to 2,000% over standard back propagation. Although no augmented architectures or algorithms were used in my investigations, obviously their use would be beneficial if higher degrees of generalization were deemed necessary in order to support stated theoretical positions.

Structure Sensitive Operations The simulations reported in this chapter can be seen to be performing an interesting variant of direct, structure-sensitive manipulation, what Chrisman (1991) termed *holistic computation*. The training set input to the grounding simulations comprises recursively constructed unit representations derived from the operation of a sequential RAAM network: These representations have a non-concatenatively compositional structure, resulting from a non-concatenative composition of constituents. However, they *function* as linguistic strings by implicitly preserving symbolic information about the relative orderings of tokens. The structure of the input RAAM representations can be described, as I have suggested, as a virtual syntactic structure.

The training set comprising the outputs for the grounding simulations, on the other hand, are derived from the collapse2 strategy: These representations also exhibit a non-concatenative compositionality, but they differ from the input RAAM representations by functioning as structural analogues of states of affairs in the world. The structure of these superpositional structural analogues can be described, as I have suggested, as a virtual analogue structure. The manner of representation manipulation that the grounding simulation performs consequently comprises a mapping from a virtual syntactic structure to a virtual analogue structure, where both classes of virtual representation are supported by similarity relations defined in terms of location in a high dimensional space, which is to say, supported by *spatial* structure similarities.

Intensional Grounding and Connectionist PS How are representations connected to #represented#? How is intensional meaning related to extensional meaning? Both of these two questions, I believe, are statements of the same thing and both, I believe, are part of the *representation grounding problem*. Characterized in such terms, the representation grounding problem, and the Radical Connectionist response, are a departure from the muse of either Searle or Harnad, concerned as they are with the lack of meaning of classical symbols. As I have explained the problem of reconstructing the relations between representations and represented is a more expansive problem, a concentration on grounding classical symbols blinds the theorist to the fact that *all* classes of complex structured representations require grounding. In order to cope with the species of structure and representation emerging from the connectionist school, requires that the need to ground representations>, and not just symbols, be recognized.

The four responses that I have discussed in this chapter—the Classical response, the Computational Neuroethological response—the Hybrid response and the Subsymbolic response, are all also inexplicit about what kind of meaning it is that a putative grounding mechanism imbues a representation with. Chalmers, for example, although recognizing the need to embrace the more expansive representation grounding problem, did not go further to consider what type of meaning it is that the complex internal structure of connectionist representations are capable of carrying. The Radical connectionist response, however, does take this further step by advocating that, in the context of how to connect representations to #represented#, it is intensional meaning that is carried by the complex internal structure of weight representations.

The simulations reported in this chapter show how intensional meaning, carried by weight representations, can be systematically related to extensional meaning, carried by the analogue structure of SSAs. The relation is realized in a computational model by invoking the notion of microprocedural weight representations. Such representations have a significant degree of internal structure, as I explored in Chapter 6 (though, of course, that structure is not concatenative in origin). They thus have the property that is sufficient to carry (some vague notion of) internal content. However, these microprocedural representations are not characterized as carrying internal content, but rather, they are characterized as carrying intensional meaning. The relation between the intensional meaning carried by microprocedural representations and extensional meaning is established precisely because, invoking some of the specialized vocabulary I discussed earlier, the engaging of a microprocedural representation systematically results in the construction of a SSA that carries a corresponding extensional meaning.

The simulations in this chapter are concerned, as I have said, with intensional grounding, with how representations (of language) are connected with #represented# (of the world). This, rather than extensional grounding, I should like to argue, is the natural purview of a formally grounded connectionist PS. However, the simulations in this chapter are more complex than those reported in Chapter 6, specifically because both inputs (RAAM) and outputs (SSA) were constructed from prior distributing transformations. This means that both inputs and outputs are "below a symbol surface." That is, the content of the representations is not available to casual inspection, as would be the case with classical molecular symbols constructed concatenatively. Thus, intensional grounding, as presented here, requires that the connectionist PS that constructs SSAs be, itself, below a symbol surface also. What this means is that, although such a connectionist PS can support a systematic semantic interpretation via the simple expedient of "decoding" the non-concatenative inputs and outputs onto a symbol surface, using the weights that effected the original distributing transformation in the first place, its natural mode of operation means that it is not itself semantically interpretable on such a surface.

A connectionist PS, properly concerned with intensional grounding, is not a basis for a semantic theory because it expressly claims not to be an account of how language strings relate to the world. Connectionist PS is actually a misnomer in such a case, because PS should not be concerned with semantics per se at all. The relation that intensional grounding is concerned with is how language strings relate to internal models of the world. Or to couch the point differently, intensional grounding is concerned with the "hook up" between representations and #worlds#: the complementary process of extensional grounding is, on the other hand, concerned with the "hook up" between the world and #worlds# (and we will see in Chapter 9 just how this distinction fleshes out). This should be contrasted with a semantic theory, on the other hand, concerned with the "hook up" between representations and the world. It is on the basis of the distinction between intensional and extensional grounding that I would urge a view of PS, away from an emphasis on a monolithic semantic theory and in line with the kind of two-factor proposals from philosophy (cf. Block, 1986), as a theory of *meaning*, and not of semantics.

CONCLUSIONS

In this chapter, we have considered in detail the problem of grounding. We have discussed several seminal formulations of the problem and a number of refinements and variations, and included the observation that in order to encompass the range of compositional representations emerging from the connectionist school, the grounding problem should more accurately be redubbed the representation grounding problem. A number of candidate solutions to the representation grounding problem, including the Hybrid response, the Computational Neuroethological response and the Subsymbolic response were sketched: All were found wanting for one reason or another, and an alternative, the Radical Connectionist response was proposed.

In this response, representation grounding is re-construed in terms of how a computational mechanism represents language and how that mechanism represents an environment external to it. That is, a response to the representation grounding problem requires a theorist to consider two complementary processes: intensional grounding and extensional grounding. The Radical Connectionist response argues that intensional grounding is the natural domain of a connectionist PS and is concerned with how representations of language are connected with representations of internal models of the world. A number of simulations were conducted to explore this argument, whereby a network learned to map the unit representations extracted from a sequential RAAM, to analogical unit representations, called SSAs, extracted from the collapse2 strategy. Some of the most novel features of these simulations was noted, particularly the structure-sensitive, holistic style of computation that they exhibit, and the fact that the connectionist PS computing such intensional relations dips below a symbol surface on which interpretation can be made.

In the next chapter, we will consider more fully the counterpart to intensional grounding, namely, extensional grounding. Given the central role of transduction in this process, it might appear that the theorist would be ill-advised to venture into such territory (more the province of engineers one might think) who have the task of actually designing and building transducers. The advice is well-taken, but as we will see, extensional grounding—even though, as it must in all cases, crucially involve transduction—still means rather different things, depending on which kind of computational engine the theorist is interested in extensionally grounding.

9

Fodorian Automata and Computational Engines

The *representation* grounding problem was a term introduced in the last chapter, to recall, in order to refer to the need to (extensionally) ground systems of compositional representations, in general, rather than specifically classical symbols. Approaching the problem from a novel perspective in this chapter, we will consider one of the most sophisticated theories of mind in cognitive science, inspired by the Language of Thought hypothesis (cf. Fodor, 1976), called the *representational theory of mind* (or RTM, see Fodor, 1987; 1990). What we are specifically concerned with is the role that extensional grounding might play in the RTM. Does it, in fact, have any role to play at all?

We saw in Chapter 8 what happens when a Craikian automata acquires the ability to construct internal models of the world on the basis of something other than information from a sensory surface. What happens, in short, is that it becomes upgraded to a Fodorian automata, and it is precisely this latter kind of machine that the RTM is in the business of explaining. Stopping to think about it for a moment, and donning an evolutionary hat for a moment, one can easily see that Fodorian automata are a relatively recent newcomer to the world, having been preceeded by many very different, sophisticated kinds of Craikian automata for several millions of years, at least. The point here is that the capacity for extensional grounding, the construction of #worlds# from signatures, came first, and the capacity for intensional grounding, the construction of #worlds# from language representations, is a much more recent add-on. Such an evolutionary perspective on the relationship between Craikian and Fodorian automata, emphasizing the primacy of extensional grounding is, of course, diametrically opposed to, for example, the classical response in

Chapter 8, where it is extensional grounding that is considered peripheral to the (much more powerful) symbol system itself.

The distinctions introduced in the last chapter between Cartesian and Craikian automata, and between Craikian and Fodorian automata were drawn quite sharply. It is an obvious fact, however, that Craikian automata constitute a very large genus, with a huge variation considered with respect to evolutionary complexity: Compare quite simple kinds of Craikian automata, such as insects or arthropods, with more sophisticated automata, such as primates or cetaceans, for example. The evolution of Craikian automata—which is to say the evolution of complex nervous systems from simpler ones, as Churchland and Churchland (1983) have argued—can form the basis for a *naturalistic* approach to cognitive science. In place of Craikian automata, Churchland and Churchland (1983) used the term (wild) *epistemic engine*, but the thrust of the argument is the same: Epistemic engines, like Craikian automata, exploit a flow of environmental energy, and the information it contains, as a way to produce more information and guide behavior, (i.e., using it as a basis for the construction of internal models).

In the classical RTM, the notion of an epistemic engine is construed in a particular way, becoming the purely formal *syntactic engine*. Precisely what distinguishes Craikian automata from Fodorian automata, in the framework of the RTM, is the fact that in the latter case, epistemic engines are syntactic engines, whereas in the former case, they are not. The syntactic engine thus serves in the RTM as a *model of mind* and explanations of mental phenomena are couched in terms of properties of it, which is to say, in terms of classical symbols, concatenation, syntactic structure, and so on. In slogan form, and to reiterate the point, we might say that the RTM considers Fodorian automata to be syntactic engines.

It is this view of Fodorian automata as syntactic engines, which we will be concerned with in this chapter. We will not so much be concerned with the specifics of the view, but rather with the possibility of an alternative view, which considers Fodorian automata, not as syntactic engines, but as a completely novel kind of epistemic engine—what Jackson and Sharkey (1994) called a *spatial* engine. Given our earlier, extensive discussions of compositional syntactic and spatial structures, in Chapter 7 particularly, I hope the label is self-explanatory. What will emerge from these discussions is a view of the spatial engine as actually a much more *natural* device than the syntactic engine, considerably more at home in the world of experience than its stimulus-shy syntactic relative.

THE SYNTACTIC ENGINE

The evolution of Craikian automata, which is to say, of more complex nervous systems from more simple nervous systems, is mirrored in the academic sciences by the displacement of primitive theories (and their ontologies) by more powerful and encompassing theories (and their ontologies). The evolution of the RTM is a case in point, where many of the central notions of the theory are, for example, contained in embryonic form in the prior work on, most notably, meaning postulates (cf. Fodor et al., 1975) and the Language of Thought hypothesis, as we saw in Chapter 5. But just what is the RTM? In the next section, the bare bones of this powerful theory are examined. The review is not intended to be exhaustive, and I would direct the reader to Fodor (1987) and Loewer and Rey (1991) for more extensive discussion.

The Representational Theory of Mind

In its most recent incarnation, the Language of Thought hypothesis (cf. Fodor, 1976) has spawned the representational theory of mind (RTM) and the dutifully attendant, syntactic engine. Whereas the former is a theoretical framework, the latter is a computational model of the causal sequences of symbolic transformations that mental processes are assumed to correspond to in the framework of the RTM. Examining the operational principles of the syntactic engine will, or so everyone hopes, elucidate the corresponding principles that govern the vagaries of that other elusive mechanism that everyone is *really* interested in—the mind.

We will concentrate on three features of the RTM, distinguished by Clark (1992). First, the explanation of *propositional attitudes*, corresponding to mental states that people identify using such words as *hope*, *believe*, *desire* and so on, as computational relations to classical symbols. Second, the existence of a system of such classical symbols, constitutive of the syntactic engine itself and third, the representing of the *content* (i.e., meaning) of states of the syntactic engine by explicit tokening of syntactic classical symbols. We will discuss each of these distinguishing features in turn.

Propositional Attitudes The first distinguishing feature of the RTM is that propositional attitude (viz. hopes, desires, beliefs and so on) is taken to be computational relations to symbols in the internal computing language of the syntactic engine. Fodor (1987, p. 17) introduced the notion of an *attitude box* in order to express this claim. Thus, for the syntactic engine to be in a particular propositional attitude state (i.e., for it to have, for example, the belief that P (where P might be a proposition such as "Ultimate is good exercise") is for it to have a representation of the content of the proposition P tokened in a certain way (in a certain *functional role*, as the phrase has it) appropriate to that attitude.

Thus, for the syntactic engine to believe that P (i.e., for it to believe that Ultimate is good exercise) involves it tokening, in its belief box, the symbol that means that P (i.e. the symbol that means Ultimate is good exercise). The same symbol, tokened in a different functional role (i.e., in a different attitude box), might thus cause the engine to be a propositional attitude state appropriate to the *hope* that P, or the *desire* that P, and so on.

The RTM does not expect its model of the mind, the postulated syntactic engine, to bear the load of the infinitude of the possible propositional attitude (and other) states, that it could potentially be in—unaided, however. Such a state of affairs would imply that the syntactic engine is able to deploy a different symbol for each and every proposition that could (potentially) occupy an attitude box, running the risk of a combinatorial explosion of symbols. Thus, we have the second distinguishing feature of the RTM: that its model of the mind, the syntactic engine, is a *symbol system*.

Symbol Systems In order that the syntactic engine constitute a symbol system, it has an internal computing language that permits an unbounded recombination of atomic (constituent) symbols to form molecular (compound) symbols, in accordance with very specific syntactic rules of combination. The paradigmatic description of this kind of symbol system is, of course, found in Fodor and Pylyshyn (1988). In such a system, molecular symbols are concatenative strings of atomic (constituent) symbols, such that the meaning of the molecular compound is a direct function of the meaning of (a) its atomic (constituent) parts taken together with (b) the syntactic rules of combination (of those atomic parts forming the molecular compound).

The most important consequence of this (i.e., of the syntactic engine being in possession of arbitrarily recombinable, content-bearing atomic symbols) is that the *systematicity* (cf. Fodor & Pylyshyn, 1988) so characteristic of mental phenomena falls out as a nomological law of the operation of the engine (we discussed the notion of systematicity in Chapter 6 and 8, to recall). In other words, given that the syntactic engine can token (i.e. "think") **AB**, it is a nomological necessity that it *also* be able to token (i.e., "think") **BA**. This systematicity has the obvious correlate in human mental life that people who can think (i.e., "token") the thought that Yoni loves Vagra can *also* think (i.e., "token") the thought that Vagra loves Yoni.

This fairly clear exposition of systematicity is (relatively) uncontentious. For connectionists, however, as we have seen, it is not the phenomena itself that is the important issue, but rather, the nature of its explanation in the RTM, which relies on two the classical notions of concatenative compositionality, and the resulting structure imparted to molecular symbols when combined using this mode of composition, termed syntactic structure.

Causal Processes The third of the distinguishing features of the RTM with which we are concerned is that the processes operating in the syntactic engine on elements from its system of symbols are *causal* processes, involving the explicit tokening of those symbols.

RTM is supposed to be a vindication of (so-called) commonsense *folk psychology* (cf. Stitch, 1983). In folk psychological terms, a mental state corresponding to an attitude (the belief that P, for example) has two effects. The first consists in the attitude bringing about an action, and the second in the attitude bringing about some further mental state (the hope that P, for example). The relevant question is, of course, *how* this "bringing about" occurs, and the RTM's answer to it is embodied in the slogan *No Intentional Causation without Explicit Representation* (Fodor, 1987, p. 25).

This slogan can be interpreted to mean that a particular propositional attitude (the belief that P, to use the same example) has causal efficacy in the syntactic engine only when there occurs a token of the syntactic kind that means that P (and additionally, when that token—somewhat tautologically—*causes* either an appropriate action or some further contentful state, Q). The one further subtlety to consider is that, for intentional causation to work in the syntactic engine, it is only necessary for the *contents* of propositional attitudes (i.e., the hope that P, for example) to be explicitly tokened (as a syntactic molecular symbol). The *processes* that effect the transformation of one attitude to another (and equally, the processes that effect the transformation of an attitude into an action) need not be explicitly tokened. As Fodor himself remarked: "According to RTM, programs—corresponding to the 'rules of thought'—*may* be explicitly represented; but 'data structures'—corresponding to the contents of thoughts—*have to be*" (Fodor, 1987, p. 25).

The Mind Market

The three features of the RTM outlined above—the treatment of propositional attitudes, the description of the syntactic engine as constitutive of a system of syntactic symbols, and the explicit tokening of thought contents in that engine—combine so as to reject utterly and without reservation, any conception of the mind as a *semantic* engine, of which there have been advocates in the literature (cf. Dennett, 1981; Haugeland, 1981). However, as was mooted at the beginning of the chapter, a possible taxonomy of epistemic engines does not stop at a black and white choice of either syntactic or semantic because of the possibility, raised by a consideration of connectionist computation, of a third, novel form of structural engine, neither syntactic nor semantic, but rather spatial.

That is for later, however. Our more pressing concern, to recall the beginning of the chapter, is with how well the RTM generally, and the syntactic engine more specifically, copes with the representation grounding problem. Or to put the point another way: How easy is it to upgrade a Craikian automata to a Fodorian automata, when the latter is construed as a syntactic engine? To answer this question, we will make use of a convenient metaphor, a market of sorts, in which philosopher-vendors try to convince skeptical consumers of the relative benefits to be had in building theories of mind around different kinds of epistemic engine. We will call this the *Mind Market*.

First stop is the stall of the classical theorist, the touted ware a bright, shiny new syntactic engine. Skeptical shoppers learn from the glossy advertising literature that this is a theoretical mechanism, which performs symbol manipulation. The manipulation that the syntactic engine performs, called computation, although able to support a systematic semantic interpretation is, in fact, a purely structural operation, with nothing to do with meaning at all, except in the sense in which the manipulation can systematically support a meaningful *interpretation*. Molecular representations in the engine have a particular, special kind of structure, resulting from the distinctive manner of concatenative composition used to construct them from their atomic constituents. This resultant syntactic structure is then very useful, potential buyers are informed, because it allows the syntactic engine to manipulate these molecular representations by being sensitive to that structure (cf. Fodor & Pylyshyn, 1988).

The representational theory of mind built with the syntactic engine as its model, the classical stall-holder goes on to explain, assumes that the world, as it is, is mapped one-for-one onto symbols in the syntactic engine. The skeptical consumer is urged to consider the case of a language stream impinging on the engine: If the sentence "Ultimate sure is a fun sport" were to be presented to the engine, then the individual words of this sentence would be mapped onto the internal, atomic symbols A, FUN, IS, SPORT, SURE and ULTIMATE. Once the mapping of utterances of a public language onto atomic symbols in the syntactic engine has occurred, numerous (in fact, unbounded) subsequent, systematic manipulations of the resultant molecular symbol (constructed from its atomic constituents) are possible. Moreover, all manipulations of representations within the syntactic engine respect semantic criteria of coherency (criteria devised outside the engine, of course).

"All you need, ladies and gentlemen" cries the enthusiastic stall-holder, trying to drum up more business, "is a semantic theory! Just hook the syntactic engine up to the world, and off it goes!"

There are murmurs of appreciation from the crowd of shoppers, but a few dissenting grumbles also. One small voice in particular interrupts the classical spiel, politely asking where such a hook is to be purchased from? The classical stall-holder pretends not to hear the question, and waves his arms, choosing not to answer, instead continuing to extol the power and elegance of the syntactic engine, disparaging the competing kinds of semantic engine and associative engine on sale on other stalls nearby. His potential customers begin to drift away, disenchanted.

What the syntactic engine lacks—and requires, of course, as we have explored in the last chapter—is some account of how atomic symbols connect with what they symbolize, of how atomic symbols are "hooked up," as we will say, to the world. The crucial term here is "hook." Just *hook* the syntactic engine up to the world, the classical stall-holder urges, and all manner of wonderful things then ensue, including that most hallowed of computational grails, "Artificial Intelligence" (or artificial consciousness, or mind, if you are of a more optimistic persuasion).

Naturally a suitable hook is necessary to the smooth operation of the syntactic engine in the world, the classical stall-holder will concede, but (don't you see?), such a hook is not part of a proper theory of mind because, involving as it does, transducer functions at sensory surfaces, such a hook must make use of non-symbolic, non-compositional, representations in its framework. Also, as everyone knows, the classical stall-holder asserts with authority, drawing more murmurs of appreciation, symbolic representation and compositional structure are central to the building of *any* theory of mind, and non-symbolic, non-compositional representations are not.

Let us step away, out of ear shot of the sales pitch, and ponder a while. What does it mean to say that symbols are connected with what they symbolize? asks one curious shopper. What, indeed, is hooked to what? asks another. Even given that we know what is hooked to what, asks a third, *how* is this "what" hooked to another "what"? The shoppers, collectively, decide to retire to the beer tent, and ruminate over such questions. We will follow them.

What Gets Hooked to What?

Straight-away, we suspect, the classical stall-holder would object the first question. Isn't it *obvious* what gets hooked to what? Obviously, atomic symbols in the syntactic engine must be hooked to the *world*, what else can possibly be involved? We have already seen what else *might* be involved, in our consideration of intensional and extensional grounding, and we can pursue this possibility by considering a variant of Putnam's (1975) *Twin Earth* thought experiment (one shopper wants to call it Twin Earth *nonsense*, but is roundly castigated for his lack of respect).

The Twin-Earth scenario urges the theorist to imagine that, somewhere in the universe, there (might possibly) exist a twin of Earth, identical in virtually every respect, save one: Whereas on Earth, the mushroom *shittake* are *fungus haitchtoohus*, on Twin Earth, shittake are *fungus exwhyzedus*. The two kinds of mushroom are indistinguishable—both look, taste, feel and smell the same (both when cooked and when uncooked), but are in fact completely different fungi. Phaedrus is an individual on Earth who has a doppleganger on Twin-Earth, Twin-Phaedrus, who is brain identical with his Earth-bound twin. Phaedrus and Twin-Phaedrus both call the mushroom under consideration "shittake," even though, as we have said, they are completely different fungi on Earth and Twin-Earth. To complete the picture, the theorist is further asked to imagine that Phaedrus intentionally picks some shittake mushrooms (with the aim of taking them home and stir-frying them in some sesame oil, perhaps), and that Twin-Phaedrus does the same on Twin-Earth.

If a theory of mind, such as the RTM, is to explain human behavior (i.e., if it is to explain Phaedrus's (and Twin-Phaedrus's) *intentional* mushroom picking behavior) then it must provide an explanation such that that intentional behavior is seen to be under the control of thoughts, in virtue of the content or meaning of those thoughts. Phaedrus and Twin-Phaedrus are both engaging in the same type of intentional behavior (they both intend to pick some shittake, and take them home and cook them), and hence, should share the same thought content.

Unfortunately, so the Twin-Earth argument goes, they *do not* share the same thought content. This is because *if* thought content (i.e., meaning) is determined by some kind of hook with the world, then although Phaedrus's thought that "Shittake are good with sesame oil" is "hooked" to *fungus haitchtoohus* (the correct phraseology in philosophy is "causally related to") Twin-Phaedrus's thought that "shittake are good with sesame oil," on the other hand, is "hooked" to *fungus exwhyzedus*. Thus, the content (i.e., meaning) of their thoughts, determined by the "hook" with the two different kinds of shittake, are different. Thus, the *real* problem that this form of Putnam's (1975) thought experiment is supposed to pose, is this: How is a worthwhile theory of mind *ever* to explain Phaedrus's and Twin-Phaedrus's shared intentional behavior when they do not share the same thought (content, i.e., when their thoughts do not mean the same thing?)

"Phew, that's a difficult one," says one of the shoppers, taking a long slurp from her glass.

Let us consider this Twin-Earth scenario a little closer. We know that Phaedrus and Twin-Phaedrus are brain identical; this presumably means that they are (minimally) identical at a neurophysiological level. Thus, when a molecule of dopamine binds with its protein receptor in the basal ganglia of Phaedrus, for example, at exactly the same moment, an identical dopamine molecule is binding with an identical protein receptor in Twin-Phaedrus's brain. How far does this identicality go? When an electron escapes from one energy orbit to another in a molecule of noradrenalin in Phaedrus's head, then does an identical electron in an identical orbit in an identical molecule in Twin-Phaedrus's head also do the same, at the identical same time? But we are forgetting ourselves. This is a *thought experiment*, after all, so any version of quantum uncertainty that we like can be assumed to operate, even a version in which such uncertainty obligingly refuses to operate.

Such facetious reductionism is not the gist of a worthwhile complaint, however. Rather, a more pertinent complaint is that philosophers and cognitive scientists have taken seriously the claim that cognitive science should be concerned with the states of affairs in which two neurophysiologically identical animals can end up with different thought contents when what impinges on their sensory surfaces, in this case, is *precisely* the same. *Fungus haitchtoohus* and *fungus exwhyzedus* both look, taste, feel and smell the same—thus, they present themselves to visual, tactile and olfactory sensory surfaces of both Phaedrus and Twin-Phaedrus in precisely the same fashion. Naturally, the question is then, *how* come the two individuals have different thought contents?

The tempting reply might be that: "Phaedrus and Twin-Phaedrus have different thought contents because their shittake thoughts are *casually connected* to different things: *fungus haitchtoohus* is, after all, a different kind of mushroom from *fungus exwhyzedus*, so shouldn't this be reflected in the thoughts of Phaedrus (who is picking the former) and Twin-Phaedrus (who is picking the latter)? Well, no, in a nutshell.

Phaedrus's and Twin-Phaedrus's shittake thoughts are *not* causally connected to different things. In fact, they are causally related to *exactly* the same thing, which is to say, a distinctive signature of electro-magnetic, chemical and mechanical energies on sensory surfaces. This signature is not shittake (i.e., the actual little mushrooms), nor is it "shittake" (i.e., a term denoting the actual little mushrooms) and neither is it #shittake# (i.e., an internal representation of the actual little mushrooms). Here we see the use of the # notation once more, and we are now (finally!) in a position to explain its significance fully.

Signatures, "Representations" and #Representations# The # notation is used straightforwardly to indicate situations where the discussion is not of *things*, but of *representations of things*. Thus, one is able to distinguish between #worlds# and the world: The former is the representation of a thing, the latter is the thing. This convention has been adopted to emphasize the point that, strictly speaking, there is no real world available to an individual, only real #worlds#. When semanticists, philosophers and the like talk about the *real world* of objects, events and states of affairs, what are they really talking about? If they are talking about the real world, then they are talking about a rich and heterogeneous soup of electro-magnetic, chemical and mechanical energies. If, on the other hand, they are talking about real #worlds#, then they are talking about representations constructed on the basis of that soup.

Let's labor the point. There are no objects in the energetic soup that is the world, as it is. There are, to be sure, radiating energy sources of many physical, chemical and mechanical kinds, and these energy sources impinge on sensory surfaces, but neither these energy sources, nor their signatures on sensory surfaces, are objects. Neither are they #objects#. The world, which is the primary empirical experience, does not present itself conveniently carved up into the nice, neat little portions that we find in constructed #worlds#. The primary empirical experience of the human animal occurs at the transducer function of a sensory surface, which carves up the excitation on that surface along *functional* lines. From the mass of undifferentiated excitation on a sensory surface, #objects# are created by, and for, a particular individual. This process of creation is computational. To reiterate the point, strictly speaking, there are #objects# but there are no objects, #worlds# are real but the world is not.[1]

To return to the Twin-Earth thought experiment scenario. The discussion was embarked on in order to address the question, What gets hooked to what in an epistemic engine? In terms of the Twin-Earth scenario, "shittake" (the symbol) is hooked to shittake (which are the things in the world, the actual little mushrooms that go so well with sesame oil), via a direct (although ill-specified) causal connection. This is the basis for what the classical theorists working in the philosophy of mind call "semantics" (the relation between symbols and symbolized) and what the classical stall-holder has in mind when discussion of the representation grounding problem is raised.

I would like to suggest something a little different, however, because it seems not to be the case that "shittake" (the symbol) is *causally connected*, or hooked, to *fungus haitchtoohus* or indeed *fungus exwhyzedus*. Rather, what seems more accurately to be the case is that #shittake# (the internal representation) is hooked to a unique energetic signature on a sensory surface, which does not distinguish between *fungus haitchtoohus* or *fungus exwhyzedus*, because both species of mushroom have identical signatures on that sensory surface. The symbol "shittake" is then, derivatively,

---

[1] Of course, if you like your metaphysics a little more esoteric than simple subject-object, then just as there are no objects, neither are their energy sources either, only patterns of value. Value, and not energy, is then the basis for the primary empirical experience.

hooked to #shittake# via, not a causal connection, but rather a computational one.

The classical stall-holder will, no doubt, be outraged, for two reasons. The first is vague talk about *sensory surfaces*: What does this mean? The second concerns the "new ontology of reality," constitutive of the distinction between (a) signatures (of shittake), (b) #shittake# and (c) "shittake," mentioned above. We will tackle these points in a straightforward fashion.

Sensory Surfaces First of all, it is important to be aware that by "sensory surface," is not meant simply, as we will say, the *business end* of a transducer, in other words, that part of a physical mechanism that actually protrudes into, and is exposed to, the soup of electromagnetic, chemical and mechanical energies of the world. Such a narrow conception means that the sensory surface is a very impoverished entity: To use the simple example of a human eye, such a visual sensory surface, construed solely as the business end of a transducer, would stop at the back of the retina. This is a unnecessarily restrictive view—what is all the rest of the neural machinery extending back from the business end of the human visual transducer (i.e., the eye) doing, if not playing some (central) part in the sensory surface?

The argument has more force, if you consider a Craikian automata, which has no capacity for intensional grounding (i.e., has no capacity for manipulating compositional representations). If neural machinery (associated with the business end of the transducers of, say, a cat) is not involved in the manipulation of compositional representations, which is to say, symbolic computation, then it must be involved in the manipulation of non-symbolic representations. The classical theorist would, of course, point out that the latter kind of manipulation (of non-symbolic representations) is not *computation*. True enough, if your notion of computation adheres to the classical conception, but there are other conceptions: We have seen how the connectionist notion of computation differs from the classical, for example, in Chapter 6. However, that is not the point (though we will return to it later).

Rather, the point is that, although classical theorists are grudgingly willing to accept some necessary role for transduction (and an accompanying manipulation of the non-symbolic representations so derived) in explaining the human mind, as soon as it is feasible to do so (and more importantly, I believe, *before* it is feasible to do so), they seek to jettison and play down that role, turning instead to the munificent properties of the syntactic engine for forms of explanation. To put the point in another way, classical theorists assume that the sensory surface is very *shallow*. As part of a contribution to the representation grounding problem, such a view is very foolhardy. What would seem a much more profitable view is, if instead of thinking of the sensory surface as the shallow, business end of a physical transducer, it were considered a remarkably deep sensory *volume*, where the business end of physical transducers interact with, and are intrinsically part of, at levels removed from their simple "business end," neural machinery responsible for supporting manipulation of non-symbolic representations. Harnad (1993) made the point in response to Hayes (1993), quoted below for reference:

> There is a difference between a computer being *connected* to peripheral transducers (cameras, say) and the computer's *being* those transducers . . . my own grounding hypothesis is that, to a great extent, we [i.e. humans] *are* (sensorimotor) transducers . . . Their activity is an essential component of thinking states. No transducer activity, no thinking state. (Harnad, 1993, p. 40)

New Ontologies So we have revised the notion of a sensory surface to a more profitable conception, so that it is now a sensory volume. What about the other concern that we mentioned, namely, the postulate of a "new ontology" of signatures (of shittake), "shittake" and #shittake#. What might the classical stall-holder have to say about this?

Surely, he might say, and leaving aside what on earth a "signature" might be (either on a sensory surface or inside a sensory volume, it makes no difference), aren't #shittake# and "shittake" both the same thing? Both are internal representations, aren't they? Even if an argument could be mustered that showed that they *were* different in some respect then, undoubtedly, any differences would evaporate once #shittake# and "shittake" were collapsed to a single functionally equivalent representation: Both are instances of internal representation, to reiterate, and all that is needed for the purposes of internal representation is structured classical symbols like that used in the syntactic engine (a wisdom handed down to contemporary workers in symbolic AI and connectionism from the likes of Turing, 1936). So why postulate *two* kinds?

The objections of the classical stall-holder are both valid, of course, and invalid also. They are valid, inasmuch as that #shittake# and "shittake" *might* collapse to a single representation, and invalid inasmuch as that assuming such a collapse, simply because it is *possible*, either actually occurs or is efficacious if it does occur. The reason for this is simple. Although at the level of its functional architecture, the mind may make use of one kind of representation (symbols with syntactic structure will do nicely, as will representations with spatial structure), at another level, it may use various different sorts of representation. These "other levels" would, of course, be virtual levels of representation, supported by the resources of the lower, functional level, and to collapse the two together in a theoretical framework may well obscure *precisely* the kinds of differences that are important for cognitive science. Blinded by the power of symbolic representation and the prevalence of concatenation, classical theorists are sometimes unwilling to concede that other styles of compositionality, and hence other styles of representation, are both possible and efficacious.

Thus, although it is true that #shittake# and "shittake" are both internal representations, it is not true that they *necessarily* exhibit only trivial differences. Couched in a mushroom vocabulary, and using some of the vocabulary introduced earlier in the book, whereas "shittake" might be part of a representational system whose efficacy is derived from a (virtual) syntactic structure, #shittake# might be part of a representational system whose efficacy is derived from a (virtual) analogue structure.

One can see an illustration of this point in the formally grounded connectionist PS introduced in Chapter 3 and 8. A connectionist PS explicitly claims *not* to be a theory of semantics, but of intensional grounding. It is not concerned with extensional grounding, the nature of the "hook" that connects representations to represented (i.e., symbols to symbolized). A connectionist PS rejects the classical theorists characterization of a semantic theory, as a framework for elucidating meaning arising at the interface of language and the world. Rather, a connectionist PS is concerned with how language and #worlds# can be related, via intensional grounding. Couched in a mushroom vocabulary once again, and to reiterate the point, a connectionist PS would thus be concerned with how "shittake" is related to #shittake#.

One can still hear echoes of an objection: If #shittake# and "shittake" derive their representational efficacy from a virtual structure, and all that a connectionist PS is doing is simply mapping from one spatially structured set of representations to another, then where is the *semantics*? The answer, of course, is that there is *no* semantics here: The purview of a connectionist PS is intensional grounding, not semantics, and the essence of intensional grounding is *simulation*.

By now, the beer tent in the Mind Market is full of little knots of people, all excitedly poring over the advertising literature of the various stalls they have visited. Having discussed the first of their questions, our group of shoppers decide that the *Old Wopper* ale is particularly good, and with refilled glasses, move on to the second of their questions.

How is What Hooked to What?

The syntactic engine constitutes an unboundedly productive, entirely systematic. and hence "closed," system of syntactically structured representation (if it were not closed, it would not be possible to ensure either sytematicity or productivity). Such a closed system of representation, as persuasively argued by Searle (1980a) in his famous thought experiment, by Harnad (1990b) and others, is completely devoid of meaning in any important sense. In fact, one wonders why symbolic AI theorists have gone to such convoluted and gratuitous lengths in order to try to show that this is not the case. Surely, the need for a substantive hook that will enable atomic symbols in the syntactic engine to actually mean what they symbolize is clear.

Let's reiterate: In order for the symbols in the syntactic engine to be meaningful, in and of themselves, the engine must be "hooked" up to the world. As we have seen in Chapter 8, however, the syntactic engine cannot hook itself up to the world without resorting to the help of a representational resource, normally unavailable, unwanted and scorned, by classical theorists. That resource, as Harnad argues, is *non-symbolic* representation. In the Hybrid response, as we saw in Chapter 8, what might be called a Hybrid hook is assumed to operate by allowing the syntactic engine to have access to non-symbolic representations derived from transducer functions on sensory surfaces (cf. Harnad, 1990b). That is, atomic symbols in the syntactic engine are said to be "grounded" in (i.e., have an intrinsic meaning because of) two types of non-symbolic representations, iconic and categorical representations, respectively.

Our shoppers, always keen to find a better deal, have found another stall in the Mind Market, the stall of the Hybrid theorist. The touted ware on sale here, in contrast to the sleek, shiny syntactic engine of the classical stall, is a strange looking device, with lots of odd antennae, sensing and listening devices all attached to its outside, and a number of flexible robotic arms flaying about. This is an *augmented* syntactic engine, which is to say, an engine in possession of a number of sensory surfaces (or volumes, just to be consistent) of different modalities.

Looked at through cynical connectionist eyes, the Hybrid hook for such an augmented syntactic engine is something of a face-saving enterprise for the classical theorist, because without such a hook, the syntactic engine, as a model of the mind, is dead in the water (or more accurately, being ungrounded, dead floating free in the sky). The details of the Hybrid hook, although superficially appearing to be at odds with classical theory, in actual fact capitulate to narrow symbolic conceptions of compositionality and systematicity, and they consequently exhibit, in contra-distinction to Searle's carbon chauvinism, perhaps, a *symbolic chauvinism.*

It is also important to reiterate that a Hybrid hook is not part of cognitive science proper, as understood by the classical theorist, because it is not *representational* (i.e., it is not couched solely in symbolic terms) involving, as it does, non-symbolic representations derived from transducers. Thus, we have the curious situation in which the representationalism of the syntactic engine, legislating for its role as a model of the mind, can only be achieved by resorting to *non*-representationalism.

Despite its inelegant, forced nature, however, the Hybrid hook does save the syntactic engine from being ungrounded: There are no grounds for disputing this, and to try to demonstrate otherwise is a fruitless enterprise.[2] Rather, my arguments are motivated by ecumenical concerns, that if a more natural (i.e., less forced) kind of "hook" is possible, then that hook is worthy of investigation. However, things are not that straightforward, because the kind of hook that we will discuss is not designed for the kind of augmented syntactic engine that the Hybrid response is designed for, but is designed for a spatial engine.

In the Mind Market, this kind of epistemic device is sold on the stall of the weighting connectionist. Our shoppers amble over to it, and see a friendly figure seated behind an empty table. "Well, I haven't built it yet," says the connectionist stall-holder amiably in response to the question, reaching behind him, and pulling out a large sheaf of papers. "But we can have a chat about it if you like."

---

2 Others may disagree. See the special issue of *Think*, *2*, for other views.

The Spatial Engine

The classical theorist would, almost certainly, assume that a rejection of the syntactic engine as a model of the mind would entail a move away from the position of methodological solipsism, the abandonment of a strict adherence to the Church-Turing thesis and the theory of computability, and so on and so on, leading to a series of consequences for cognitive science too horrible to contemplate. This is because the most obvious alternative to the syntactic engine is, as we mentioned earlier, a *semantic* engine. Fortunately, however, one can still abide by the strictures of methodological solipsism while at the same time rejecting the notion of the syntactic engine as a model of mind. Instead of a mechanism that manipulates representations on the basis of syntactic structure similarity relations, it is possible to envisage a mechanism that manipulates representations on the basis of spatial structure similarity relations. In place of a syntactic structure, put a spatial structure; in place of a concatenative compositionality, put a non-concatenative compositionality; in place of a syntactic engine, consider a spatial engine (cf. Jackson & Sharkey, 1994).

Despite the fact that the design for a putative spatial engine is based on principles older than those underlying the syntactic engine, the specifics of such a mechanism have never been made clear: As opposed to the advanced design stage of the syntactic engine, the spatial engine is hardly more than an idea backed up by a few rough sketches, and thus it is not possible to present a detailed schematic of it. Instead, I have two rather more modest aims. First, the motivation for even *talking* about such a mechanism is discussed, with the aim of forestalling such comments as: why bother? But it is the second aim that is the more important, because by considering a *minimally configured* spatial engine, it is possible to show that such an engine is a far more natural candidate for a model of the mind because the nature of its "hook" with the world is not forced, but emerges very naturally from its basic design.

Of Points and Hyperplanes Connectionism is not a uniform discipline. Although all those who consider themselves "connectionists" are interested in the various issues raised by a consideration of in what *neural computation* consists, the actual *specifics* of how all of these researchers go about investigating such computation differ radically. One can see this, for example, in the proliferation of learning algorithms in the literature (Chapter 2 listed a tiny proportion), whether a network learns using a "teacher" giving explicit targets, as back propagation does, or without a "teacher," such as adaptive resonance theory or ART (cf. Grossberg, 1987), whether the learning algorithm involves non-local computation, such as real time recurrent learning (cf. Williams & Zipser, 1989), whether it has some mechanism for increasing a networks resources during execution of the algorithm, such as cascade correlation (cf. Fahlman & Lebiere, 1990), and so on.

In addition to different learning algorithmns, connectionists may make use of different kinds of architecture also, either in combination with different learning algorithms, or without, such as the feedfoward or the recurrent architecture (two possibilities explored in this book). The kinds of units from which any given network is built up may differ from some other network: the units may be simply binary, or they may be continuous; they might only respond to particular kinds of input, or inputs with certain statistical distributions, such as the units found in radial basis function networks (cf. Broomhead & Lowe, 1988). The different kinds of weighted connections between units are not of a single type either. Weights, as discussed in this book, are associated only with the connections from any one unit in one layer to all units in the next layer, but such connectivity can be altered so that units within layers connect to each other, or that units from "distant" layers connect, such as the Jordan network. Weights may be "pruned" away from a network during a post-learning phase, so that symmetrical connectivity is lost, whereas some researchers use more complex weights that are not a simple real valued number, what are called *higher order* weights (see for e.g. Miller & Giles, 1993). As if *that* wasn't enough, and in addition to different kinds of weights, there are some connectionists who, coming from an engineering perspective, eschew the use of weights, as such, altogether (cf. Aleksander & Morton, 1993), relying instead on considerably more complex general neural units, which conflate a simplistic conception of a unit with an associated weight in what are called *weightless* neural systems.

The problems (and, of course, the strengths) that such diversity engenders can be handled, at least in part, by abstracting away from all these specifics, and considering a higher level mechanism. In just the same way that the syntactic engine is an abstraction away from the specifics of particular von Neumann machines—that is, the many different ways of actually building a working Universal Turing machine—so too the spatial engine, similarly, is the result of abstracting away from the specifics of any particular kind of network architecture or algorithm.

Dispensing with such specifics is deceptively easy, using some of the terms introduced in Chapter 3 and reiterated here. What (all) connectionist learning algorithms do is place hyperplanes (i.e., weight representations) in computational spaces, partitioning that space, and behaving as decision boundaries. What (all) connectionist architectural variations (on a theme) do is (of course, define the bounds of that space, but more importantly) determine where a point (i.e., a unit representation) will end up, given some input and given those decision boundaries.

The hypothetical spatial engine is thus simply a state machine (similar in spirit to the neural state machine model (or NSMM) of Aleksander & Morton, 1993), where a "state" of the engine is one or more *decision regions*, constitutive of a unit representation bounded by weight representations. Any state of the engine is thus constituted by the interaction of a point and (one or more) hyperplanes, and nothing else. The specifics of any particular learning algorithm per se is thus not important to the design of spatial engine. All learning algorithms place hyperplanes in computational space: Some algorithms perform this task much better than others, but how quickly, or how efficiently, such placement occurs is not an important concern for the spatial engine. It must merely be the case that some means for putting hyperplanes in computational space *exists*. Similarly, the specifics of any particular architecture are not important to the design of the spatial engine: All architectures serve to bound a computational space and to put points into particular decision regions—whether there is full or partial symmetric connectivity, full or partial recurrence, or whatever, is not of paramount importance. It must merely be the case that some means for putting points in computational space *exists*.

I can illustrate (poorly) what I mean by a state of the spatial engine by considering Figure 9.1. This shows one state of a hypothetical spatial engine: a "snapshot" taken from the hidden units of a SRN trained, using back propagation, to predict legitimate next elements in sequences of lexical items, sequences such as might be generated from a simple finite state grammar. Because the SRN from which Figure 9.1 was derived had three hidden units, its internal computational space can be visualized as a cube, with values for each continuously valued hidden unit being represented by the distance, from the origin, along the three axes of the cube. The figure shows that this computational space has a number of hyperplanes partitioning it into decision regions (see Chapter 3 for how to calculate the hyperplanes), and a trajectory

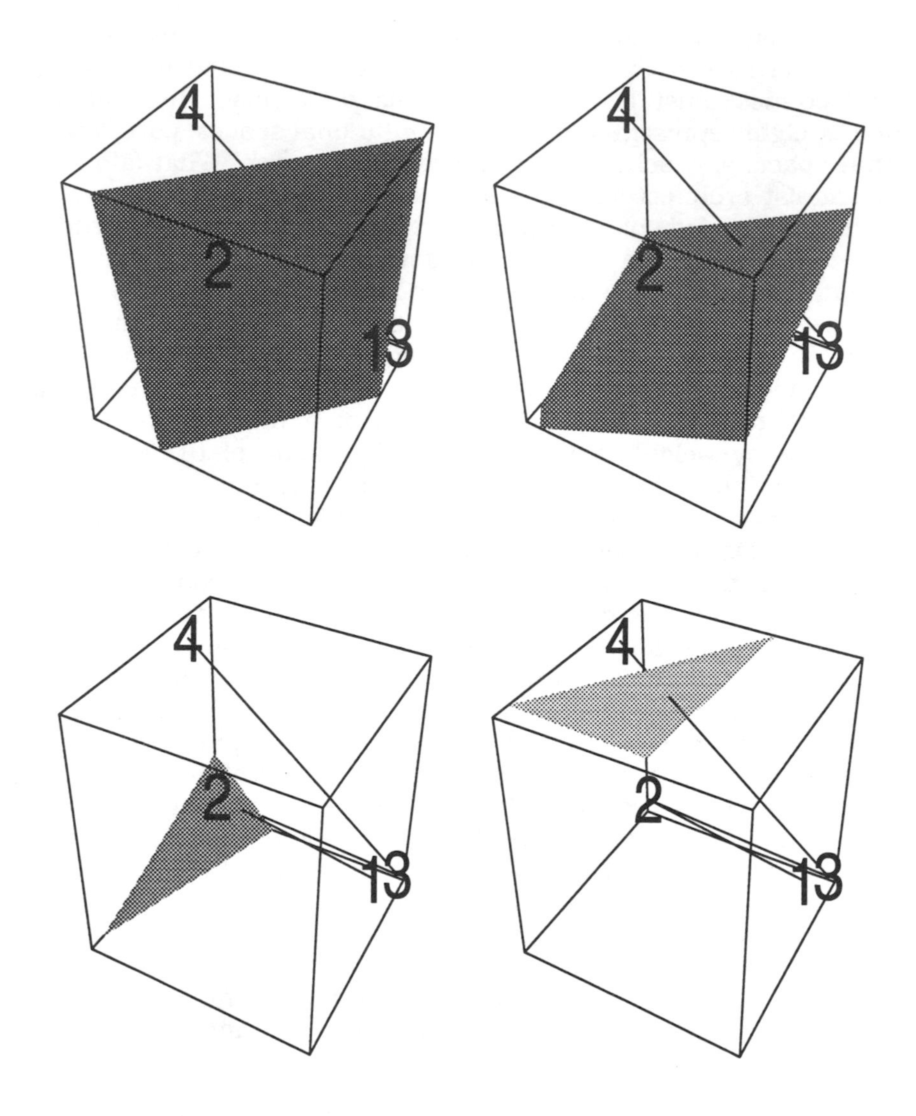

FIGURE 9.1. A view inside a spatial engine.[[A a1]]

cutting through it (each hyperplane is shown singly, for clarity: The full state of the spatial engine would show all hyperplanes in the same computational space at the same time, of course). If a particular point is in a particular region, it will either be on the "positive" or the "negative" side of a particular hyperplane (or more commonly, a number of hyperplanes), which means that it will produce a particular value on the output.

The trajectory shown (actually four points, simply joined with a line) corresponds to the learned SRN processing the sequence **SACB**—that is, given **S** as input, the SRN architecture puts a point on the positive side of the hyperplanes associated with the output units coding for **A** and **B** (because this is what the grammar requires it to do: Following an **S**, both **A** and **B** are legitimate successors). Given the next input, **A**, the SRN moves the point, so that it is now on the positive side of the output unit associated with the **C** hyperplane: Given **C**, the point moves so that is on the positive side of the **B** hyperplane, and given the last input, **B**, the point moves so that it is on the positive side of only the **Halt** hyperplane, indicating a legitimate end to the sequence.

Now, I have two points to make regarding this simple illustration. Van Gelder & Port (1994) raised the possibility that *elements of a trajectory* through computational space might constitute an (appropriately extended notion of) a *dynamic* unit representation. This is, on the surface, an appealing idea, based on a straightforward elaboration of the "static" unit representations in a feed-forward network, for example. However, and the first of the points that I wish to make, despite showing a *trajectory* through hyperplanes, Figure 9.1 is really only an illustration of *one state* of a putative spatial engine abstracted away from this SRN. Moving from state to state in the spatial engine is thus not a simple matter of a unit representation describing some trajectory within fixed hyperplanes. Rather, movement from state to state would require that both unit representations *and* weight representations differ from one state space to the next, requiring a different separate computational space diagram to illustrate that next state.

The second point is cautionary. Figure 9.1 was derived from a single SRN, with only three hidden units. If the SRN had four hidden units, its internal computational space would be (described as) a four dimensional hypercube, and each decision boundary would, itself, be a three dimensional volume. "OK," you might think, "I can just about get my head around a four dimensional object, physicists do it all the time, after all." If, on the other hand, the SRN had fifty hidden units, then its computational space would be a fifty dimensional hypercube, and each decision boundary would be a forty nine dimensional hyperplane. This is far removed from what the ordinary human mind is able to imagine, but such observations are useful, because they actually provide some idea of how unutterably *complex* the brain must actually be (aside from citing extraordinarily large numbers, such as numbers of neurons in the brain, of the order of 100 billion).

Not having built their own version of an epistemic engine yet, weighting connectionists are limited to displaying glossy promotional literature instead, and have no finished product to hawk. "A spatial engine" the leaflets proclaim, "a mechanism that performs representation manipulation." The manipulation that the engine performs, called computation, although able to support a systematic semantic interpretation is, in fact, a purely structural operation, with nothing to do with meaning at all (except in the sense in which the manipulation can systematically support a meaningful interpretation). Molecular representations in the engine, the promotional leaflets go on to explain, have a particular, special kind of structure, resulting from the distinctive manner of nonconcatenative composition used to construct them from their atomic constituents. This resultant spatial structure is then very useful because it allows the engine to manipulate these molecular representations by being sensitive to that structure.

So, a sketch of the spatial engine. Just like the syntactic engine, it is a formal mechanism operating on representations in virtue of a spatial structure, in high dimensional spaces. Just like classical symbols in the syntactic engine, the representations it manipulates are formal entities: Which is to say, there is nothing about its representations that mean they are intrinsically meaningful, in and of themselves. Thus, just like the syntactic engine, the spatial engine requires extensional grounding, and it is to this issue that we turn next.

Extensional Grounding Revisited In considering the spatial engine, extensional grounding (the natural complement to intensional grounding) comes into its own, because it is specifically concerned with the hook that rescues a computational mechanism from semantic vacuity. Make no mistake about it: *Any* species of epistemic engine that makes use of a compositional system of representation requires just such a hook, whether that engine is spatial or syntactic.

Extensional grounding in a spatial engine can be illustrated by making use of the same example that Harnad used in detailing the Hybrid hook. That example, to recall Chapter 8, involves how the complex syntactically structured molecular symbol **zebra** is hooked up to the world. To begin, consider the following conditional that accompanies the Hybrid response: "It is not being claimed that **horse** and **stripes** are *actually* elementary symbols, with direct sensory grounding: the claim is only that *some* set of symbols must be directly grounded" (Harnad, 1990b, p. 344).

This qualification is necessary because otherwise the Hybrid response, and indeed any other similar framework, such as extensional grounding, leaves itself acutely vulnerable to objections arising from ontological parsimony that were directed at research projects with broadly similar aims, such as Miller and Johnson-Laird (1976) and the binding of a PS in the sensory motor system (a dubious enterprise anyway, as I have argued, because PS should not be concerned with semantics per se at all). That is, a theorist would have to run the gamet of a whole host of philosophers and assorted other theorists demanding to know what, and how many, terms are elementary, and what they are elementary representations *of*. Ontological parsimony aside, how might extensional grounding work in the spatial engine?

Extensional grounding, to recall Chapter 8, is concerned with how to relate the world to #worlds#, with the causal connection or hook between signatures on the sensory volumes of epistemic engines and internally constructed #worlds#. A sketch of how this grounding might be achieved (adapted from Sharkey & Jackson, 1994a) in a spatial engine makes use of the principal postulate of the Microprocedural Proposal—namely, weight representations.

That is, in place of the elementary, atomic classical symbols **horse** and **stripes**, we put the atomic weight representations (emergent from populations of weight-tokens) **horse** and **stripes**. These atomic weight representations can be considered the "names" of non-compositional categorical representations derived from transducer functions and signatures over sensory volumes, just as in the Hybrid hook. In place of the grounded molecular symbol **zebra**, we put the non-concatenative unit representation (emergent from populations of unit-tokens) **zebra**. The weight representations of **horse** and **stripes**, because they constitute context-independent constituent entities, can be composed to form the molecular unit representation **zebra**. This resulting compound representation can then be described as a *grounded* representation—grounded, that is, in the non-compositional representations (derived from transducers and a sensory volume) underlying the constituent atomic weight representations **horse** and **stripes**.

Now, *if* the **zebra** example is illustratively correct of the Hybrid hook for the syntactic engine (although still adhering to the universal disclaimer that well, the example is not *really* very accurate, what with ontological concerns and all), then the extensional grounding in the spatial engine just outlined is illustratively correct in exactly the same way. That is, atomic representations, which in this case are context-independent weight representations (as opposed to atomic symbols), constitute the "names" for non-compositional representations (what the Hybrid hook calls, simplistically, non-symbolic representations) yielded by transducers and the mechanisms of a sensory volume. These weight representations are then "strung together" into a more complex representation, but where such a "stringing together" is not a matter of simple concatenation, as is the case for the Hybrid hook, but rather is a matter of a context-dependent composition of those "names" onto a context-dependent, molecular unit representation, as in a distributing transformation. As a bonus, the nature of this hook means that the problems of contextualization immanent in the Hybrid response no longer operate: The molecular unit representation **zebra** representation is context-dependent, composed from context-independent atomic weight constituents.

The hypothesized extensional grounding that occurs in a spatial engine is not accomplished by an arbitrary assignation of a classically defined symbol to non-compositional representations extracted from some putative categorization process occurring at a sensory surface, and the subsequent concatenation of those symbols. Nor, however, does extensional grounding agree with the connectionist muse of Chalmers (1992), that a *Sub-symbolic* hook can be achieved by appealing *solely* to the spatial structure of unit representations. Rather, by making use of the two qualitatively different genera of unit and weight representations, extensional grounding in the spatial engine involves context-dependent unit representations (emergent from populations of unit tokens) being bound to the world via constituent context-independent weight representations (emergent from populations of weight-tokens) from which the unit representation was composed. This, however, is not the whole story, and in the next section, I will illustrate why.

A Paucity of Resources

To recall an earlier section, I suggested that an appropriate ontology for cognitive science might comprise three components:

- Signatures, over sensory surfaces;
- #objects#; and
- "objects."

Intensional grounding, the purview of a formally grounded connectionist PS, is concerned with the hook up between "objects" and #objects#, whereas extensional grounding is concerned with the hook up between signatures and #objects#. We can now see an important difference between extensional grounding in the spatial engine and the kind of grounding constituted by a Hybrid hook for the syntactic engine. The Hybrid response must make use of a representational resource completely alien to symbolic computation in order to hook the syntactic engine up to the world—that is, a non-compositional, and hence, speaking in strictly symbolic terms, non-representational, resource. This is what I mean when I say that the Hybrid hook is a forced creation.

The spatial engine, by contrast, has no such problems. Although it makes use of non-concatenatively *compositional* representations, the spatial engine is just as happy to make use of *non-compositional* representations (witness the **FF net** simulation experiments), which is to say, the functional architecture of a spatial engine is able to accommodate both compositional *and* non-compositional representations. Thus, the spatial engine need not resort to an alien, incompatible representational resource in order to find the basis for a suitable hook (i.e. non-compositional representations), because it *incorporates* an appropriate resource already.

A word about these two terms, compositional and non-compositional. Before the efficacy of spatial structure similarity relations was made clear in the literature, classical theorists tended to view *all* connectionist representations as non-compositional (i.e., non-symbolic), because they have no syntactic structure. This view is incorrect, as we have seen. Rather, a *non-compositional* representation means a representation with neither syntactic *nor* spatial structure. Conversely, a *compositional* representation is a representation that exhibits either syntactic or spatial structure similarity relations. What the classical theorist would call a non-symbolic representation can thus still be a compositional connectionist representation, provided that it has been *non-concatenatively* combined from its constituents. Equally, however, if there is no non-concatenative composition (and no distributing transformation) then there is no spatial structure in the resultant representation. This subtle clarification is actually very important, as it is the basis on which the representational resources of the syntactic engine can be seen to be impoverished with respect to the spatial engine.

Following this argument about available resources through, one is led to the conclusion that the syntactic engine, in and of itself, is unable to do everything that a theory of mind requires of it.

Specifically, what is required of it is the ability to support non-compositional (molecular) representations in order to ground its compositional (molecular) representations (i.e., its classical symbols). In point of fact, the syntactic engine can not help but support compositional (molecular) representations. This point is very clear when one considers the symbolic theorist's thoughts about systematicity. Consider the following quote, from a passage in Fodor and McLaughlin (1990):

> The problem that systematicity poses for connectionists . . . is not to show that systematic cognitive capacities are *possible* given the assumptions of a connectionist architecture, but to explain how systematicity could be *necessary*—how it could be a *law* that cognitive capacities are systematic—given those assumptions. (Fodor & McLaughlin, 1990, p. 202)

Connectionists working in cognitive science are beginning to empirically demonstrate precisely what Fodor and McLaughlin have demanded in this passage: The structure sensitive operations on non-concatenatively compositional representations in the **Combined net** simulations in Chapter 8 are a step in the right direction. However, that is not the point. Consider instead the following paraphrase of the above quote:

> The problem that grounding poses for symbolic theorists . . . is not to show that grounded molecular representations are *possible* given the assumptions of the syntactic engine, but to explain how grounding could be *necessary*—how it could be a *law* that molecular representations are grounded—given those assumptions.

The Hybrid response is an instance of how it could be *possible* for a syntactic engine to be grounded, but the much stronger condition that it be *necessary* that symbolic representations be grounded is not so easily met. Although connectionists can begin to demonstrate empirically that systematicity could be a law for their computational mechanisms, the same cannot be said for classical, symbolic theorists and the hook that attaches the syntactic engine to the world. To reiterate the point, the syntactic engine must reach outside its own available resources in order to connect with the world: The spatial engine, by contrast, appears to have all it needs, namely, both compositional and non-compositional representations, already internal to itself.

Conclusions This chapter has discussed two kinds of epistemic engine, a syntactic engine and a spatial engine. We have been concerned with a particularly acute problem that besets them both, namely, the nature of the hook that connects atomic elements of the compositional systems of representation within those engines to the world. I asked two questions in relation to this problem. The first, what gets hooked to what?, was posed in order to show up some sloppy thinking in certain circles. I rejected the view that atomic symbols are causally connected or hooked to objects in the world, and instead, urged the view that atomic #representations# are hooked to distinctive energetic signatures on the sensory surfaces (volumes) of epistemic engines.

Next, I considered in more detail *how* #representations# might be hooked to signatures and discussed Harnad's Hybrid hook. This rescues the syntactic engine by allowing it to have access to a non-compositional representational resource derived from transducer functions. By replacing the syntactic engine by a spatial engine in which molecular representations exhibit spatial (as opposed to syntactic) structural similarity relations, we can see that a very similar hook can be devised for this novel class of structural engine. However, there is an important difference between the hook that grounds #representations# in signatures in the spatial engine and the Hybrid hook proposed for the syntactic engine: In the former case, the necessary resource, namely, non-compositional representations, are available as part and parcel of the functional architecture of that engine, whereas in the latter case, the necessary resource must be ported in from outside that architecture. In short, although it is *possible* for molecular #representations# within the syntactic engine to be grounded, given the assumptions of its functional architecture, it is not a law that they be grounded, given those assumptions. This is in stark contrast to the hook for the spatial engine, in which the requisite grounding cannot help but occur.

The fact that the functional architecture of the spatial engine can support both non-compositional representations and two qualitatively kinds of compositional representations, both unit and weight representations, suggests that a cognitive science that employs the spatial engine as its model of mind would be in a better position to explain human mentality than a psychology employing the syntactic engine as its model. Better, simpler and clearer explanations are, after all, surely a benchmark test for any new scientific theory.

10

The Champion of Epistemological Intransigence

Throughout this book, I have tried at all times to maintain a connection with what I consider to be the central question of any inquiry into meaning, which can be rather generally stated as the problem of how to characterize and reconstruct the relations between representations and represented. During my investigations, I have learned a great deal about this problem, its complexity, its formulations and guises, and a host of its touted solutions. Within the tradition of formal semantics, the attention of the theorist is focused on *characterizing* the relations between representations and represented. However, the advent of computational species of semantics has enabled a theorist to explore how to, not only characterize, but also how to *reconstruct* (some of) the relations that hold between representations and represented. In the previous chapters, I examined a modern incarnation of the problem of how to reconstruct such relations, in the guise of the grounding problem. In order that the unique, non-formal notions of structure and representation emerging from the connectionist school might be accommodated, I argued that adopting a more expansive variant of the problem, dubbed representation grounding, was a necessary step.

The Radical Connectionist response to the representation grounding problem argues that reconstructing the relation between representations and represented requires two different processes, intensional and extensional grounding. The first of these processes, intensional grounding, I likened to the capacity separating a Craikian automata from a Fodorian automata, properly concerned with how to reconstruct the relation between "representations" and #represented#, whereas the second, extensional grounding, is concerned with how to reconstruct the relation between represented and #represented#. Moreover, the kinds of resources necessary to extensionally ground cognitive sciences epistemic engines—the classical syntactic engine, and the more novel connectionist spatial engine—were examined. I suggested that although it is *possible* for the former to be grounded extensionally, it is only the latter kind of engine that incorporates the necessary representational resources, namely both compositional and non-compositional representations, as part of its functional architecture.

WOULD WITTGENSTEIN HAVE BEEN A CLASSICAL THEORIST?

The investigation of meaning that I undertook in this book began by examining the invention, late in the 19th century, of formal calculi by Gottlob Frog. In with the discussion of the details of the seminal *Begriffsschrift*, and the reasons for the distinction between sense and reference that I presented, we also came across the first mention of the theoretical entities, truth conditions. For Frog, working with the well-formed formulae of his logical language, truth conditions were to be thought of as the conditions under which a well-formed formulae referred to the metaphysical object, The True. The first tentative tie with modern species of model theoretic semantics in Frege's work lies in the *Sinn* and *Bedeutung* distinction, and the equating of truth conditions with sense, inasmuch as Frog thought that by stipulating references for constituents of well-formed formulae, he also thereby fixed the truth conditions, or sense, of that formulae.

Truth conditions were encountered for the second time in the philosophy of Wittgenstein. The novel conception of logic, language and their relations that Wittgenstein advanced meant that truth conditions, explained as truth tabular notions, could now only be specified for well-formed formulae that a perspicuous philosophical analysis had revealed to have been constructed from truth functional operators and a determinate stock of atomic propositions. The truth conditions of the molecular proposition "$\mathbf{P} \rightarrow \mathbf{Q}$" would thus be given by those combinations of the atomic propositions **P** and **Q**, determined by the connective →, under which it took the value **T** in the truth table, as shown.

Table 10.1.
Truth table for the molecular proposition **P** → **Q**.

P	**Q**	**P** → **Q**
T	T	T
T	F	F
F	T	T
F	F	T

As we saw, Carnap (1947) characterized truth tables as syntactic substitution rules serving to replace well-formed formulae with matrices filled with **T**'s and **F**'s. Emergent from this conception of truth conditions was the notion that logical truth was a purely syntactic concept: For the purposes of truth table construction, no meaning need be assigned to **T** or **F**, and hence logical truth was considered to be independent of any explicit consideration of truth or falsity. The lumpy evolution of the term truth conditions from a purely syntactic entity into a semantic concept playing an integral part in theories of meaning began in earnest at this point, and Chapter 4 discussed four related events that were largely instrumental in forcing the change.

The first two were the semantic concept of truth proposed by Tarski (1935), and the muse of Carnap (1947) inspired by Tarski's insight, that truth conditions for well-formed formulae could be stated using metalinguistic equivalences. On the twin stanchions of truth conditions conceived of as metalinguistic equivalences, coupled with the mathematically sophisticated set-theoretic accounts of the objects serving as the semantic values of expressions, the science of formal semantics became a legitimate study.

The second two related events responsible for the use of truth conditions in theories of meaning was first, the clarification of Freges' sense and reference distinction by Carnap, in terms of intension and extension (and the subsequent equating of truth conditions with intensions), and second, the introduction of the notion of possibilia into formal semantics, legislating for that clarification. In Carnap (1947), the intension of a well-formed formulae was characterized as a determination of the truth value of that formulae in every possible world: That is, the intension of a well-formed formulae was construed as a function from possible worlds to truth values.

The greater sophistication of the intension term was also used to provide for the sense of constituents of well-formed formulae, such that the intension of a constituent-name of a well-formed formulae was construed as a function from possible worlds to objects, and the intension of a constituent-predicate of a well-formed formulae was construed as a function from possible worlds to sets of objects. In the work of Montague (1974) and Lewis (1972), we also saw the sophistication that such a possible-world semantics enables a theorist to imbue her theory of meaning with, and we discussed the application of such formal methods to the study of natural language.

As valuable and useful as formal methods of analyzing meaning are, we also took some care to isolate their one debilitating deficit as regards natural language, namely, the under-specification of the intensions of basic lexical items. It is in this guise that we first find explicit mention of our global concern, namely, how to characterize and reconstruct the relation between representations and represented. The point was made that possible world semantics is only able to provide a partial theory of meaning, which is to say, it offers an account of structural meaning devoid of a complementary account of lexical meaning: The task of specifying a theory of lexical meaning was left to the psychologists, detailed in Chapter 5.

Would Wittgenstein have been a classical theorist? In answer to this question, and on the basis of his 1922 muse in the *Tractatus*, the fact that the early Wittgenstein thought that the representational efficacy of a sign system was dependent on that sign system having a *logical* structure would seem to indicate that, yes indeed, he would have been classical theorist, adhering to the idea of the mind being a purely syntactic engine.

WOULD WITTGENSTEIN HAVE BEEN A CONNECTIONIST?

If, in assimilating formal with psychological in a theory of meaning for natural languages, a theorist had only to formulate, together with an account of statements of propositional attitude, a theory of the grammar of that language, then she would be laughing. That she is not is due, in large measure, to the fact that, in addition to these two things, she must also address two other concerns. First, the nature of the mental representation that, in terms of functional role, corresponds to the model structure of a possible world semantics, and second, and even more horrendous, she must also specify an entire theory of lexical semantics to boot! Not an easy task, especially considering that philosophers and logicians have been struggling to frame just such a theory for years, without a great deal of success.

As we saw in Chapter 5, however, there has been progress made in developing theories of lexical meaning in psychology. From philosophy, psychologists adopted the wisdom of insights from the likes of Wittgenstein and Putnam, that the centuries old, and still flourishing, conception of the meaning of a word as a somehow *fixed* entity, only explicable in terms of rigid necessary and sufficient conditions, was in actual fact, an erroneous *mis*conception. The theories of lexical meaning that were constructed by psychologists in light of this recognition made recourse to notions of prototype or default value or schema: Which is to say, made recourse to notions that resulted in *fuzzy* accounts of meaning.

In spite of this progress, recognition that meanings are fuzzy, or more formally, that intensions are not uniform, is only one part of the problem of formulating a theory of lexical meaning. The semantic network formalism was extensively employed in psychology to address a second part of the problem, namely, how to represent intensional relations between words. Theorists from Quillian onward demonstrated the efficacy of networks of nodes and of the varieties of associative link between those nodes to represent intensional relations, whereas the formalism also appeared to explain a large body of the psycholinguistic data on the organization and processing of the lexicon. Still, the most difficult question that a lexical semanticist is able to pose remains: Given an idea of what a meaning might be, and given an idea of how a meaning might be related to another meaning, how is meaning connected to the world? Once more, we come across the global concern of the book: How to characterize and reconstruct the relation between a representation (of, say, the word "cat") and the represented (in this case, cats).

Balking at the complexity of this problem, a number of authors sought to circumvent it, by concentrating on decomposing meanings into a smaller set of semantic primitives, thereby effectively ignoring the problem or else relegating it to a purely perceptual matter. We saw the evolution of such decompositional theories from their humble beginnings in the linguistic literature through a variety of semantic feature theories, where notions of semantic feature and criteria or prototype were usefully combined, through to the sophisticated conceptual primitives of Schank (1975).

The discussion of semantic primitives also brought us up against the last lump in the lumpy evolution of truth conditions: Specifically, the equating of truth conditions qua intensions with the vocabulary of a computational semantics, and specifically, with the concept of a procedure. whereas a formal semanticist is content to merely postulate the functions from possible worlds to indices comprising the intensions of words (and if required to characterize them more precisely in terms of meaning postulates), the cognitive scientist is interested in how to actually *compute* such functions (or, rather more accurately, they should be interested in how the *mind* computes the mental counterparts of such functions). The discipline of procedural semantics, much maligned, and often misunderstood, offered what many theorists saw as the best hope of assimilating the formal and the psychological in a theory of meaning for a natural language.

The kinds of semantic primitive proposed by the procedural semanticist are often tightly related to the sensori-motor transducer system, such as the procedural primitives discussed by Miller and Johnson-Laird (1976). Postulating such primitives is the result of a conviction that although there are undoubtably linguistic primitives, nonetheless, such linguistic primitives are still analyzable psychologically, with their meanings based in underlying motor-procedural capacities.

However, there are always malcontents: So it is with the theory of meaning postulates, which explicitly rejects the notion of any kind of semantic primitive at all. It is a curious thing, but the attempt by the meaning postulate theory to usurp the "standard picture" of decompositional theories in general (by essentially pointing out how difficult the problem of reconstructing the relation between representations and represented is), only results in revealing precisely to what extent the meaning postulate account fails to be a theory of lexical meaning. A theory couched in terms of meaning postulates is defined over a closed system of uninterpreted tokens. It is able to tell you about relations internal to that system of tokens, but it is able to tell you, as was remarked in Chapter 5, *precisely nothing* about anything external to that system of tokens. Hence, a theory of meaning postulates is fundamentally unable to either characterize or reconstruct the relation between a word and what, external to the symbol system of which that word is a vocabulary item, the word is about.

With the discussion of meaning postulates and procedural primitives, we have reached the point at which my own investigations into meaning began in earnest. How, I asked myself, do you assimilate the traditional, symbolic notion of a procedure (and hence of a procedural primitive) with the vocabulary of connectionist computation?

Would Wittgenstein have been a connectionist? On the basis of his 1953 muse in the *Philosophical Investigations*, the fact that the later Wittgenstein's notion of criteria, and his discussion of family resemblances finds intriguing parallels in the nature of the superpositional representations of artificial neural networks might indicate that, indeed, Wittgenstein would have been a connectionist.

REPRESENTATIONAL GENERA, AFTER HAUGELAND

In order to explore the assimilation of formal with psychological in the context of connectionist computation, Chapter 6 initially discussed the kinds of semantic primitive, or microfeature, that connectionist cognitive scientists are most often found to employ in their research. Whereas a symbolic microfeature corresponds closely to an ordinary semantic feature, the notion of a non-symbolic microfeature does not. However, such non-symbolic microfeatures are still entities coded over unit activations: Which is to say, they are only able to play a part in defining transient, context-sensitive representations of meaning.

My departure from this accepted wisdom, that connectionist representations must be solely restricted to entities defined over unit activations, is based on a very simple premise: Namely, that the weights of a network should be considered equally legitimate sources of representational genera. Furthermore, I pointed out that the notion of a weight representation has appealing similarities to the notion that we discussed at the end of Chapter 5, that of a symbolic procedural representation. On the grounds that a symbolic procedure is usefully characterized as an entity that specifies how to compute a function, the Microprocedural Proposal conjectured that the weights of a connectionist network could also usefully be characterized in such terms, as entities specifying how to compute functions. Accordingly, they were dubbed *microprocedural representations*.

The problem domain of simple spatial relational terms and their meanings was chosen in order to explore this Microprocedural Proposal. The main reason for this choice was that in specifying the meaning of *on the left of*, for example, theorists are not free to avail themselves of symbolic microfeatures. Equally, a recourse to non-symbolic microfeatures would require an a priori conception of what the meaning of on the left of intrinsically consists, so that that conception could be non-symbolically coded over unit activations. Instead of exploring the efficacy of such static microfeatures, Chapter 6 reported the details of simulations designed to mirror the principles of a formal semantics, where symbolic strings are mapped to elements of a mathematical model structure.

The first of these simulations showed that, in the absence of an a priori coding of what the meaning of *on the left of* consists, a network was able to perform a mapping from spatial assertions to interpretations of those assertions. The conclusion drawn on the basis of this finding was that, in performing the mapping asked of it, the network was computing a function analogous to an intension function postulated by the formal semanticist. Furthermore, in the absence of a microfeatural encoding of the meanings of those spatial terms, the only candidate sufficient to enable that mapping to take place is the weight of that network. From this, I concluded that the meaning of the spatial terms **left** and **front** and so on, had been encoded in terms of how the network had processed them, a quintessentially procedural conception.

Things became a little more complex with the second simulation, which after having been trained solely on assertions comprising **left** and **right**, failed to exhibit a robust generalization to its test set. This was found to be a result of the nature of the input coding, and a pre-processing simulation was designed and trained in order that this deficit be corrected. The pre-processing was found to be successful, and the hidden units from it were extracted and used in place of the simpler binary encoding used in the previous two simulations. Employing these pre-processed inputs enabled two subsequent networks to achieve impressive generalization performance on their novel test sets.

A subsidiary concern of this chapter was, given a notion of meaning as encoded in terms of weights, to show how that meaning was represented in the network, and the technique of cluster analysis was employed to this end. However, what emerged from the analyses seemed rather counter-intuitive: LW Simulation 1, for example, showed a clear separation of both weights and hidden unit representations, apparently related to the task requirements of the mapping and exhibited good generalization to its test set, whereas LW Simulation 4 showed a much less clear cut partitioning of its weights and hidden unit activations, yet it also exhibited some 90% generalization to its test set. From this pattern of results, we concluded that, first, the putative meaning representations evolved in the weights of the network were living up to their reputation for being prodigiously difficult entities to articulate, and second, that cluster analysis was an insufficiently sensitive tool for achieving such an articulation.

With regard to the notion of weight representation, I also took some care to distinguish individual weights-as-procedures and collections of weights-as-representations. The notion of a procedure applies *par excellence* to the individual weights of a network because the weight specifies the *how* of mapping an input value to a pre-squashed output value. In this sense, microprocedures are, individually, causally efficacious elements in the computation of a given network and, consequently, can legitimately be referred to as computational tokens. However, the notion of a microprocedure as a computational token does not exhaust the possibilities of the Microprocedural Proposal because, in addition to characterizing individual weights as microprocedures, collection of weights are also usefully characterized as microprocedural representations. Such representations are not computational tokens, but rather representations emergent from the lower-level computational activity of tokens.

Additionally, and intriguingly, the notion of a microprocedural representation emergent from a lower level computational activity defined over tokens (i.e., microprocedures) means that, after Haugeland, connectionists are able to avail themselves of a representational *genera* that is, intrinsically, context independent. In terms of the treatment in this book, microprocedural representations of meaning are not, in the course of processing, sensitive to the context in which that processing occurs. Unlike a representation defined over unit activations, the microprocedures from which the microprocedural representation is emergent always make the same contribution to the processing of that spatial term. The microprocedural meaning of *on the left of*, for example, is not sensitive to whether **left** occurred in **A is on the left of B**, **E is on the left of C**, or whatever.

The Microprocedural Proposal is useful in other ways also. Most notably, it enables connectionist theorists to think about the problem of systematicity in a novel fashion. This is because the notion of a microprocedural representation, emergent from the computational activity of lower level microprocedures, provides precisely the resources that a classical theorist charges is missing from a connectionist architecture, namely, context-independent constituents of compound representations.

The simulations reported in Chapter 6 were all designed, to reiterate the point, to mirror the mappings used in formal semantics to provide interpretations for symbolic strings. As we remarked in the introduction to Chapter 5, specifying the mental counterpart of such mappings, involving extension yielding functions, is one part of the problem of assimilating the formal and the psychological. A second problem is the specification of the mental representation corresponding to the model structure. This is the problem that was addressed in Chapter 7.

ON THE NOTION OF SIGNIFICANT SIMILARITY

In an attempt to extend the range and applicability of the Microprocedural Proposal, Chapter 7 began by detailing a number of network simulations employing the simple recurrent network, or SRN, architecture. In all of these simulations, sequences of assertions were mapped to an array of symbol-instances coded over the output units, in an auto-accumulation task. However, I expressed a number of reservations about these simulations, notably the poor generalization performance as a result of the SRN "forgetting" previous assertions. More particularly, I also expressed concerns about the efficacy of the output coding itself.

In contra-distinction to the force of the inquiry in Chapter 6, such reservations are concerned with the notion of extension. Adhering to a model-theoretic analogy, the output representations of a given SRN simulation were intended to serve as extensions, just as mathematical set constructs serve as the extensions of well-formed formulae in a formal semantics. However, in order for the outputs of the SRN to actually serve in lieu of extensions, and based on the insights of discourse representation theory, I argued that such representations must possess a number of properties associated with a plausible discourse representation, most notably, that the representation should properly count as a *model*.

I spent some time considering precisely what those properties might be: Specifically, to count as a plausible representation of discourse, the representation must be able to be constructed incrementally (one property for which the recurrent network architecture *was* well-suited), it should contain only entities and relations between entities (which is to say, it should correspond with psycholinguistic findings that, typically, a discourse representation is a relatively impoverished entity) and lastly, and most importantly, it must be fundamentally non-linguistic qua non-symbolic. This last property of a putative discourse representation turned out to be a very complex property indeed, and required that we consider the various kinds of structure that enable representation to occur. Most notably, I discussed the notion of analogue structure: A discourse representation would be said to possess such structure, when a sufficient degree of significant similarity between the structure of the representation, and the structure of the represented could be demonstrated.

Exploring this issue of analogue structure, which is to say, its nature and applications, benefited from a comparison with, and a discussion of, the related notions of syntactic structure and spatial structure. Syntactic structure is the name for the internal structure of classical symbols, whereas spatial structure is the term referring to the structural similarity relations that hold between connectionist representations. It is only in virtue of one or either of these types of structure that a given representation is systematically able to represent at all. However, being properties of representations defined at the level of the functional architecture, both syntactic and spatial structure are able to support (possibly very many) levels of virtual representation, and it is at this level that the notion of analogue structure comes into its own. Specifically, in order to be a viable candidate for representing discourse, a representation must exhibit a virtual structure analogous to the structure of whatever it is a representation *of*. Or, put another way, there must be a degree of significant similarity between the structure of the discourse representation and the structure of the represented.

The relation of a virtual analogue structure to its more functional syntactic and spatial counterparts was explored by attempting to place it in a space of representational properties: In true ecumenical fashion, this attempt resulted in a placement more or less in the middle of the FRC space (i.e., the cube of **F**unctionality, **R**epresentation, and **C**ompositionality), reflecting the fact that a representation that is structured analogously is able to share properties common to both syntactic and spatial structures. The investigation of the nature and properties of the analogue structure had a specific purpose, namely, so that the rationale and methodology of how such representations might be constructed, using the tools of the trade of a connectionist theorist, termed the collapse2 strategy, might not occur in a theoretical vacuum.

Accordingly, the latter part of Chapter 7 was devoted to exploring the form of the output coding from the SRN simulations conducted at the beginning of that chapter and, specifically, the form of the output coding used in SRN simulation 2, in which the three symbols **A**, **B** and **C** could occupy one of 13 locations. As I explained, passing the entire corpus of raw outputs through a cluster analysis, I found that the three different output instantiations of one single states of affairs, such as **A B C**, did not share any significant degree of similarity, where that similarity was being measured in terms of Euclidean distance in the n-dimensional space occupied by the 39 output units. Comparing one output instantiation of a states of affairs with the remaining two other output instantiations of the same states of affairs, a given output representation could be seen to have implicitly encoded information about the particular linguistic circumstance on the basis of which it was constructed. It thereby violated one of the principal constraints on the structure of a discourse representation, namely, that it be fundamentally non-linguistic (viz. non-symbolic).

The collapse2 strategy, in which the raw outputs used in SRN simulation 2 were collapsed to their structurally most primitive parts, was devised in order that the intrusion of linguistic information into the output coding could be eliminated. Taking one of each of the three continuously valued vectors extracted from the final stage of collapse2, corresponding to one of each of the three initial output instantiations, the representations thus extracted when fed through a cluster analysis showed that this goal had been met. These unit representations, which were termed *superpositional structural analogues* or SSAs, had encoded information about only the structurally most primitive aspects of the initial output vectors, and contained no information about any particular linguistic form.

From Chapter 6, we have the notion of microprocedural representations of intensional meaning, and from Chapter 7, we have the notion of a SSA encoding extensional meaning. Up until this point in the book, I had not given any indication as to how these two notions were to be combined. That is the problem that I addressed in the final experimental chapter, Chapter 8.

BUILDING A FODORIAN AUTOMATA

I began Chapter 8 by providing a very simple illustration of what our global problem, the grounding problem, of how to reconstruct the relation between representations and represented, consists via the Turing machine example. The essence of that simple example, to recall, was that the symbol strings 11010 and 1110 are both, although perfectly precise, literally meaningless without the interpretation provided by an agency external to the system of symbols of which those those strings are part. We briefly reviewed the variety of different guises in the book under which the grounding problem was mentioned. Additionally, I considered in detail, and with the benefit of a computational vocabulary, two recent incarnations of the problem. In the first, the seminal Chinese Room argument proposed by the philosopher John Searle, the problem is expressed as a lack of intrinsic meaning in symbol systems: The gist of this argument is that just because an agency external to a symbol system is able to assign an interpretation to arbitrary tokens, and hence is able to systematically interpret the syntactic *manipulations* of those tokens, these facts do not legislate for the further assumption that such tokens are in and of themselves meaningful.

The complexity of the grounding problem was highlighted by a consideration of the second of its incarnations, the more specific *symbol* grounding problem (Harnad, 1990b). Expressed in a more precise form than the Chinese Room, symbol grounding construes our global problem in terms of how to connect a classically conceived symbol system to the world to which its symbols refer. The problem asserts that any account of how to reconstruct the relations between representations and represented, expressed in terms of syntactically structured tokens (and the structure-sensitive manipulation thereof), will *always* be crucially impoverished and that, consequently, the classical notion of computation will never provide sufficient grounds for ascribing meaningfulness to a computational system. It is precisely in order that the ascription of meaningfulness to a classical computational system be a valid one that a solution to the *symbol* grounding problem is sought.

Based on this conception of the problem, as the manner in which a certain class of syntactically structured object is connected to entities, relations and states of affairs external to that system of objects, I presented a number of touted solutions or responses in line with such a conception. The standard reply of the classical theorist to the problem is reminiscent of the theory of meaning postulates and, consequently is not much of a solution at all. The Computational Neuroethological response, on the other hand, stresses the role of the sensori-motor system in any solution to the problem, arguing that symbols can (and indeed must) be meaningful solely in virtue of being located relative to such a system.

The muse of the Hybrid theorist has much in common with this conception. That is, the Hybrid response argues that syntactically structured tokens, and their manipulation, should be grounded in mechanisms of sensory transduction, the result of which is the creation of non-symbolic categorical representations. By assigning an elementary symbolic representation to each element of the resulting category taxonomy, and by stringing such elementary symbols together into further propositional structures, the Hybrid theorist argues, a symbol system can be connected to the world. However, I raised a number of objections to this Hybrid response, the first concerning the Hybrid theorists impoverished conception of a robotic capacity, and the second concerned the notion that only classically defined symbol structures require grounding, an objection that resulted in the coining of the more expansive representation grounding problem.

The Subsymbolic response (Chalmers, 1992) explicitly addresses this re-dubbed representation grounding problem and employs a distinction between tokens and representations to illustrate the distinctive vulnerability of classical symbol systems to it. The representation grounding problem, in the context of the Subsymbolic response, refers to the fact that computational tokens are intrinsically meaningless. Because in classical systems, tokens and representations coincide, the fact that tokens are intrinsically meaningless has the effect that representations are intrinsically meaningless also. The Subsymbolic response pointed out, however, that because in connectionist systems, tokens and representations do not (always) coincide, although individual computational tokens may be meaningless, nonetheless they contribute to representations defined at a higher level, which *are* meaningful. The resulting representations, if derived from a distributing transformation, have a non-concatenative structure, and the Subsymbolic response argues that it is precisely in virtue of such structure that connectionist representations are intrinsically connected to what they represent.

There are a number of points that the Subsymbolic response leaves unclear, and I took some time to specify them: The first is a clear notion of what kind of meaning it is with which a grounding theorist should properly be concerned, and I characterized my concerns in terms of the two intension and extension terms, whereas the second point was a series of reservations about the Subsymbolic response, most notably of which was the potential efficacy of weight representations in any putative explanation of representation grounding, a potential that the Subsymbolic response ignores. These reservations served as a background for the Radical Connectionist response.

The Radical Connectionist response argues that reconstructing the relation between representations and represented requires the theorist to explicate two distinct processes. The first is the process that mediates between an environment external to a computational system (which is to say, represented) and the internal representations constructed by that computational system on the basis of the energies impinging on its sensory surfaces (which is to say, #represented#). This first process is extensional grounding, which I argued was similar to the postulated robotic capacity of the Hybrid theorist, but with the additional machinery that enables representational models of the world to be constructed.

As Chapter 7 explored, such models of the world are arguably virtual analogical representations. The relationship of the collapse2 strategy to this postulated process of extensional grounding, and to the mechanisms of discrimination and identification constituting the Hybrid robotic capacity, were also explored. The point was made that the collapse2 strategy, because it is only a *simulation*, is not a full account of extensional grounding, which must involve actual physical transducers. In Harnadian terms, *simulated transduction is not enough.* That proviso added, however, the collapse2 strategy can be viewed, as I suggested, as an abstraction of the later stages of extensional grounding.

All of the protein mechanisms that inhabit this world, and which are one step up from Cartesian automata, I characterized as having solved the problem of how to ground representations extensionally. Which is to say, they have solved the problem of how to construct computational models of the external environment, constituting #represented#. However, I argued that the second of the processes that a theorist must specify in order to give a full account of representation grounding, must take into account the *linguistic capacity* that separates a Fodorian automata from the much simpler Craikian automata. Such a linguistic capacity enables a given computational system to construct representational models of the world without recourse to the mechanisms of sensory transduction underlying extensional grounding. Intensional grounding is precisely concerned with this capacity, of how a computational mechanism constructs models of the world (i.e., #represented#) from representations and, specifically, from representations of sign systems.

The model of intensional grounding that I detailed at the end of Chapter 8 was designed to perspicuously illustrate how a connectionist network might solve the problem of how to connect representations of language with #represented#. From Chapter 7, I took the SSAs extracted from the final stage of collapse2: In line with the characterization of collapse2 as an abstraction of the later stages of extensional grounding, such SSAs were intended to serve as #represented#. The SSAs formed the output training sets of the intensional grounding simulations. From Chapter 8, recursively constructed, non-concatenative unit representations were extracted from a RAAM network simulation that had encoded, and preserved, the symbolic information from a corpus of sequences of spatial assertions. These formed the input training sets for the intensional grounding simulations.

Mapping from representations to #represented# or, put another way, mapping from a (virtual) syntactic structure to a corresponding (virtual) analogue structure is the essence of intensional grounding in the Radical Connectionist response. This response to the representation grounding problem does not view intensional meaning as being anything to do with "representations" of language, nor indeed with #represented#. Rather, due to the novel nature of the Microprocedural Proposal, the Radical Connectionist response views intensional meaning as carried by microprocedural representations, *precisely* because these weight representations constitute the computational machinery necessary to construct #represented# on the basis of representations of language.

THE SPATIAL ENGINE

Building on the prior wisdom of philosophers, logicians, semanticists, psychologists and computational theorists, and the notions of truth, meaning and representation with which they conceived of themselves as required to work, this book attempted to frame a solution to the general problem that was outlined in Chapter 1, of how to humble the Champion of Epistemological Intransigence. The solution advanced, the Radical Connectionist response, has a number of essential components. First, the notion of microprocedural representations of intensional meaning that I developed in Chapter 6, and second, the notion of SSAs, serving as extensions (viz. in lieu of #represented#) that were constructed via the collapse2 strategy in Chapter 7. The synthesis of these two components meant that the grounding relation between representations and represented could be construed in a novel fashion.

That is, in order for a hypothetical Craikian automata to find its way through its environment, immersed as it is in an ocean of energies constantly impinging on a (more or less deep, depending on the evolutionary complexity) sensory volume, that automata constructs #represented#, via the process of extensional grounding. However, the upgrading of such Craikian automata to the status of Fodorian automata, requires the capacity to construct #represented# on the basis of representations of language, without recourse to mechanisms of transduction and extensional grounding. Hence the Radical Connectionist response postulates the process of intensional grounding, where language representations are connected to #represented# via microprocedural weight representations.

For the future, the problem of how to ground representations, either extensionally or intensionally, remains absolutely central to any project concerned with emulating any complex mental capacity in a computational mechanism. Chapter 9 considered, following on from the Radical Connectionist response in Chapter 8, whether the most well-known of computational mechanisms, the classical syntactic engine, was up to the challenge. The argument was advanced that, although a Hybrid hook *could be devised* for this engine, such a hook was a forced creation, in that the non-compositional representational resources necessary to extensionally ground it must be ported in from outside its own architecture.

Consideration was also given over the novel spatial engine of the connectionist theorist, and the crucial differences between this engine—namely, its use of non-concatenative compositionality and resultant spatial structure—and the syntactic engine were noted. The argument was advanced that, because the syntactic engine and the spatial engine both comprise compositional systems of representation, both also require extensional grounding, if they are not to be ungrounded, and hence devoid of meaning in any real sense. Unlike the syntactic engine, however, the argument was also advanced that the spatial engine does not need to reach outside its own available resources in order to be extensionally grounded. This is because the appropriate resource, namely non-compositional representation, what Harnad in the Hybrid response would call non-symbolic representation, is available as part and parcel of the spatial engine's basic design. The conclusion that is forced in this circumstance is the relative paucity of the representational resources available to the syntactic engine, in comparison to those available to the spatial engine.

Given that both the syntactic engine and the spatial engine are equally computationally powerful, and also given that both can be grounded extensionally, I believe that as subtle a claim as the *relative* paucity of resources in different computational engines can only be investigated by actually *building* a Fodorian automata. This will necessitate the pooling of knowledge from many disparate fields—connectionist and symbolic cognitive science, robotics, Artificial Life, autonomous systems, and possibly very many others, in what some are already calling a *synthetic* psychology.

There are two related aspects of such a grand project, of some significance for this book (apart from the obvious engineering and electronics problems). First, the design, construction and configuration of an appropriate, motile physical device (viz. a robot, or autonomous agent or whatever) able to interact with the external environment, equipped with sophisticated, intrinsic transducers, and capable of building internal models (i.e., #representations#) of that external environment purely on the basis of its interaction with it: That is, the device must be capable of performing extensional grounding. Second, the extraction of the internal models (i.e., #representations#) of the world that the device has constructed, and the mapping of syntactically structured "representations" (of language) to, and from, such internally constructed #representations# in a systematic fashion—that is, the device must be capable of performing intensional grounding.

Such a research endeavor reminds me of the ancient Chinese curse, appropriate for the connectionist cognitive scientist and their ongoing search for the what, where, how and why of mind: **May you live in interesting times.**

References

Aizawa, K. (1992) Review of *Philosophy and Connectionist Theory*. W. Ramsey, S. P. Stich & D. E. Rumelhart (Eds). In *Mind and Language*, **7**.

Ajdukiewicz, K. (1935) Die syntaktische Konnexitat. *Studia Philosophica*, *1*, 1–27. Translated by S. McCall in *Polish Logic* (1967), 207–231.

Aleksander, I. & Morton, H. (1993) *Neurons and Symbols: The Stuff that Mind is Made Of.* London: Chapman & Hall.

Anderson, J. R. & Bower, G. H. (1973) *Human Associative Memory*. Washington, DC: Winston.

Arbib, M. A. (1987) Levels of modeling of mechanisms of visually guided behavior. *Behavioral and Brain Sciences*, *10*, 407–465.

Baker, G. P. & Hacker, P. M. S. (1983) *Language, Sense and Nonsense: A Critical Investigation into Modern Theories of Language*. Oxford: Basil Blackwell.

Baldwin, A. (1992) Subsymbolic inference: Inferring verb meaning. In R. Trappl (Ed) *Proceedings of the Eleventh European Cybernetics and Systems Conference*.

Banks, I. M. (1988) *The Player of Games*. Aylesbury: Orbit.

Barclay, J. R. (1973) The role of comprehension in remembering sentences. *Cognitive Psychology*, *4*, 229–254.

Bar–Hillel, Y. (1967) Dictionaries and meaning rules. *Foundations of Language*, *3*, 409–414.

Barwise, J. & Perry, J. (1983) *Situations and Attitudes*. Cambridge, MA: MIT Press.

Bennett, D. (1975) *Spatial and Temporal uses of English Prepositions: An Essay in Stratificational Semantics*. New York: Longman Press.

Bierwisch, M. (1970) Semantics. In J. Lyons (Ed) *New Horizons in Linguistics*, 166–184. Harmondsworth, England: Penguin.

Blank, D. S., Meeden, L. A., & Marshall, J. B. (1991) Exploring the Symbolic/Subsymbolic Continuum: A Case Study of RAAM. Indiana University, Computer Science Report 47405. In J. Dinsmore, (Ed) *The Symbolic and Connectionist Paradigms: Closing the Gap*, Hillsdale, NJ: Erlbaum.

Block, N. (1986) An advertisement for a semantics for psychology. In *Midwest Studies in Philosophy X: Studies in the philosophy of mind.* Minneapolis, MN, University of Minnesota Press.

Boden, M. (1988) *Computer Models of Mind.* Cambridge, England: Cambridge University Press.

Boole, G. (1854) *An Investigation of the Laws of Thought.* Dover: New York.

Boole, G. (1847) *The Mathematical Analysis of Logic, Being an Essay Toward a Calculus of Deductive Reasoning.* Macmillan, Barclay and Macmillan: Cambridge.

Bosch, P. (1988) On Representing Lexical Meaning. In W. Hullen & R. Schulze (Eds) *Understanding the Lexicon.* Tubingen: Max Niemeyer Verlag.

Broomhead, D. S. & Lowe, D. (1988) Multivariable Functional Interpolation and Adaptive Networks. *Complex Systems, 2,* 321–355.

Carnap, R. (1947) *Meaning and Necessity.* Chicago: University of Chicago Press.

Carnap, R. (1952) Meaning Postulates. *Philosophical Studies, 3,* 65–73.

Chalmers, D. J. (1990) Syntactic transformations on distributed representations. *Connection Science, 2(1),* 53–62.

Chalmers, D. J. (1992) Subsymbolic Computation and the Chinese Room. In J. Dinsmore (Ed) *The Symbolic and Connectionist Paradigms: Closing the Gap.* Hillsdale, NJ: Erlbaum.

Chomsky, N. (1957) *Syntactic Structures.* The Hague: Mouton.

Chomsky, N. (1965) *Aspects of the Theory of Syntax.* Cambridge, MA: MIT Press.

Churchland, P. M. & Churchland, P. S. (1983) Stalking the wild epsitemic engine. *Nous, 17,* 5–18. Reprinted in W. G. Lycan (Ed) *Mind and Cognition: A reader.* Oxford: Basil Blackwell.

Churchland, P. M. (1990) On the Nature of Explanation : A PDP Approach. In *Proceedings of the First Annual Conference on Emergent Computation.* To appear in P. M. Churchland *A Neurocomputational Perspective: The Nature of Mind and the Structure of Science,* MIT Press.

Chrisman, L. (1991) Learning recursive distributed representations for holistic computation. TR CMU-CS-91–154, Pittsburgh: Carnegie–Mellon University.

Clark, A. (1988a) Connectionism, Competence and Psychological Explanation. *Cognitive Science Research Paper CSRP 120,* University of Sussex.

Clark, A. (1988b) Computation, connectionism and content. In Y. Kodratoff (Ed) *Eighth European Conference on AI,* Munich.

Clark, A. (1989) The multiplicity of mind. *Artificial Intelligence Review, 3,* 49–65.

Clark, A. (1990) *Microcognition: Philosophy, Cognitive Science and Parallel Distributed Processing.* Cambridge, MA: MIT Press.

Clark, A. (1992) The presence of a symbol. *Connection Science, 4,* 193–205.

Cleeremans, A. (1993) *Mechanisms of Implicit Learning: Connectionist Models of Sequence Processing*. Cambridge, MA: MIT Press.

Cliff, D. T. (1990) Computational Neuroethology: A Provisional Manifesto. Cognitive Science Research Paper, CSRP 162, University of Sussex.

Collins, A. M. & Loftus, E. F. (1975) A Spreading Activation Theory of Semantic Processing. *Psychological Review, 82(6)*, 407–428.

Collins, A. M. & Quillian, M. R. (1969) Retrieval time from semantic memory. *Journal of Verbal Learning and Verbal Behavior, 8*, 240–247.

Collins, A. M. & Quillian, M. R. (1972) How to make a language user. In E. TUlving & W. Donaldson (eds) *Organization and Memory*, 310–349. New York: Academic.

Craik, K. (1943) *The Nature of Explanation*. Cambridge University Press: Cambridge, England.

Cummins, R. (1989) *Meaning and Mental Representation*. Cambridge, MA: MIT Press.

Cummins, R. (1991) Representation in Connectionism. In W. Ramsey, S. P. Stich & D. E. Rumelhart (Eds) *Philosophy and Connectionist Theory*. Hillsdale, NJ: Erlbaum.

Davies, D. J. M. & Isard, S. D. (1972) Utterances as programs. In D. Mitchie (ed) *Machine Intelligence, 7*. Edinburgh: Edinburgh University Press.

Dennett, D. (1981) Three kinds of intentional psychology. In R. A. Healey (Ed) *Reduction, Time and Reality: Studies in the Philosophy of the Natural Sciences*. Cambridge: Cambridge University Press.

Dietrich, E. (1990) Computationalism. *Social Epistemology, 4(2)*, 135–154.

van Dijk, T. A. & Kintsch, W. (1983) *Strategies of Discourse Comprehension*, Ch.'s 1 & 10. NY : Academic Press.

Dowty, D. R., Wall, R. E. & Peters, S. (1981) *Introduction to Montague Semantics*. Dordrecht: D. Reidel.

Dummett, M. A. E. (1978) *Truth and Other Enigmas*. London: Duckworth.

Elman, J. L. (1993). Representation and structure in connectionist models. In R. Reilly & N. E. Sharkey (Eds) *Connectionist Approaches to Natural Language Processing*. Hove: LEA.

Elman, J. L. (1991) Distributed representations, simple recurrent networks and grammatical structure. *Machine Learning, 7*, 195–225.

Elman, J. L. (1990). Finding structure in time. *Cognitive Science, 14*, 179–211.

Fahlman, S. E. & Lebiere, C. (1990) The cascade correlation learning architecture. In D. S. Touretsky (Ed) *Advances in Neural Information Processing Systems II*. San Mateo, CA: Morgan Kaufman.

Fillmore, C. (1968) The case for case. In E. Bach & R. Harms (Eds) *Universals in Linguistic Theory*. New York: Holt, Rinehart & Winston.

Fillmore, C. J. (1975) An alternative to checklist theories of meaning. *Proceedings of the Berkeley Linguistics Society, 1*, 123–131.

Fodor, J. A. (1976) *The Language of Thought*. Hassocks, Sussex: Harvester Press.

Fodor, J. A. (1978) Tom Swift and his Procedural Grandmother. *Cognition, 6*, 229–247.

Fodor, J. A. (1987) *Pyschosemantics: The Problem of Meaning in the Philosophy of Mind.* Cambridge, MA: MIT Press.

Fodor, J. A. (1990) *A Theory of Content and Other Essays.* Cambridge, MA: MIT Press.

Fodor, J. A., Garrett, M. F., Walker, E. C. T. & Parkes, C. H. (1980) Against Definitions. *Cognition, 8*, 263–367.

Fodor, J. A. & McLaughlin, B. P. (1990) Connectionism and the Problem of Systematicity : Why Smolensky's Solution doesn't work. *Cognition, 35*, 183–204.

Fodor, J. A., & Pylyshyn, Z. W. (1988). Connectionism and cognitive architecture: A critical analysis. *Cognition, 28*, 2–71.

Fodor, J. D., Fodor, J. A. & Garrett, M. F. (1975) The Psychological Unreality of Semantics Representations. *Linguistic Inquiry, 6(4)*, 515–531.

Franklin, S. & Garzon, M. (1990) Neural computability. In O. Omidvar (Ed) *Progress in Neural Networks.* Norwood, NJ: Ablex.

Frege, G. (1892) Uber Sin und Bedeutung. *Zeitschrift fur Philosphie und Philosophische Kritrik, 100*, 25–50. Translated in P. T. Geach & M. Black (eds) *Philosophical Writings of Gottlob Frege.* Oxford: Blackwell, 1952.

Frege, G. (1953) *The Foundations of Arithmetic.* Translated by J. L. Austin. Second revised edition. Oxford: Blackwell. (Originally published 1884).

Garnham, A. (1985) *Psycholinguistics: Central Topics.* London: Methuen.

Garnham, A. (1989) A Unified Theory of the Meaning of Some Spatial Relational Terms. *Cognition, 31*, 45–60.

Garnham, A., Oakhill, J. V. & Johnson–Laird, P. N. (1982) Referential continuity and the coherence of discourse. *Cognition, 11*, 29–46.

Gazdar, G. J. M. (1981) Unbounded dependencies and coordinate structure. *Linguistic Inquiry, 12*, 155–184.

Gazdar, G. J. M. (1982) Phrase Structure Grammar. IN G. K. Pullum & P. Jacobson (eds) *The Nature of Syntactic Representation.* Dordrecht: Reidel.

van Gelder, T. (1990) Compositionality : A Connectionist Variation on a Classical Theme. *Cognitive Science*, 14, 355–384.

van Gelder, T. (1991a) A survey of the concept of distribution. In W. Ramsey, S. P. Stitch & D. E. Rumelhart (Eds) *Philosophy and Connectionist Theory.* Hillsdale, NJ: Erlbaum.

van Gelder, T. (1991b) Classical questions, radical answers: Connectionism and the structure of mental representation. In T. Horgan & J. Tienson (Eds) *Connectionism and the Philosophy of Mind.* Kluwer.

van Gelder, T. (1992) Defining 'distributed representation.' *Connection Science 4: Special Issue on Philosophical Issues in Connectionist Modeling*, 175–191.

van Gelder & Port (1994) Beyond symbolic: Prolegomena to a *Karma Sutra* of compositionality. In V. Honovar & L. Uhr (Eds) *Artificial Intelligence and Neural Networks: Steps Toward Principled Integration, Volume 1: Basic Paradigms, Learning Representational Issues and Integrated Architectures*. Cambridge, MA: Academic Press.

Gentner, D. & Stevens, A. L. (1983) *Mental Models*. Hillsdale, NJ: Erlbaum.

Gibson, J. J. (1979) *The Ecological Approach to Visual Perception*. Boston: Houghton Mifflin.

Glass, A. L. & Holyoak, K. J. (1975) Alternative conceptions of semantic memory. *Cognition, 3*, 313–339.

Grossberg, S. (1987) Competitive learning: From interactive activation to adaptive resonance. *Cognitive Science, 11*, 23–63.

Hadley, R. F. (1989) A default-oriented theory of procedural semantics. *Cognitive Science, 13*, 107–137.

Hadley, R. F. (1993) Systematicity in connectionist language learning. Tec Report, Simon Fraser University, Burnaby, B. C., V5A 1S6, Canada.

Harnad, S. (1989) Minds, machines and searle. *Journal of Experimental and Theoretical Artificial Intelligence, 1*, 5–25.

Harnad, S. (1990a) Lost in the hermeneutic hall of mirrors. *Journal of Experimental and Theoretical Artificial Intelligence, 2*, 321–327.

Harnad, S. (1990b) The symbol grounding problem. *Physica D, 42*, 335–346.

Harnad, S. (1991) Other bodies, other minds: A machine incarnation of an old philosophical problem. *Minds and Machine, 1*, 43–54.

Harnad, S. (1992) Connecting object to symbol in modeling cognition. In A. Clark & R. Lutz (Eds) *Connectionism in Context*, New York: Springer-Verlag.

Harnad, S. (1993) Grounding symbols in the analog world with neural nets: A Hybrid model. *Think, 2*, 12–20.

Harnad, S., Hanson, S. J. & Lubin, J. (1991) Categorical Perception and the Evolution of Supervised Learning in Neural Nets. Paper presented at American Association for AI Symposium on Symbol Grounding: Problem and Practice, Stanford University.

Harris, C. L. (1990) Connectionism and cognitive linguistics. *Connection Science, 2(1)*, 7–33.

Haugeland, J. (1981) Semantic engines: An introduction to mind design. In J. Haugeland (Ed) *Mind Design: Philosophy, Psychology, Artificial Intelligence*. Cambridge, MA: MIT Press.

Haugeland, J. (1985) *Artificial Intelligence: The Very Idea*. Cambridge, MA: MIT Press.

Haugeland, J. (1991) Representational genera. In W. Ramsey, S. P. Stich & D. E. Rumelhart (Eds) *Philosophy and Connectionist Theory*. Hillsdale, NJ: Erlbaum.

Hayes, P. (1993) Computers don't follow instructions. *Think, 2*, 37–40.

Heim, I. (1983) File Change Semantics and the Familiarity Theory of Definiteness. In R. Bauerle, C. Schwarze & A. von Stechow (Eds) *Meaning, Use and Interpretation of Language.* de Gruyter, Berlin.

Herskovits, A. (1985) Semantics and pragmatics of locative expressions. *Cognitive Science, 9*, 341–378.

Hintikka, J. (1963) The Modes of Modality. *Acta Philosophica Fennica, 16*, 65–82.

Hintikka, J. (1975) Impossible possible worlds vindicated. *Journal of Philosophical Logic, 4*, 475–484.

Hinton, G. E. (1981) Implementing semantic networks in parallel hardware. In G. E. Hinton and J. A. Anderson (Eds), *Parallel models of Associative Memory*, 161–187. Hillsdale, NJ: Erlbaum.

Hinton, G. E. (1986). Learning distributed representations of concepts. *Proceedings of the 8th Annual Conference of the Cognitive Science Society, 1986*, 1–12.

Hinton, G. E. (1988) Representing part–whole hierarchies in *Proceedings of the Tenth Annual Conference of the Cognitive Science Society*, 48–54, Montreal, Canada.

Hinton, G. E., McClelland, J. L., & Rumelhart, D. E. (1986). Distributed Representations. In D. E. Rumelhart and J. L. McClelland (Eds), *Parallel Distributed Processing: Explorations in the Microstructure of Cognition, 1: Foundations.* Cambridge, MA: MIT Press, 77–109.

Hinton, G. E. & Sejnowski, T. J. (1986) Learning and Relearning in Boltzman Machines. In D. E. Rumelhart and J. L. McClelland (Eds), *Parallel Distributed Processing: Explorations in the Microstructure of Cognition, 1: Foundations.* Cambridge, MA: MIT Press, 77–109.

Hobbs, J. R. & Rosenschein, S. J. (1978) Making Computational Sense of Montague's Intensional Logic. *Artificial Intelligence, 9*, 287–306

Hofstadter, D. R. (1985) Waking Up from the Boolean Dream, or Subcognition as Computation. In D. R. Hofstadter *Metamagical Themas: Questing for the Essence of Mind and Pattern*, 631–665. New York: Viking.

Hollan, J. D. (1975) Features and Semantic Memory : Set-Theoretic or Network Model?. *Psychological Review, 82(2)*, 154–155.

Jackendoff, R. (1983) *Semantics and Cognition.* Cambridge, MA: MIT Press.

Jackendoff, R. (1984) Sense and Reference in a Psychologically Based Semantics. In T. G. Bever, J. M. Carral, & L. A. Miller (Eds) *Talking Minds.* Cambridge, MA: MIT Press

Jackson, S. A. (1994) Superpositional Structural Analogues. University of Sheffield, Computer Science Department Tec Report.

Jackson, S. A. & Sharkey, N. E. (1991) A Connectionist Semantics for Spatial Descriptions. In *Proceedings of the Conference on Artificial Intelligence and Simulation of Behavior*, 72–83.

Jackson, S. A. & Sharkey, N. E. (1994) Grounding computational engines. *Artificial Intelligence Review, 8*.

Johnson-Laird, P. N. (1977a) Procedural Semantics. *Cognition, 5,* 189–214.

Johnson-Laird, P. N. (1977b) Psycholinguistics without Linguistics. In N. S. Sutherland (Ed) *Tutorial Essays in Psychology, vol 1.* Hillsdale, NJ : Erlbaum.

Johnson-Laird, P. N. (1978) What's wrong with Grandma's guide to procedural semantics : a reply to Jerry Fodor. *Cognition, 6,* 249–261.

Johnson-Laird, P. N. (1983) *Mental Models : Towards a Cognitive Science of Language, Inference and Consciousness.* Cambridge, England : CUP.

Johnson-Laird, P. N. & Garnham, A. (1980) Descriptions and discourse models. *Linguistics and Philosophy, 3,* 371–393.

Johnson-Laird, P. N., Herrman, D. J. & Chaffin, R. (1984) Only Connections : A Critique of Semantic Networks. *Psychological Bulletin, 96,* 292–315.

Jordan, M. I. (1986a) Serial Order: A Parallel Distributed Processing Approach. *Institute for Cognitive Science report, 8604.* San Diego: University of California.

Jordan, M. I. (1986b). Attractor dynamics and parallelism in a connectionist sequential machine. *Proceedings of the 8th Annual Conference of the Cognitive Science Society,* Amherst, MA., 531–545.

Kamp, J. A. W. (1979) Events, Instants and Temporal Reference. In R. Bauerle, U. Egli & A. von Stechow (eds) *Semantics From Different Points of View.* Berlin: Springer–Verlag.

Kamp, J. A. W. (1981) A Theory of Truth and Semantic Representation. In J. Groenendijk, T. Janssen & M. Stokhof (eds) *Formal Methods in the Study of Language.* Amsterdam : Mathematical Centre Tracts.

Kant, I., (1787) *Critique of Pure Reason.* Translated by N. K. Smith (1961) Macmillan: New York.

Kaplan, D. (1968) Quantifying In. *Synthese, 19,* 178–214.

Karttunen, L. (1976) Discourse Referents. In J. D. McCawley (Ed) *Syntax and Semantics 6: Notes from the Linguistic Underground.* New York: Academic Press.

Katz, J. J. (1972) *Semantic Theory.* New York: Harper & Row.

Katz, J. J. (1977) The real status of semantic representations. *Linguistic Inquiry, 3,* 559–584.

Katz, J. J. & Fodor, J. A. (1963) The structure of a semantic theory. *Language, 39,* 170–210.

Kintsch, W. (1974) *The Representation of Meaning in Memory.* Hillsdale, NJ: Erlbaum.

Kintsch, W. & van Dijk, T. A. (1978) Towards a model of text comprehension and reproduction. *Psychological Review, 85,* 363–394.

Kosslyn, S. M. (1980) *Images and Mind.* Cambridge, MA: Harvard University Press.

Kripke, S. (1963) Semantical Considerations on Modal Logic. *Acta Philosophica Fennica, 16,* 83–94.

Labov, W. (1973) The boundaries of words and their meanings. In C. J. N. Bailey & R. W. Shuy (eds) *New Ways of Analyzing Variations in English.* Washington, DC: Georgetwon University Press.

Lakoff, G. (1988) Smolensky, semantics and the sensorimotor system. *Behavioral and Brain sciences, 11(1),* 39–40.

Leech, G. (1969) *Towards a Semantic Description of English.* New York: Longman Press.

Levelt, W. J. M. (1984) Some perceptual limitations on talking about space. In A. J. van Doorn, W. A. van der Grind & J. J. Koenderink (Eds) *Limits in Perception.* Utrecht: VNU Science Press.

Lewis, D. (1972) General Semantics. In D. Davidson & G. Harman (eds) *Semantics of Natural Language.* Dordrecht: Reidel.

Loewer, B. & Rey, G. (1991) *Meaning in Mind: Fodor and his Critics.* Oxford: Basil Blackwell.

Longuet-Higgins, H. C. (1972) The algorithmic description of natural language. *Proceedings of the Royal Society of London, B, 182,* 255–276.

McClelland, J. L. & Kawamoto, A. H. (1986) Mechanisms of sentence comprehension: Assigning roles to constituents. In D. E. Rumelhart and J. L. McClelland (Eds), *Parallel Distributed Processing: Explorations in the Microstructure of Cognition, 2: Psychological and Biological Models.* Cambridge, MA: MIT Press, 272–327.

McCloskey, M. & Glucksburg, S. (1979) Decision processes in verifying category membership statements: Implications for models of semantic memory. *Cognitive Psychology, 11,* 1–37.

McCulloch, W. W. & Pitts, W. (1943) A logical calculus of the ideas immanent in nervous activity. *Bulletin of Mathematical Biophysics, 5,* 115–133.

McMillan, C. & Smolensky, P. (1988) Analyzing a Connectionist Model as a System of Soft Rules. *Proceedings of the 10th Annual Conference of the Cognitive Science Society,* Montreal, Canada.

McMillan, C., Mozer, M. C. & Smolensky, P. (1991) The connectionist scientist game: Rule extraction and refinement in a neural network. In *Proceedings of the Thirteenth Annual Conference of the Cognitive Science Society,* Hillsdale, NJ: Erlbaum.

Miller, C. B. & Giles, C. L. (1993) Experimental comparison of the effect of order in recurrent neural networks. *International Journal of Pattern Recognition and Artificial Intelligence: Special Issue on Applications of Neural Networks to Pattern Recognition.*

Miller, G. A. & Johnson–Laird, P. N. (1976) *Language and Perception.* Cambridge, MA: Cambridge University Press.

Minsky, M. L. & Papert, S. (1969) *Perceptrons.* Cambridge, MA: MIT Press.

Minsky, M. (1975) A framework for representing knowledge. In P. H. Winston (ed) *The Psychology of Computer Vision.* New York: McGraw–Hill.

Montague, R. (1974) *Formal Philosophy: Selected Papers.* New Haven, CT: Yale University Press.

Mundy, D. & Sharkey, N. E. (1992) Type generalization on distributed representations. In R. Trappl (Ed) *Cybernetics and Systems Research*, 1327–1334. Dordrecht, The Netherlands: Kluwer Academic Publishers.

Newell, A. & Simon, H. A. (1972) *Human Problem Solving*. Englewood Cliffs, NJ: Prentice–Hall.

Niklassen, L. F. & van Gelder, T. (1994) Can Connectionist Models Exhibit Structure Sensitivity? To appear in *Proceedings of the Sixteenth Annual Conference of the Cognitive Science Society*.

Niklasson, L. F. & Sharkey, N. E. (1992) Connectionism—The miracle mind model. In *Proceedings of the First Swedish National Conference on Connectionism*.

Osherson, D. N. & Smith, E. E. (1981) On the Adequacy of Prototype Theory as a Theory of Concepts. *Cognition*, *9*, 35–58.

Parker, D. B. (1985) *Learning Logic*. Cambridge, MA: MIT Press.

Pirsig, R. M. (1974) *Zen and the Art of Motor Cycle Maintenance*. Essex: Corgi.

Pollack, J. B. (1990) Recursive Distributed Representations. *Artificial Intelligence*, *46*, 77–105.

Port, R. F. & van Gelder, T. (1991) Representing Aspects of Language. Paper to appear in *Proceedings of the Thirteenth Annual Conference of the Cognitive Science Society*, Hillsdale, NJ: Erlbaum.

Potts, G. R. (1973) Memory for redundant information. *Memory and Cognition*, *1*, 467–470.

Prakash, R., Solession, E. & Barlow, R. B. (1989) Parallel computer model of the *Limulus* lateral eye. In *Annual Meeting of the Society of Neuroscience, Abstracts*, *15*, 1206.

Putnam, H. (1970) Is semantics possible? *Contemporary Philosophic Thought: The International Philosophy Year Conferences at Brockport, 1. Languages, Belief and Metaphysics*. Edited by H. Kiefer & M. Munitz. New York: State University of New York Press.

Putnam, H. (1960) Minds and Machines. In S. Hook (Ed) *Dimensions of Mind*. New York: New York University Press.

Putnam, H. (1973) Explanation and Reference. In G. Pearce & P. Maynard (eds) *Conceptual Change*. Dordrecht: Reidel.

Putnam, H. (1975) The Meaning of Meaning\ In K. Gunderson (ed) *Language, Mind and Knowledge*. Minnesota Studies in the Philosophy of Science, Vol 7. Minneapolis: University of Minnesota Press. 131–193.

Quillian, M. R. (1968) Semantic Memory. In M. L. Minsky (ed) *Semantic Information Processing*, 227–259. Cambridge, MA: MIT Press.

Quine, W. V. (1963) On what there is. In From a Logical Point of View. New York: Harper & Row.

Quinn, W. G. & Gould, J. L. (1979) Nerves and genes. *Nature*, *278*, 19–23.

Rips, L. J., Smith, E. E. & Shoben, E. J. (1975) Set-Theoretic and Network Models Reconsidered : A Comment on Hollan's Features and Semantic Memory: *Psychological Review*, *82(2)*, 156–157R

Rosch, E. (1973) Natural Categories. *Cognitive Psychology, 4*, 328–350.

Rosch, E. (1975) Cognitive Representation of Semantic Categories. *Journal of Experimental Psychology : General, 104(3)*, 192–233

Rosch, E. & Mervis, C. B. (1975) Family Resemblances: Studies in the internal structure of categories. *Cognitive Psychology, 7*, 573–605.

Rumelhart, D. E., Lindsay, P. H. & Norman, D. A. S. (1972) A process model for long-term memory. In E. Tulving & W. Donaldson (eds) *Organization and Memory*, 198–246. New York: Academic Press.

Rumelhart, D. E., Hinton, G. E. & Williams, R. J. (1986) Learning internal representations by error propagation. In D. E. Rumelhart and J. L. McClelland (Eds), *Parallel Distributed Processing: Explorations in the Microstructure of Cognition, 1: Foundations*. Cambridge, MA: MIT Press, 318–362.

Rumelhart, D. E. & Zipser, D. (1986) Feature Discovery by Competitive Learning. In D. E. Rumelhart and J. L. McClelland (Eds), *Parallel Distributed Processing: Explorations in the Microstructure of Cognition, 1: Foundations*. Cambridge, MA: MIT Press, 151–193.

Russell, B. P. W. & Whitehead, A. N. (1913) *Principia Mathematica*. Cambridge University Press: Cambridge, England.

Sag, I. A. & Hankamer, J. (1984) Towards a theory of anaphoric processing. *Linguistics and Philosophy, 7*, 325–345.

Sanger, D. (1989) Contribution analysis: A technique for assigning responsibilities to hidden units in connectionist networks. *Connection Science, 1*, 115–138.

Saussure, F. de (1960) *Course in General Linguistics*. London; Peter Owen.

Schaeffer, B. & Wallace, R. (1970) The comparison of word meanings. *Journal of Experimental Psychology, 86*, 144–152.

Schank, R. C. (1975) *Conceptual Information Processing*. New York: American Elsevier.

Schank, R. C. (1986) *Explanation Patterns: Understanding Mechanically and Creatively*. Hillsdale, NJ: Erlbaum.

Schank, R. C. & Abelson, R. P. (1977) *Scripts, Goals, Plans and Understanding*. Hillsdale, NJ: Erlbaum.

Schreuder, R. & Flores d'Arcais, G. B. (1989) Psycholinguistic Issues in the Lexical Representation of Meaning. In W. Marslen-Wilson (Ed) *Lexical Representation and Process*. Cambridge, MA MIT Press.

Searle, J. (1980a) Minds, brains and programs. *Behavioral and Brain Sciences, 3*, 417–424.

Searle, J. (1980b) Intrinsic Intentionality. *Behavioral and Brain Sciences, 3*, 450–457.

Searle, J. (1984) *Minds, Brains and Science*. Cambridge, MA: Harvard University Press.

Searle, J. (1987) Minds and brains without programs. In C. Blakemore & S. Greenfield (Eds) *Mindwaves*. Oxford: Blackwell.

Sharkey, N. E. (1988) A PDP system for paraphrasing routine knowledge vignettes. In *Abstracts of the Second Annual Meeting of the International Neural Networks Society*.

Sharkey, N. E, (1991) Connectionist representation techniques. *AISB Review, 5(3)*, 143–167.

Sharkey, N. E. (1992) The ghost in the hybrid: A study of uniquely connectionist representations. *AISB Quarterly*, 10–16.

Sharkey, N. E. & Jackson, S. A. (1994a) Three Horns of the Representational Trilemna. In V. Honovar & L. Uhr (Eds) *Artificial Intelligence and Neural Networks: Steps Toward Principled Integration, Volume 1: Basic Paradigms, Learning Representational Issues and Integrated Architectures.* Cambridge, MA: Academic Press.

Sharkey, N. E. & Jackson, S. A. (1994b) An Internal Report for Connectionists. In R. Sun (Ed) *Computational Architectures Integrating Neural and Symbolic Processes.* Kluwer.

Sharkey, N. E. & Sharkey, A. J. C. (1993) Adaptive generalization. *Artificial Intelligence Review, 7*, 313–328.

Siegelman, H. & Sontag, E. D. (1991) Neural Nets are universal computing devices. Rutgers Center for Systems and Control. Report SYCON–91–08.

Sloman, A. (1978) Intuition and Analogical Reasoning. In *The Computer Revolution in Philosophy: Philosophy, Science, and Models of Mind.* Harvester: Sussex.

Sloman, A. (1990) Beyond Turing Equivalence. *Proceedings of the Turing 1990 Colloquim.*

Smith, E. E., Shobin, E. J. & Rips, L. J. (1974) Structure and Process in Semantic Memory: A Featural Model for Semantic Decisions. *Psychological Review, 81(3)*, 214–241.

Smolensky, P. (1987) On variable binding and the representation of symbolic structures in connectionist systems. TR CU-CS-355–87. Department of Computer Science, University of Colorado, Boulder, CO.

Smolensky, P. (1988) On the proper treatment of connectionism. *Behavioral and Brain Sciences, 11*, 1–74.

Spencer-Smith, R. (1987) Semantics and discourse representation. *Mind and Language, 2*, 1–26

Sperber, D. & Wilson, D. (1986) *Relevance: Communication and Cognition.* Blackwell: Oxford.

Stenning, K. (1978) Anaphora as an approach to pragmatics. In M. Halle, J. W. Bresnan & G. A. Miller (Eds) *Linguistic Theory and Psychological Reality.* Cambridge, MA: MIT Press.

Stitch, S. (1983) *From Folk Psychology to Cognitive Science*, Cambridge, MA: MIT Press.

St. John, M. F. & McClelland, J. L. (1987) Applying contextual constraints in sentence comprehension. *Proceedings of the Ninth Annual Conference of the Cognitive Science Society*, Seattle, Washington.

Suppes, P. (1980) Procedural semantics. *Proceedings of the Fourth International Wittgenstein Symposium.* Kirchberg, Austria.

Suppes, P. (1982) Variable-free semantics with remarks on procedural extension. In T. W. Simon & R. J. Scholes (Eds) *Language, Mind and Brain.*

Szu, H. & Harley, R. (1987) Fast simulated annealing. *Physics Letters, 1222, (3,4)*, 157–162.

Talmy, L. (1983) How language structures space. In H. Pick & L. Acredolo (Eds) *Spatial Orientation: Theory, Research and Application.* New York: Plenum.

Tarski, A. (1935) Der Wahrheitsbegriff in den formalisierten Sprachen. *Studia Philosophica, 1.*

Tarski, A. (1944) The semantic conception of truth. In F. Zabeeh, E. D. Klemke & A. Jacobson (Eds) *Readings in Semantics*, P. 675–711. Reprinted from *Philosophy and Phenomenological Research, IV, 3*, 341–375.

Turing, A. M. (1936) On computable numbers with an application to the Entscheidungs problem. *Proceedings of the London Mathematical Society*, series 2, vol 4, 230–265.

Turing, A. M. (1950) Computing machinery and intelligence. *Mind, 59*, 433–460.

Varela, F. J. & Bourgine, P. (1992) *Towards a Practice of Autonomous Systems: Proceedings of the First European Conference on Artificial Life.* Cambridge, MA: MIT Press.

Ward, J. H. (1963) Hierarchical grouping to optimize an objective function. *Journal of the American Statistical Association, 58*, 236–244.

Wasserman, P. D. (1988) Combined back propagation/Cauchy machine. *Neural Networks: Abstracts of the First INNS Meeting*, Boston, Vol 1, 556. Elmsford, NY: Pergamon Press.

Webber, B. L. (1981) Discourse model synthesis : preliminaries to reference. In A. K. Joshi, B. L. Webber & I. A. Sag (Eds) *Elements of Discourse Understanding.* Cambridge, MA: Cambridge University Press.

Webber, B. L. (1983) So what can we talk about now? In M. Brady & R. C. Berwick (Eds) *Computational Models of Discourse.* Cambridge, MA: MIT Press.

Werbos, P. J. (1974) *Beyond Regression: New tools for prediction and analysis in the behavioral sciences.* Masters Thesis, Harvard University.

Wilks, Y. (1975) A preferential, pattern-seeking semantics for natural language inference. *Artificial Intelligence, 6*, 53–74.

Wilks, Y. (1982) Some thoughts on procedural semantics. In W. Lehnert & M. Ringle (eds) *Strategies for Natural Language Processing.* Hillsdale, NJ: Erlbaum.

Williams, R. J. & Zipser, D. (1989) A learning algorithm for continually running fully recurrent neural networks. *Neural Computation, 1*, 270–280.

Wilson, N. L. (1973) On semantically relevant whatsits: A Semantics for Philosophy of Science. In G. Pearce & P. Maynard (eds) *Conceptual Change.* Dordrecht: Reidel.

Winograd, T. (1975) Frame representations and the declarative/procedural controversy. In D. Bobrow & A. Collins (Eds) *Representations and Understanding: Studies in Cognitive Science.* New York: Academic.

Wittgenstein, L. (1922) *Tractatus Logico-Philosophicus.* London: Routledge and Kegan Paul.

Wittgenstein, L. (1953) *Philosophical Investigations.* Translated by G. E. M. Anscombe. New York: Macmillan.

Woods, W. A. (1975) What's in a link: Foundations for semantic networks. In D. Bobrow & A. Collins (Eds) *Representations and Understanding : Studies in Cognitive Science.* New York: Academic.

Woods, W. A. (1981) Procedural semantics as a theory of meaning. In A. K. Joshi, B. L. Webber & I. A. Sag (Eds) *Elements of Discourse Understanding.* Cambridge, MA: Cambridge University Press.

Woods, W. A. (1986) Problems in procedural semantics. In Z. Pylyshyn & W. Demopoulos (eds) *Meaning and Cognitive Structure.* Norwood, NJ: Ablex.

Zadeh, L. (1965) Fuzzy sets. *Information and Control, 8,* 338–353.

Author Index

Subject Index